Conserving Living Natural Resources

Conserving Living Natural Resources provides students, managers, and general readers with an introduction to the principles of managing biological resources. It presents the historical and conceptual contexts of three seminal approaches to the management of living natural resources: utilitarian management for harvest of featured species and control of unwanted species, protection and restoration of populations and habitats to maintain biodiversity, and management of complex ecosystems to sustain both productivity and biodiversity. The book shows how the first two approaches were grounded in the belief that nature is "in balance" and that people are outsiders, and then goes on to show how the "flux-of-nature" viewpoint suggests new strategies for conservation grounded in a view of nature as dynamic, and people as participants in the natural world.

Rather than endorsing a single approach as the only correct one, this book investigates the historical and philosophical contexts, conceptual frameworks, principal techniques, and limitations of each approach.

BERTIE JOSEPHSON WEDDELL teaches principles of conservation in the Distance Degree Program of Washington State University, and is principal of *Draba*, a natural resource management consulting business she founded.

To Jim, Wes, and Angie

Conserving Living Natural Resources

in the context of a changing world

Bertie Josephson Weddell

Washington State University

CAMBRIDGE
UNIVERSITY PRESS

CAMBRIDGE UNIVERSITY PRESS
Cambridge, New York, Melbourne, Madrid, Cape Town, Singapore, São Paulo

Cambridge University Press
The Edinburgh Building, Cambridge CB2 2RU, UK

Published in the United States of America by Cambridge University Press, New York

www.cambridge.org
Information on this title: www.cambridge.org/9780521782708

First published 2002
Reprinted 2006
Digitally reprinted 2006

A catalogue record for this publication is available from the British Library

Library of Congress Cataloguing in Publication data

Weddell, Bertie, J., 1948–
Conserving living natural resources in the context of a changing
world / Bertie Josephson Weddell.
   p.   cm.
Includes bibliographical references (p.)
ISBN 0 521 78270 8 (hb) – ISBN 0 521 78812 9 (pb.)
1. Conservation biology.   2. Conservation of natural resources.   3. Natural
resources – Management.   I. Title.
QH75.W42   2002
333.7´2–dc21   2001037355

ISBN-13 978-0-521-78270-8 hardback
ISBN-10 0-521-78270-8 hardback

ISBN-13 978-0-521-78812-0 paperback
ISBN-10 0-521-78812-9 paperback

Contents

Preface

I wrote this book to introduce students to and to review for managers three different approaches to natural resource management. Until the early 1970s, college courses and texts in natural resource management dealt primarily with a utilitarian approach to resources, with a little preservationist management thrown in (for example, the story of Yellowstone National Park and parks in general). Then endangered species and nongame species began to attract the attention of managers. During this period, a number of ecologists in North America also undertook studies of ecosystem functions and processes. (In Europe ecosystem studies had been receiving attention since late in the nineteenth century.) By the 1980s, courses and texts in conservation biology began to appear. At first these emphasized the management of small, fragmented populations; later, management to maintain fundamental ecosystem processes became a more prominent theme.

I wanted to present these different strands of thought to students who were not necessarily majoring in the natural sciences or in management and to show how these threads interweave in the fabric of natural resource management. This book attempts to do that, by presenting the historical and conceptual contexts of different approaches to resource management. It begins with the utilitarian approach to harvesting featured species, proceeds to recent responses to the biodiversity crisis, and culminates in efforts to manage ecosystems sustainably.

In writing this book, I started from the premise that it is more useful for students and managers to learn about the historical conditions that gave rise to different strategies for resource stewardship and the strengths and weaknesses of each, than to study a single approach as the only correct one. The book is organized into three main sections, which cover three approaches to conservation, more or less in chronological order: management for products (the utilitarian approach), preserves (the preservationist approach), or processes (the sustainable-ecosystem approach). Utilitarian management focuses on the harvest of featured species to provide desired products, preservationist management stresses the protection and restoration of populations and habitats to maintain biodiversity, and the sustainable-ecosystem approach seeks to conserve both productivity and biodiversity by maintaining healthy ecosystems. Each section describes the historical and philosophical context, the conceptual framework, the principal techniques, and the limitations of the approach in question. By showing how each period in natural resource management has operated within a particular world view, made important contributions, and had definite limitations, I hope this volume will encourage readers to view science and resource management as ongoing processes rather than as static entities. In keeping with its historical bent, the book includes substantial amounts of quoted material from bygone decades. Some of these are from seminal thinkers; others are simply examples that reflect the thinking of the day. It is hoped that this will give students some appreciation of the flavor of the different mind-sets that are discussed.

Many books present the philosophical context of conservation as a dichotomy between anthropocentric and biocentric approaches, or use versus preservation. A number of scholars have recently pointed out, however, that *both* the utilitarian and the preservationist perspective share a similar philosophical context: the idea that people are outside of the balance of nature. At the same time, some scientists have suggested that the flux of nature is a more appropriate metaphor for nature. This leads to novel ways of thinking about the natural world and our role in it.

Parts I and II of this book deal with management strategies that are grounded in the balance-of-nature view. Part I shows how the academic disciplines of forestry, wildlife management, and range management developed in response to the unregulated exploitation of wild plant and animal populations after Europeans colonized the New World, Africa, the Australian region, and parts of Asia. Professionals in these disciplines approach natural resources from a utilitarian perspective, that is, they seek to regulate the exploitation of economically valuable resources such as timber, game species, and livestock forage. To accomplish their objectives, utilitarian resource managers attempt to enhance populations and habitats that provide economic benefits and to reduce or eliminate processes (such as fire) and species (such as predators) that are viewed as detrimental. They focus primarily on a small number of natural processes, such as density-dependent population growth and the development of stable plant communities. The underlying assumption is that managers can maximize the flow of useful products by controlling or compensating for forces that upset the balance of nature.

Part II covers efforts to preserve natural places and living things regardless of their economic values. Whereas utilitarian management attempts to conserve natural resources *for* people, this type of management seeks to protect those resources *from* people. The roots of this movement go back to nineteenth-century efforts to preserve wild places for their intrinsic beauty and spiritual value. A century later, awareness of environmental problems and accelerating losses of species led to a new goal, the protection of biodiversity. Preservationist resource managers apply insights into processes such as extinction and colonization and the genetic consequences of small population size to this challenge. Like utilitarian resource managers, preservationists envision nature as tending toward equilibrium. In this view, the activities of people upset the balance of nature, and the goal of conservationists should be to reinstate that balance by protecting and restoring populations and habitats. The assumption underlying this approach is that the natural world often needs to be protected from the degrading and disturbing influences of people because these influences upset the balance of nature.

Part III investigates an alternative view – that nature is in a state of flux which people are part of. This approach, termed the sustainable-ecosystem approach, draws on insights from both utilitarian and preservationist management, but it suggests new ways of thinking about our place in nature and of managing natural resources to sustain ecosystems. It was fostered by a variety of practical and theoretical considerations that highlight the need for an approach to resource management which emphasizes the variability of nature, addresses social issues of equity and power, and includes the activities of people as part of the natural world. Recent insights into disturbance dynamics and heterogeneous environments suggest new techniques for managing to conserve ecological processes at a variety of temporal and spatial scales. The underlying assumption here is that by preserving ecological processes and natural variability we can maintain both productivity and diversity.

The first chapter in each part describes the historical conditions that set the stage for that particular type of resource management and concludes with a discussion of historical perceptions about fundamental problems in resource management and how they should be solved. I suggest that readers, especially those who are not resource managers or ecologists, begin by reading those chapters (1, 7, and 12) to get some idea of the issues being addressed by the different kinds of resource management. These chapters are followed by one or two chapters that outline the central concepts of each approach, and two or three chapters dealing with its principal techniques.

The sequence presented in this book, of three stages in resource management – utilitarian management, preservationist management, and management to sustain ecosystems – is not, strictly speaking, a chronological one. Although in a general sense utilitarian management led to biodiversity protection and subsequently to management for sustainable ecosystems, there is plenty of temporal overlap between these strategies. Furthermore, all three approaches are thriving and producing useful insights in the twenty-first century. Many resource managers hold views that are a combination of the different approaches covered in this book and practice a type of management that synthesizes elements of the different styles. But even though reality is complicated, and managers don't really fit into pigeon-holes, understanding the different schools of thought that have influenced the theory and practice of resource management during the past century and a half can help us to put current challenges in context. If we understand the assumptions underlying various resource management policies, we are in a better position to evaluate them. Organizing those ideas into three categories – utilitarian, preservationist, and sustainable-ecosystem – is a heuristic device that helps us do that.

Throughout the discussion of these different approaches, I point out that each approach is appropriate under certain circumstances and that each has its limitations. For example, after Aldo Leopold suggested that game species tend to prefer edges, wildlife biologists set out to create habitats with large amounts of edge. This resulted in small patches of forest surrounded by fields. In terms of the objectives of game managers, this has been quite appropriate. Later, however, it became apparent that some species prefer forest interiors, and these species do not do well in highly fragmented landscapes. Most of them are not game species. Their needs were overlooked by game managers, and now some are now threatened or endangered. If our objective is to maximize populations of game species, creating edges is a good idea. But if we seek to preserve enough habitat to maintain viable populations of all species, maximizing the amount of edge is not a good way to do this. Changing objectives often lead to changing strategies.

Students may well ask "If managers no longer believe that maximizing edge is a good strategy, then why should I bother to learn about it?" There are several reasons why learning about "outmoded" ideas is essential. First, there is no single correct way to manage natural resources. As noted above, managing to maximize edge is still appropriate under some circumstances, that is, when the goal is to benefit edge-dependent species. Second, our generation has no special corner on the truth. To act as if we do seems arrogant and only invites our successors to **wonder** how we could possibly have been so naive. Every generation, in every **cultural** setting, focuses on certain things and develops insight about those **things**. Likewise, every generation and culture has its own particular blinders **and** prejudices. Our predecessors did, and so do we. Third, it is important to understand how we got where we are. If mistakes were made in the past, can we learn from them? Although we know that we too have a particular slant on reality, perhaps we can be a little less shortsighted if we understand past shortcomings. Fourth, sometimes without realizing it, we hold on to ideas that are out of date. Many of the ideas that are discussed in this book are widely held, even though scientists are now questioning them. This is partly because science reporters and science teachers themselves do not always keep abreast of the latest developments. But it is also because old ways die hard. That may put us in the position where we are reasoning from contradictions that we do not see. For example, if we try to manage to sustain ecosystem processes, an approach that is based upon the flux-of-nature perspective, yet we continue to envision nature as tending toward balance, the contradictions in our approach are likely to undermine our efforts. With a good understanding of historical context, however, we can understand and disentangle the different threads of thought contributing to our thinking.

A great many people contributed to this book, often in ways that were not apparent at the time. It is not possible to name them all here, but some of the most important deserve special thanks. My mother, Lucy Ellen Wishart Josephson, made sure that I was able to spend my childhood summers away from our apartment in New York City, so that I had some opportunities to be around wild things. My grandmother, Mabel Bradshaw Wishart, taught me the names of plants and animals and ignited my desire to know more. My father, Leon Josephson, taught me the importance of ideas and of history. Nancy Reed Kykyri's enthusiasm and love of people, Cape Cod, and life was infectious. In junior high school, Leah Wallach's intellectual curiosity, honesty, and imagination opened many doors for me and Mrs. Dorothy Young taught me how to organize my writing. In graduate school Richard E. Johnson's meticulous editing trained me to pay attention to the details of written presentation. More recently, my husband Jim has always been supportive and believed that a city girl like me could become a field biologist, even when I had my doubts, and my children, Wes and Angie, were fun to be around and put up with the demands on my time and the piles of books and papers in our home. I also thank Yoram Keyes Bauman, John Donnelly, and John Lawrence for reading and commenting on portions of the manuscript; the other three members of the *Sage Notes* Gang of Four (Karen Gary, Juanita Lichthardt, and Sarah Walker) and Alan Busacca, Jack Connelly, Michael Dexter, Jean Gorton,

Bill Lipe, and Kerry Reese for the stimulating discussions and helpful suggestions they provided; Bob Greene and Betsy Dickow of Bookpeople for suggesting important readings and for ordering obscure books for me; the staff of Washington State University's Owen Science and Engineering Library for helping me find useful references; Charlie Robbins for moral support at many points along the way; Jim and Zoe Cooley for their encouragement; Berta Herrera-Trejo for assistance with translating, and Dan Bukvich for making the University of Idaho's Jazz Choir I the ultimate antidote to stress. In addition, I am grateful to Melissa Rockwood and Julie Flynn for doing some of the illustrations, to Emily Silver for working closely with me to develop landscape illustrations that elegantly convey the different approaches to managing living natural resources, to Susan Vetter at Washington State University's Distance Degree Program and Carol Borden for editorial assistance early on, to an anonymous reviewer whose thoughtful and thorough comments were invaluable, and especially to Ward Cooper at Cambridge University Press for believing in this project, and Anna Hodson, for her meticulous copy-editing, and Carol Miller for her fine assistance with proofreading.

Introduction: Balance and flux

Apart from the hostile influence of man, the organic and the inorganic world are . . .
bound together by such mutual relations and adaptations as secure, if not the absolute permanence and equilibrium of both, a long continuance of the established conditions of each at any given time and place, or at least, a very slow and gradual
succession of changes in those conditions. But man is everywhere a disturbing agent.
Wherever he plants his foot, the harmonies of nature are turned to discords. The
proportions and accommodations which insured the stability of existing arrangements are overthrown.

(Marsh 1874:34)

This statement, by George Perkins Marsh – a nineteenth-century American
diplomat, conservationist, and writer – expresses a concept that can be traced
in western thought as far back as ancient Greece: the idea that nature in the
absence of human intervention is in a state of balance which changes little
over long periods of time. In the nineteenth century, this view became a credo
for the young science of ecology.

The concept of balance figures so prominently in discussions about natural
resource management that it is worth looking at in more detail. In scientific
formulations, balance – or equilibrium – is defined as a state in which there is
no net change in a system. For example, in a chemical reaction, substances A
and B might join to form compound AB, but AB also breaks down to form
A and B. This is denoted by arrows going in two directions:

$$A + B \rightleftharpoons AB$$

The amounts of reagents on the two sides of the equation do not have to be
equal, they just have to be stable. In the example above, equilibrium occurs

when most of the system consists of compounds A and B. This is indicated by a longer arrow going to the left, and we say the equilibrium is to the left. If equilibrium occurs when most of the system is in the form of compound AB, we say the equilibrium is to the right; this is indicated by a long arrow pointing to the right. Either system is in equilibrium. Notice that change continues (A and B combine to form AB, and AB breaks down to form A and B). The important point is that there is no *net* change in the composition of the system; at equilibrium the ratios of AB to A and B do not change.

A similar type of equilibrium can occur in natural communities, if the composition of a system is stable. For example, suppose that 55% of a forest is old growth. From time to time, fires burn some patches of the old-growth forest, converting them to open fields. At the same time, however, young forests are getting older. Eventually they become old growth. If the rate at which old-growth forest is created equals the rate at which it is destroyed, the amount of old-growth forest will remain constant, like compound AB in the chemical equation, and the forest as a whole can be said to be in a state of equilibrium.

The idea of equilibrium is closely connected with the idea of self-regulation. This is an important concept for resource managers, because many ecological processes are considered self-regulating. If a self-regulated system is truly at equilibrium and something happens to cause it to deviate from that equilibrium, then we would expect to see a compensatory change that moves the system back to its equilibrium state. A thermostat is a familiar example. If a room's temperature is regulated by a thermostat set at 20°C, then the thermostat should cause the heater to shut off when the room becomes warmer than 20°C. When the heater is shut off and heat production ceases, heat loss exceeds production and the room's temperature falls, restoring the temperature after a while to 20°C. With the heater off, the room continues to cool, until its temperature drops so far below 20°C that the thermostat causes the heater to turn on again, thereby initiating a compensatory production of heat designed to return the room's temperature to 20°C (Figure 1.1). (Thermostats vary in the precision with which they do this. A very sensitive thermostat will turn off the heater when room temperature rises just slightly above 20°C; a less sensitive thermostat will not respond until the temperature rises several degrees. But regardless of whether temperature fluctuates a lot or a little, the thermostat maintains temperature near a set level.) This type of regulation, in which change in a variable in one direction sets in motion a compensatory change that causes the value of the variable to change in the opposite direction, thereby tending to return it to its original level, is termed negative feedback.

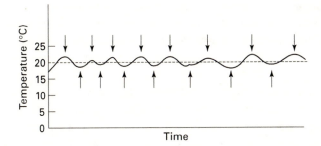

Figure I.1. A thermostat is a device that regulates temperature at a set point. In this example, the set point is 20 °C. Arrows pointing down indicate points at which the thermostat turns the heater off. Arrows pointing up indicate points at which the thermostat switches the heater back on.

It is easy to conceive of a population that is regulated at a set level, or carrying capacity, in this way. If such a population increases above carrying capacity, then there will be a decrease in reproduction and/or an increase in mortality until the population declines. If it drops below carrying capacity, then reproduction will increase or mortality will decrease, or both, allowing the population to grow until it reaches carrying capacity. If there is not a long time-lag between the changes in population size and compensatory adjustments, then this hypothetical population will remain fairly stable.

Until very recently, the prevailing scientific theories about populations and communities hinged on the idea of balance or equilibrium. George Perkins Marsh's idea of a harmoniously balanced natural world that people perturb permeates much scientific and popular writing about the nonhuman world. In this view, the nonhuman world is like a pendulum, characterized by a tendency to return predictably to its starting-point. Ecologists grounded in equilibrium theories focus their attention on populations that are in equilibrium with their resource base (Chapter 2), plant communities that return to an equilibrium or climax state after they are disturbed (Chapter 3), and species assemblages in which rates of extinction and colonization are in balance (Chapter 8). From this perspective, the activities of people are outside of and disturbing to the balanced natural world. We will see in Chapter 12, however, that scientists are now questioning the idea that most of the natural world is in a state of equilibrium most of the time. The idea that people should consider themselves outside of nature has also been called into question, for a variety of reasons.

The idea that the natural world without people is in a balanced state can lead to two quite different strategies for management. Either we can leave it alone and protect it (a preservationist approach), or we can take a utilitarian approach, manipulating it. In the words of ecologist Daniel Botkin, if nature is like a watch, we can "appreciate the beauty of the watch" or we can "attempt to take the watch apart and improve it" (Botkin 1990:156). The examples below, from North America and Africa, illustrate these two approaches.

Historically, American resource managers have tended to fall into one of two groups: utilitarian managers, who strive to maximize the amount of economically valuable products obtained from the natural world, and preservationist managers, who seek to preserve a substantial fraction of the natural world by protecting it from human use. In reality, these are two extremes on a continuum; most people's views fall somewhere in the middle, but some managers are much closer to the use end, while others are closer to the preservation end.

In North America, the roots of this controversy go back over 100 years. The controversy over the building of Hetch Hetchy dam in the scenic Yosemite Valley was one of the first of many conflicts in North America between those who want to preserve resources and those who want to use them for the benefit of people. The Yosemite Valley was set aside for public use by President Lincoln in 1864. Until 1890, when it became a national park, it was administered by the state of California. In 1901, San Francisco city officials proposed damming the Tuolumne River in the park to provide power and water for the residents of San Francisco. A bitter and emotional controversy ensued.

John Muir, champion of wilderness and founder of the Sierra Club, argued that the sublimely beautiful canyon should be left in its natural state for people to appreciate. He described the falls as "harmonious and self-controlled," "without a trace of disorder" (Muir 1912:251–252), and compared the inundation of the canyon to the destruction of the garden of Eden:

Our magnificent National parks . . . Nature's sublime wonderlands . . . have always been subject to attack by despoiling gainseekers and mischief-makers of every degree from Satan to Senators. . . . Thus long ago a few enterprising merchants utilized the Jerusalem temple as a place of business instead of a place of prayer . . .; and earlier still the first forest reservation, including only one tree, was likewise despoiled.

Their arguments are curiously like those of the devil, devised for the destruction of the first garden – so much of the very best Eden fruit going to waste; so much of the best Tuolumne water and Tuolumne scenery going to waste. (Muir 1912:256,257,260)

The dam's proponents believed the energy of the water going over the falls was wasted and should be harnessed for the benefit of people. They argued that controlling the machinery of nature would enhance, not degrade, the value of the canyon. This viewpoint was summed up by Representative Ferris of Oklahoma, Chair of the House Public Lands Committee:

These patriotic earnest men believe it is a crime to clip a twig, turn over a rock or in any way interfere with Nature's task. I should be grieved if I thought practicality should completely drive out of me my love of nature in its crude form, but when it comes to weighing the highest conservation, on the one hand, of water for domestic use against the preservation of a rocky, craggy canyon, allowing 200,000 gallons of water daily to run idly to the sea, doing no one any good, there is nothing that will appeal to the thoughtful brain of a commonsense, practical man. (Quoted in Ise 1961:92)

The argument that people are entitled to use nature's resources was put even more emphatically by Representative Martin Dies of Texas, who stated "God Almighty has located the resources of this country in such a form as that His children will not use them in disproportion," and implied that to utilize them was to follow "the laws of God Almighty" (quoted in Ise 1961:92).

In 1913 Congress passed a bill authorizing the project, and the valley was dammed.

These two views of our relationship to nature might seem to be light years apart, but they have a lot in common. They are grounded in the same world view – the idea that people are separate from a natural world that tends toward a stable equilibrium. In one case people are superior to nature and entitled to manipulate it, dominate it, control it, use it, and improve upon it; in the other the natural world is pure and good, while people are morally tainted but long to be reunited with nature. In either case, we are outside of that which is in balance without us; humanity is either better than or worse than the non-human world. The primary difference between the two viewpoints is that utilitarian managers see themselves as manipulators of nature, while preservationists see themselves as nature's protectors.

The dam's proponents were utilitarian in their approach to resource management; they advocated the utilization of economically valuable natural resources. The dam's opponents took a preservationist stance; they argued that conservation should involve the protection of natural places from exploitation by people. These two positions exemplify two approaches to the management of resources.

It might at first seem odd that those who wished to dominate nature accepted the view that nature is in balance and people are outside of that balanced world. Yet if we return to the writings of George Perkins Marsh, who so clearly articulated the idea that nature is in balance, it becomes evident that this perspective is quite compatible with the idea that people are separate from nature and entitled to manipulate it. In Marsh's view, humanity was "of more exalted parentage" than "physical nature" and belonged "to a higher order of existence" (Marsh 1874:34). Consequently,

man [and domesticated animals and plants] . . . cannot subsist and rise to the full development of their higher properties, unless brute and unconscious nature be effectually combated, and, in a great degree, vanquished by human art. Hence, a certain measure of transformation of terrestrial surface, of suppression of natural, and stimulation of artificially modified productivity becomes necessary. (Marsh 1874:37)

Marsh felt that by changing nature, people had "effected . . . changes which . . . resemble the exercise of a creative power" (Marsh 1874:10,37). Although Marsh argued that civilization had gone too far in transforming the natural world, he saw no contradiction between the idea that nature is harmonious and the idea that people should manipulate nature's harmonies.

If we turn our attention outside North America, we can again find examples of utilitarian and preservationist management plans that are rooted in the balance-of-nature perspective. Kenya's Tsavo National Park was set aside as a preserve by colonial authorities in 1948. At the time of its creation, most of the park was densely vegetated with trees, and the premier attraction was its elephant and rhinoceros populations (Sheldrick 1973). When the park was formed, people who had lived in the area were evicted and prevented from using the land for hunting or grazing. Wildlife viewing became the only permitted land use.

Throughout most of the park's history, managers took a hands-off, let-nature-take-its-course approach, with the expectation that the park would continue to support trees, elephants, and rhinos. By the late 1950s, however, it had become clear that elephants were destroying the trees and preventing their regeneration, and widespread elephant mortality seemed imminent because of this habitat degradation. Wildlife researcher Ian Parker reported that "many visitors who saw the ravaged woodlands were appalled. The acres of dead and battered wood were likened repeatedly to the Somme battlefields of the First World War" (Parker and Amin 1983:71).

The park's management argued that such die-offs were part of a natural cycle, and they continued to follow a strategy of minimum intervention. Things got worse instead of better, however. The effects of habitat alteration were compounded by severe droughts in the 1960s and early 1970s, and as a result thousands of rhinos and elephants died (Sheldrick 1973). By 1973, grassy areas had replaced woody vegetation throughout the park. Elephants, rhinoceroses, and other wildlife species associated with trees had declined, whereas populations of grazing species such as zebras and gazelles had increased markedly. Ironically, the policy of eliminating people and letting nature take its course led to a dramatic alteration in the landscape and its wildlife, instead of perpetuating a stable community as managers had envisioned (Botkin 1990; Rogers 1999).

In South Africa's Krueger National Park, managers pursued a different strategy of preserve management. In a decidedly hands-on program, they intervened to control the balance of nature by culling lions, elephants, and ungulates; constructing deep wells and dams; burning vegetation; and controlling diseases. These actions were designed to maintain the habitats and species that were prevalent at the time the park was created. In other words, a highly manipulative management style was used to keep a "natural" area in a particular state. The connection between equilibrium thinking and this type of management is less obvious than in the Tsavo example, but it is equally strong. In fact, the balance-of-nature viewpoint was explicitly accepted in the proclamation setting aside the area in 1898 (Rogers 1999).

Tsavo and Krueger were managed in strikingly different ways, yet both these strategies are grounded in the idea that nature tends toward balance and stability. If both hands-on and hands-off management are rooted in the equilibrium viewpoint, one might ask if any other alternative is possible. But if we stop assuming that nature tends to be in balance, new possibilities emerge.

As a result of several high-profile controversies about resource management, most people are aware of the tension between preservationist and utilitarian approaches to resource management, even if they do not use those terms to describe the situation. Unfortunately, the popular media have presented the debate as if these were the only two alternatives, posing questions like: Do we want owls or jobs? In reality, this is not a helpful dichotomy. There is a third approach.

The third approach, which I call the sustainable-ecosystem approach, seeks to integrate resource preservation and use. Instead of focusing on products or preserves, this approach focuses on conserving the processes that sustain healthy ecosystems. It is grounded in a different view of nature, which is sometimes termed the flux-of-nature viewpoint. From this perspective, natural systems are often in a state of flux and people are an integral part of that flux.

In Chapter 1, we will see how unregulated exploitation set the stage for the development of utilitarian management of natural resources. First, however, let us consider how we use information to manage living natural resources.

References

Botkin, D. B. (1990). *Discordant Harmonies: A New Ecology for the Twenty-First Century.* New York, NY: Oxford University Press.

Ise, J. (1961). *Our National Park Policy: A Critical History.* Baltimore, MD: Johns Hopkins University Press.

Marsh, G. P. (1874). *The Earth as Modified by Human Action.* New York, NY: Scribner, Armstrong, and Co.

Muir, J. (1912). *The Yosemite.* New York, NY: Century Co.

Parker, I. and M. Amin (1983). *Ivory Crisis.* London: Chatto and Windus.

Rogers, K. H. (1999). Operationalizing ecology under a new paradigm: an African perspective. In *The Ecological Basis of Conservation: Heterogeneity, Ecosystems, and Biodiversity*, ed. S. T. A. Pickett, R. S. Ostfeld, M. Shachak, and G. E. Likens, pp. 60–77. New York, NY: Chapman and Hall.

Sheldrick, D. (1973). *The Tsavo Story.* London: Harvill Press.

Methodology: Getting the information we need to manage living natural resources

The scientific method

Resource managers need scientific information on which to base their decisions about conservation. The scientific method is a mode of inquiry in which testable propositions, termed hypotheses, are formulated and information is gathered to test them. The investigator makes predictions about what will happen under certain circumstances if a particular hypothesis is true and then determines whether or not those predictions are fulfilled. If the predictions are not fulfilled, the hypothesis is falsified.

Controlled experiments

There are many ways to test hypotheses. One is through a controlled experiment, in which a scientist compares a test group with a control group. Controlled experiments are not the only way to do science, however. The real world is more complex than the laboratory. It does not always lend itself to and sometimes it cannot tolerate experimental manipulation. I will return to this point below, but first, let us consider how controlled experiments can be used to provide the sorts of information resource managers need.

The process of hypothesis testing must begin, obviously, with the formulation of a hypothesis. The more we know about our subject, the more likely we are to come up with a plausible hypothesis. Reading what others have reported can help, but there is no substitute for the insight which comes from having observed the experimental system and becoming thoroughly familiar with it. This is one place where intuition plays an important role in scientific inquiry. Barbara McClintock, who made the revolutionary discovery of transposable genetic elements – or "jumping genes" – in maize, says that it is essential to have "a feeling for the organism," to understand "how it grows, understand its parts, understand when something is going wrong with it" (Keller 1983:198).

Suppose you are conducting research on the nutritional requirements of white-tailed deer, and you want to find out whether the amount of weight the deer gain depends upon dietary protein. You might begin by reading the accounts of some early naturalists. If you found a description written by a nineteenth-century rancher stating that mule deer at a particular site with "high-quality forage" (food plants) were in good physical condition, would that prove that protein is what controls weight gain in white-tailed deer? Not really, for several reasons. First, the rancher's observations pertain to mule deer, not white-tailed deer. Second, the observer presented no data on either weight gains among deer or what exactly was meant by "high-quality forage." Third, you have no way of checking the accuracy of this reported observation. Fourth, there is no indication of the sample size represented by this observation, so even if events were recorded accurately, the weight difference might just be a fluke stemming from something unusual about those particular individuals, or the difference might be so slight that it really is not meaningful. Fifth, you have only this one "study" on which to base your conclusions.

Although this type of anecdotal account cannot provide conclusive evidence, it can nevertheless suggest areas of fruitful inquiry. The hypothetical description referred to above would suggest that there *might* be a relationship between food quality (and, perhaps, protein content) and weight gain in mule deer (and, perhaps, in white-tailed deer as well). So you decide to design a controlled experiment to find out if this is the case.

You should begin by stating a hypothesis. In order to facilitate statistical analyses of your results (see below), your hypothesis should be framed as a statement of no difference, which is termed a null hypothesis. This is the proper procedure even if you think there will be a difference. If your intuition is correct and there is a difference, then the hypothesis of no difference will be falsified. This could be stated as the following hypothesis: *The amount of protein in the diet of white-tailed deer has no effect on weight gain.*

To do a controlled experiment, you compare one or more groups that receive a particular treatment to a control, a group that does not receive the treatment. If you have two similar groups of deer, you could weigh the deer in each group and then feed a high-protein diet (Diet A) to one group (Group A) and a diet with normal levels of protein (Diet B) to the other (Group B)

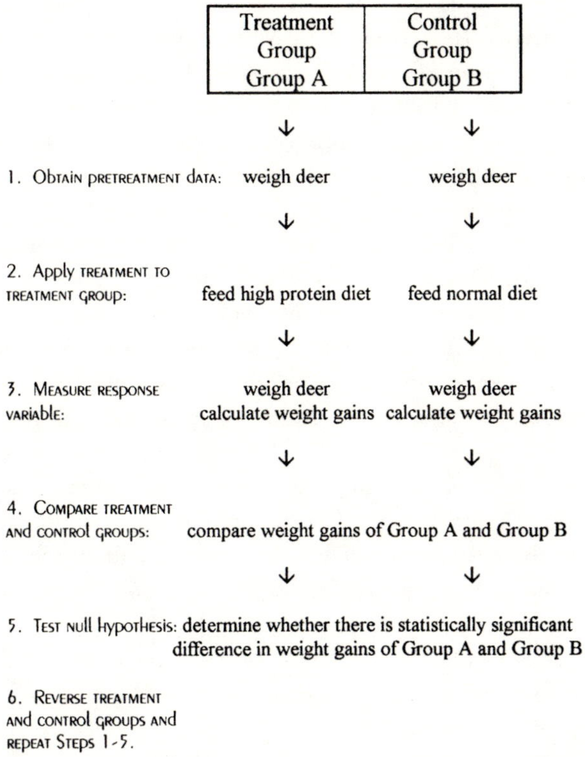

	Treatment Group Group A	Control Group Group B

↓   ↓

1. Obtain pretreatment data: weigh deer   weigh deer

↓   ↓

2. Apply treatment to
treatment group: feed high protein diet   feed normal diet

↓   ↓

3. Measure response   weigh deer   weigh deer
variable: calculate weight gains calculate weight gains

↓   ↓

4. Compare treatment
and control groups: compare weight gains of Group A and Group B

↓   ↓

5. Test null hypothesis: determine whether there is statistically significant
                   difference in weight gains of Group A and Group B

6. Reverse treatment
and control groups and
repeat Steps 1-5.

Figure F.1. Procedure for an experiment designed to test the hypothesis that the amount of weight gained by white-tailed deer depends on the amount of protein in their diet.

(Figure F.1). In this experiment, protein content is the treatment. The experiment is set up to test whether the treatment is responsible for changes in a response variable. Group A is the test group, which receives the treatment, and Group B is the control group, which serves as a standard for comparison with the treatment group. The use of a control group allows the investigator to assess the effects of the treatment. Weight gain is the response variable, that is, the variable you are testing to see if its value depends upon the treatment. After a specified amount of time, you can weigh the deer in each group. If the deer in Group A have gained significantly more weight than the deer in Group B, this is evidence that Diet A caused the deer to gain more weight than Diet B. (I will elaborate on what is meant by the term "significantly" below.)

Ideally, the treatment and control should be applied to more than one group of experimental subjects. This is termed replication.

Even if you have a large sample size and a carefully controlled study, you should reverse your treatments after a while, if possible, so that the group that had been receiving Diet A is given Diet B and vice versa. If you *still* find that the group receiving the high-protein diet gains more weight than the group on the control diet, then you have an even stronger case. Although you should try to insure that Groups A and B are identical at the start of your study, it is always possible that there are some differences you are unaware of between your treatment and control groups. The reversal of treatment and control groups serves as a safeguard against this possibility. If there are undetected differences between Groups A and B that are influencing the results of your study, this should show up when you switch the treatment group and the control group. Finally, after performing the experiment with the treatments reversed, you can strengthen your case even further by repeating the entire set of experiments.

It is important to realize that the experimental design assumes that the *only* difference between Group A and Group B is the amount of protein they receive. This amounts to three requirements for your experiment: (1) initially the two groups of deer should be the same in all respects (such as weight, area of origin, age, sex ratio, and any other variables you can think of), (2) the two groups should be kept under identical conditions except for the different diets they receive, and (3) the only difference in the two diets should be the difference in the treatment, protein content.

If the deer in the group that receives Diet A are initially different from the deer that receive Diet B, then any difference in weight gain that you observe might not be due to the amount of protein in their diets after all. For instance, if one group contains all males and the other group is all females, then you will not be able to attribute differences in weight gain solely to diet; they might be due to the sex difference between the two groups. Similarly, if the two groups differ in age structure or genetic constitution or in any other ways, your conclusion will be suspect. In other words, if you want the results of your experiment to be convincing, you must have properly designed controls, so that the treatment and control groups really are identical in all respects before the treatment is applied. If that is the case, then any observed differences after the treatment can be assumed to be due to the treatment itself.

In addition, if the two groups are not maintained under identical conditions or if the diets differ in fat content or some other variable as well as protein content, then problems might arise in interpreting the results. Again, any differences in weight gain observed at the end of the experiment might not be caused by the amount of dietary protein; instead, other factors could have caused the observed differences.

You must also be sure that you have accurately assessed any changes in the response variable. The deer must be weighed accurately, and their weights must be recorded correctly. If you use different techniques, different observers, or different scales to weigh the individuals in Groups A and B, then you may introduce sources of error into your experiment. Bias, a consistent tendency to deviate in one direction from a true value, could be introduced by using an inaccurate scale, or it might result if an observer allows his or her perceptions to be influenced by preconceived notions.

Even honest scientists are potentially influenced by their expectations; therefore, the results of your study are more likely to be objective if the investigators do not even know which group a deer is in when they weigh it. This precaution should guard against any tendency to unconsciously inflate the weights of deer in the high-protein group. (For an example of how observers' biases influenced the data they reported, see Chapter 2.) Although it is unethical to bias one's results deliberately, it is important to realize that no one is completely free from bias either. What we perceive and how we interpret our perceptions are colored by our past experiences and cultural context.

You will be in a better position to test your hypothesis if you have a large number of individuals in your treatment and control groups. If you have only two deer in each category, the results of your study will be much less persuasive than if each group contains 20 or 100 or 1000 deer. This is because with small groups your results may be unduly influenced by idiosyncrasies or unusual circumstances affecting one or a few individuals.

The use of statistics allows us to evaluate rigorously any differences between treatment and control groups. The term "significant" has a specific meaning in the discipline of statistics. To use statistics, we test a null hypothesis, which states that there is no difference between the treatment and the control groups in the response variable being tested. If there is a statistically significant difference in the average value of the response variable in the treatment and the control groups, then the null hypothesis of no difference is rejected and the alternative hypothesis, that there is a difference, is supported.

A statistically significant difference in the average amount of weight gained by the deer in Groups A and B means that there is a low probability (which is stated as a specified "P level") that the differences we observe are due to chance alone rather than to the different treatments. In other words, if the only difference between the two groups of deer is the amount of protein they receive, a significant difference in weight gain means it is probable that the observed difference was caused by the difference in dietary protein. (Sample size is taken into consideration when significance levels are calculated, to account for the greater likelihood that differences in small groups are due to chance alone.) Statistics thus make it possible for one to judge whether or not observed differences between treatment and control groups are likely to be due to the treatment used in an experiment.

If you have done all these things, then you are in a good position to answer the question you originally asked. If you repeatedly perform a rigorous experiment – with adequate controls, sample sizes, and replication as well as unbiased methods of collecting and recording data – and you observe a statistically significant difference each time you do the experiment, then you can feel reasonably confident in your results. On the other hand, if the results are not what you expect then you should consider alternative explanations.

As you can see, there are a number of pitfalls that must be avoided in an investigation such as this. Nevertheless, well-designed controlled experiments can shed light on many important questions. They are particularly useful when questions of policy (What level of pesticide application should be permitted? Should predators be killed? Should naturally started fires be put out?) are at stake, because in matters where we need to evaluate alternative courses of action, this methodology defines a standard for objective evidence.

Controlled laboratory experiments are not the only valid way to test hypotheses, but they do appeal to our desire to reduce the workings of the world to a small number of understandable variables that can be explained by a scientist "in a white coat twirling dials in a laboratory – experiment, quantification, repetition, prediction, and restriction of complexity to a few variables that can be controlled and manipulated" (Gould 1989:277). In western culture, the respect that scientists enjoy and the premium placed on "objectivity" mean that poorly done experiments may be given more weight than they deserve, however. For this reason, it is extremely important that findings that are presented as resulting from controlled experiments be carefully examined and evaluated. I shall return to this point below.

So far, we have been discussing laboratory experiments, but controlled experiments can also be used to test hypotheses in the field. The natural world is more complicated than the laboratory, however, and studying it presents some daunting logistic challenges for the experimenter.

Suppose you want to find out whether predation limits densities of ring-necked pheasants in a given area. You might select several study sites, divide each into two plots, A and B, remove all predators from the A plots, leave predators on the B plots, and then measure and compare pheasant densities on the sites with and without predation. In this experimental design, the A plots are the treatment group, the B plots are the control group, predator removal is the treatment, and pheasant density is the response variable. Of course, you will also need to prevent predators from re-entering the experimental sites. After you have completed this portion of your experiment, you should follow the same procedure described for the deer diet experiment: reverse the treatment and control plots, so that predators are removed from the B plots but not from the A plots. (Be sure to allow enough time for predators to recolonize the A plots, from which they were removed in the initial phase of the experiment.)

This is a good plan for your field experiment, but you will run into challenges not encountered in the laboratory. In field studies it is very difficult to be sure that the only difference between the control group and the experimental group is the treatment; it is often difficult to obtain even moderate sample sizes; and the measurement of variables is not always straightforward. Furthermore, because field studies are expensive and time-consuming, it may be impossible to reverse the treatment and control groups if the treatment has irreversible effects, and replication is not always possible.

In laboratory studies, investigators can use genetically identical strains of mice or other experimental subjects to lessen the chance that observed differences in results will be due to differences within treatment groups or between the treatment groups and the control group. If this is not possible, one way to minimize the problem of variability is to have a very large sample size. This is often done in medical studies, which can sometimes involve hundreds of thousands of subjects. The larger the group, the more likely it is that individual differences between people in the groups will cancel each other out. Usually neither of these options – genetically uniform populations or huge samples – is practical in studies of wild organisms, though.

Natural variability presents another problem: the lack of uniformity in the test environment. When you divide each study site into experimental and control plots, you cannot be sure that the two halves are identical unless you start with study sites that are absolutely homogeneous. This is virtually impossible in a natural setting; it is far more likely that the two sections of each study site will differ in topography, slope, soil fertility or depth, type of bedrock, moisture regime, vegetation, or some other factors.

One thing you can do to address some of these problems is obtain pretreatment data from all your study sites *before* the experimental treatment begins. You are then in a position to compare these data on pheasant density before predator removal with the data you obtain after you begin the treatment. The pretreatment data enable you to assess differences between the treatment and control groups. By doing this you are, in effect, adding another control – the pretreatment data – to use as a standard in assessing the effects of the treatment. The use of before-and-after data allows you to compare the magnitude of the *change* in pheasant densities on the treatment and control plots. This is a more meaningful comparison than a comparison of absolute values. Suppose that in the initial stages of your study, pheasant density is 15 pheasants/km² on the A plots and 20 pheasants/km² on the B plots, but after predator removal both the A and the B plots average 20 pheasants/km². If you had no data on densities before predator removal, you would conclude that predators had not affected pheasant density, but when you compare densities before and after removal you find that productivity on the A plots, from which predators were removed, has actually increased. (Recall that in your study of weight gain among deer you used a similar procedure. You incorporated pre-treatment data by weighing the deer before you began administering the diets and subsequently calculating weight gain – the difference between initial and final weights – for each individual.)

In your laboratory study of weight gain in deer, it was fairly easy to measure changes in the response variable, weight, if you had a good scale and careful observers, but the measurement of pheasant density is more complex. To calculate the density of a population, you need to know the number of individuals per unit of area. The most thorough way to determine population size is to count all the pheasants in your plots; however, in studies of wild organisms it is rarely possible to count every individual in a population. Consequently, you may need to estimate the population, that is, to make an approximation on the basis of a sample, and you will need to choose from among several possible sampling methods – each with particular advantages and disadvantages – to do this. In some instances, you may not even be able to observe individuals directly. Instead, you may have to rely on counting signs of animal activity, such as calls, feces, nests, or burrows. If you know how many individuals produce a given number of signs, then you can convert your counts to a population estimate. Obviously, these difficulties present additional opportunities for errors to be introduced into your experiment.

Sometimes when we cannot answer a question through field experiments, we can get part of the answer we want by doing laboratory experiments and applying our results to field conditions. For instance, populations of the American black duck have been declining in eastern North America in recent decades. Suppose that you want to find out if this decline has been caused by acidification of the ponds on which black ducks breed. There are substantial obstacles to doing the sort of controlled field experiment that might definitively answer this question. You would not want to acidify a large number of ponds and monitor black duck production on them in comparison to untreated control lakes, because this would involve substantial risk to the treated habitats. Even if two groups of very similar, pristine lakes were available, most scientists would be reluctant to acidify one group deliberately. On the other hand, you could create artificial ponds in which to do such an experiment. You might treat some of the ponds with acid and keep others as untreated controls. Then you could use the results of these "laboratory" experiments to interpret conditions in nature (Luoma 1987).

The principal advantage of this approach is that, in contrast to field studies, you have the ability to regulate many variables in your experiment. You should be able to create identical ponds that differ only in acidity and maintain them at identical conditions. On the other hand, the experimental ponds will differ in numerous ways from natural conditions, and you will have to take this into consideration when you interpret and report your results. Critics of your study are likely to claim that your results do not apply to the wild because of factors you did not take into consideration.

To summarize, both laboratory experiments and field experiments can shed light on the workings of the natural world, provided that: (1) they are carefully designed and executed so that there are adequate controls, sample sizes, and replication, (2) accurate estimates of the relevant variables can be obtained, and (3) the data are objectively recorded. But, there are many examples where we cannot carry out controlled experiments to answer the questions we would like to answer. In the next section we will consider some other ways of testing hypotheses about the world around us.

Comparative studies

Controlled experiments are not the only way to do science, however. The real world is more complex and less predictable than the laboratory, and it does not always lend itself to experimental manipulation. That is part of its appeal, but because of this quality, natural phenomena do not always lend themselves to controlled studies. There are many situations where experiments are either not feasible or not desirable. This is particularly true when we are studying rare or sensitive species and habitats or past environments.

For instance, the ozone layer in the earth's atmosphere reduces the amount of harmful ultraviolet radiation entering the atmosphere. This layer has become thinner in recent decades, and it has been suggested that this thinning of the ozone layer is responsible for a variety of ecological changes ranging from altered food chains to cataracts and blindness in some species. Suppose that we wanted to design a controlled experiment to find out the effects of ozone thinning on wildlife populations. Using the model of the controlled experiment described above, we would need to have a large number of identical planets, divide them into two groups, reduce the ozone layer in one group, and monitor the results. Clearly this is preposterous. It is totally impractical to carry out an experiment on this scale, and even if it were possible it would not be ethical because it would involve deliberately risking serious and irreversible alterations to the experimental planets.

There are many other examples of situations in which it is not appropriate or feasible to carry out rigorously controlled experiments to answer questions in resource management. If we want to know the effects of potentially harmful treatments – such as pesticides, oil spills, or acid rain – on populations of wild organisms, there are substantial practical and ethical obstacles in the way of doing large-scale, controlled experiments like the ones we described above.

Thus, the problems facing ecologists using controlled experiments to study the natural world are sometimes daunting, but fortunately there are other ways of getting information about natural systems. One approach is the comparative study, in which conditions are compared in two or more situations that differ in place, time, or another variable but are alike in many other respects. For instance, we might look at similar events among closely related organisms or in similar habitats or in the same place at different times. In the discussion of the importance of pretreatment data in field experiments, I emphasized the importance of comparing conditions before and after the application of the treatment. The same approach is often used to assess the effects of inadvertent environmental perturbations, such as the thinning of the ozone layer. To do this, scientists analyze data on the composition of the atmosphere over a period of time. If a pronounced change in the value of a variable, such as a decrease in the amount of ozone, is observed after a certain date, investigators search for events that preceded and might have caused the change. Again, a profound understanding of the system plays an important role in suggesting what might be responsible for the observed changes.

This approach can shed some light on what might have caused thinning of the ozone layer, but it has some limitations. First of all, in most studies of this type, the early data were not gathered in the same way as the more recent data. Second, the difference in the thickness of the ozone layer is not likely to be the only variable that changed during the period of interest, and this will complicate the interpretation of the data. Third, there are usually time-lags between a cause and its effect, but most often we don't know how long those lags are. Whatever caused ozone thinning might have begun changing a few years or a hundred years before the resulting change in atmospheric composition was noticed. Finally, we have a sample size of one. There is only one planet earth, and the thinning of its ozone layer is a singular event.

Often we do not foresee the consequences of our actions, so we do not plan before-and-after comparisons ahead of time. If we had known one hundred years ago that the ozone layer might wane, we could have tried to gather data on conditions before this phenomenon began (although we wouldn't have had the technology to do this very well) and compared it to data we gathered subsequently. (On the other hand, if we had foreseen the thinning of the ozone so far in advance, perhaps we would have taken steps to prevent it or at least to slow its course.)

So, we are often left scrambling to conduct the first phase of these unplanned comparative studies. To do this, we may scour historical records and earlier studies to glean information about prior conditions. This kind of information is very useful, and it underscores the importance of keeping accurate records, because you never know what use data will be put to in the future. But frequently this type of information is gathered using methods that differ from the ones we would choose if we designed our study from scratch. We might have to use data from many different sources, or data that were gathered using different methods and by workers with different degrees of expertise and training. Furthermore, since this type of information was rarely compiled with the question that concerns us in mind, in many cases the relevant information was not recorded. Consequently, we are left having to make inferences from scraps of information, rather like a detective trying to solve a case with only a few clues.

Like controlled experiments, the results of comparative studies can be used to test hypotheses. We can state hypotheses, make predictions derived from our hypotheses, and then look at evidence from the past to find out if our predictions are correct. Care must be taken, however, in interpreting the results of this type of study. When trying to disentangle cause and effect in the past, we often face situations where several variables changed simultaneously. *Correlation* (the association of variables with each other) does not equal *causation*. Comparative studies can identify certain factors that occur together in time, and statistical techniques can be used to evaluate the significance of these associations (that is, the likelihood that they are due solely to chance), but this does not prove a causal relationship between them. There may be other factors that changed at the same time and that actually caused the effects we are interested in. (In Chapters 2 and 10 we will encounter examples of before-and-after studies in which it was difficult to determine causation.)

As in the case of the ozone example, ecologists often look to the past to get information that will help them understand the present or plan for the future. If we want to know how much the climate varied in the last 2 million years, how often forest fires occurred in California before policies of fire suppression were instituted, where wetlands used to occur in China, what was the former geographic range of the red kite, or what the extinction rate of native Australian mammals was before Europeans arrived, we must turn to the past for answers. The principal sources of information about the past are discussed briefly in the Box below.

Sources of information about the past

We learn about the past by consulting natural records and historical documents. These are described in more detail below. Additional information can be found in Swetnam *et al.* (1999), Norris (2000), and U.S. Army Corps of Engineers (2001).

Natural records

Natural records include fossils and artifacts preserved in sedimentary rocks, ice, soils, or sediments; packrat middens (piles of accumulated objects); and the tissues of long-lived individuals. Processes that occur in pulses often produce layered records that are especially useful. Examples include periodically deposited sediments as well as rings or layers that result from variations in the growth rates of wood, bone, or coral. In living tissues, this phenomenon occurs where alternating cold and warm seasons produce marked differences in seasonal growth rates. For instance, in temperate climates trees produce distinct annual rings, hibernating mammals deposit bone in annual rings, and fish have layered scales. The position of material in a sequence of layers provides information on its relative age. By combining inferences about the relative age of deposited materials and the conditions present at the time of deposition, we can reconstruct a sequence of past environments. By comparing growth rings from trees with overlapping life spans, scientists can date tree rings over periods that are longer than the life span of an individual tree. In this way, chronologies can be constructed that cover thousands of years. Within that time period, it is possible to determine the exact year when an event such as a fire occurred.

Plants and animals are useful indicators of environmental conditions because every species has a specific range of environmental conditions it can tolerate (see Chapter 3). The presence of muskrats or cattails in the past indicates that surface water was present, and cacti indicate a hot, dry environment. (This only works if we find them where they lived, not if they were transported somewhere else after they died.)

Natural records are a valuable source of information about the past, but it must be kept in mind that some materials are more likely to be preserved than others. In other words, the samples passed down to us by natural records are biased. For example, packrat middens are found only in rocky terrain. In addition, the record of the past that natural processes provide is often too short or too fragmented to tell us what we want to know. Or the record may be extensive but not provide information for the places and time periods we are interested in.

The time-span covered by natural records ranges from years to millennia. For some kinds of natural records that cover long time-spans, however, the resolution is fairly poor. Fossils can give us information about what was going on thousands of years ago, but we cannot distinguish a given year or even decade within the fossil record. On the other hand, tree-ring chronologies allow us to pinpoint when an event occurred quite precisely.

Historical documents and oral traditions

This category includes written and oral records, such as drawings made by surveyors, settlers, explorers, naturalists, ethnographers, etc.; tabular data; photographs; genealogies; oral histories; and maps. Documents are an inexpensive, easy-to-use source of information about the past; however, the value of historical documents depends in part on whether the information that is preserved is a good sample, that is, whether it is representative of the past. Historical documents provide valuable windows to the past, but the viewpoint of the observer must be taken into consideration. The decision about what to record is always subjective, and historical documents reflect the recorder's biases about what was important. The usefulness of historical documents also depends on the accuracy of the recorded information, which in turn depends on the observational skills, memory, meticulousness, and honesty of the person who recorded the information and also on the technical capabilities of the cameras or other equipment used.

The time-span covered by historical documents is relatively short (usually decades or centuries), but such documents often allow us to pinpoint when events occurred to the nearest day, week, or month.

Simulations and models

Scientists often seek to understand the behavior of systems under conditions that cannot be observed directly. This may be because the system is too small (an atom) or too large (the earth's atmosphere) to observe directly or because the phenomena of concern took place in the past. Clearly, controlled experiments are of little use where we are dealing with phenomena that we cannot observe. In such situations, scientists sometimes construct models. A model is a concrete or abstract representation of a system that can be used to predict how the system behaves under specified conditions. A scientific model may take many forms – a physical structure, a description, an equation, an analogy, or a theoretical projection.

A simulation is a type of model that predicts the changes a system undergoes given certain starting conditions and assumptions. Computers are very useful for this type of modeling because they allow researchers to manipulate the values of variables in their models and to perform calculations rapidly under a wide range of scenarios.

Models are used a great deal in the management of natural resources. In cases where we want to predict what effects a proposed policy will have on habitats or populations, models are extremely useful. If we introduce six wolves into an area of suitable habitat, what size will the wolf population attain in 20 years? How long will the world's tropical forests last if we continue clearing them at current rates? What will the average summer temperature be in London in AD 2050 if we continue producing greenhouse gases at current rates? If we spray a marsh with an insecticide, what concentration will keep damage to fish populations at an acceptable level? How long will a marine community take to recover from an oil spill? We cannot answer these questions directly through experimental studies, but we can measure responses under certain conditions and use this information to predict the behavior of a system under other conditions.

Although models usually represent systems that are not amenable to experimentation, experiments may be useful for examining how certain parts of a system work. For example, you might wish to conduct experimental tests of a variety of treatments for cleaning up spilled oil. The information obtained in this way could then be used to modify or refine your model of recovery time. A model's predictions should be tested against reality repeatedly, and the information generated in this way should be used to modify the model in order to make it more realistic.

Models are particularly useful where the risks of doing experimental studies are unacceptable, as in the case of research on rare organisms. Field studies

inevitably involve a degree of disturbance to wild populations, while laboratory experiments require the removal of some individuals from the wild. Both
these outcomes should be avoided when dealing with highly sensitive populations. Models are one way to avoid these negative impacts. Sometimes the
urgency of a situation is used as a justification for basing management decisions on models rather than data, however. While it is true that models can be
useful in situations where we do not have time to gather data, we should never
confuse models and data and never neglect the testing of models. (See
Chapter 8 for an example where a prediction stemming from a model was
treated as finding rather than as hypothesis.) In the words of Graeme
Caughley, "We are faced not by a shortage of models but by a shortage of the
data needed to adjudicate between them" (Caughley 1985:13).

Simulations and other models have their limitations. A model always incorporates certain assumptions about how a system behaves. One reads a great
deal these days about debates over models that predict global changes in environmental conditions, population growth, and resource availability. Much of
the debate focuses on differences in assumptions about how the system in
question behaves.

Furthermore, models necessarily oversimplify the behavior of the systems
they portray. A model that incorporates most of the important factors influencing a system and contains realistic assumptions about how the system
changes will do a good job of predicting that system's behavior. A very oversimplified model with unrealistic assumptions will not.

Experiments, models, and comparative studies are not mutually exclusive
approaches to finding out about nature. They can be used to complement one
another, each suggesting fruitful areas of inquiry that can be pursued using
other tools. Regardless of which tools the scientist uses, however, evidence
that is presented as "scientific" should be evaluated carefully.

Evaluating evidence with a critical eye

Advertisers and the popular press often distort the results of scientific
research. Statements that you will live longer if you eat more of Product Y are
completely insupportable. At best, one might be able to claim that if you eat
more of Product Y you will increase the probability that you will live a long
life (but you still might get hit by a truck tomorrow, and if you have a certain
genetic constitution, Product Y may do you no good whatsoever).
Furthermore, correlation is often mistakenly reported as causation. If a study
finds that people who exercise more have fewer heart attacks, that only tells

us that those two variables – exercise frequency and heart attack rate – tend to change in a parallel fashion. It tells us nothing about whether one variable is affecting the other, and if so, which is the cause and which is the effect: does exercise lessen one's chance of having a heart attack, or does heart disease cause one to exercise less?

Because of these problems, it is very important to evaluate carefully all claims of scientific findings. I have already identified a number of pitfalls to watch out for, such as poorly controlled studies, anecdotal evidence that is presented as conclusive proof, data that are subjectively interpreted, and studies that confuse correlation with causation. Another problem arises when unsubstantiated claims are published and then cited without being examined critically. We are all aware that the popular media sometimes sensationalize news in order to generate interest. Claims made by scientists are exaggerated or taken out of context, without regard to a host of qualifications that were specified in the original study. Sometimes an informed reader can spot this type of sensationalism, sometimes not. This is bad enough in itself, but when the reported "findings" are accepted as fact and then repeated, the problem is compounded.

Professional publications are not immune to this problem. Journal articles and texts sometimes repeat an unsubstantiated conclusion, or the findings of a poorly designed study, or a theoretical prediction, as if it were a demonstrated fact. Once the "finding" has been cited in this way, the process tends to be self-perpetuating, with additional papers citing the supposed finding until it works its way into the scientific literature and acquires a sort of authority of its own. We will see an example of this in Chapter 2.

The point here is that we must not accept everything that is presented as scientific evidence without questioning it. Whenever you encounter material that is presented as scientific evidence, you should ask yourself the following questions:

1. How was this evidence obtained? Was sound methodology used? If the evidence comes from a scientific experiment, were there good controls? Was the sample size adequate? Were all relevant factors considered? What types of biases might have influenced the collection and reporting of data?

2. Who is the author? Is the author well qualified to discuss this topic? Does he or she have any professional credentials? (Don't make the opposite mistake, of being overly influenced by whether or not a person has credentials or fame, either. It is unwise to assume that someone who is well known should always be believed or that an unknown person is never reli-

able.) Is the author unusually biased? Does the author have a strong interest in advocating a particular measure or in defending a certain school of thought? Is the author associated with a group that is defending or attacking a particular position? If the evidence comes from historical documents, what was the perspective of the author? What interests and biases might have colored the information that was recorded?

3. Where is this evidence presented? Is it in a professional journal? If not, is it in a publication that is trying to push a particular viewpoint or perhaps to sensationalize news? (Remember that information on the World Wide Web is not subject to any review process.)

4. When was this evidence presented? Is it out of date? Has new information come to light that might elucidate the subject or suggest other possible interpretations?

5. How well are the conclusions supported by the data? Does the article contain contradictions? Do the author's arguments make sense? Are they logical and consistent?

6. How is the material presented? Are the data presented in a misleading way? Does the author use emotional language to try to influence the reader? Is the article well documented? (To determine this, you will need to look at the list of references and evaluate whether they are likely to be reliable sources. It is a good idea to consult some of the references that are cited, so that you can evaluate them and can determine whether they were used in a misleading fashion.)

7. Is the evidence consistent with what you know? Here you will need to use your knowledge to evaluate the claims that are being made. If the evidence is not consistent with what you know, you may be able to identify some problems with the study's design or with the author's reasoning. On the other hand, if you cannot identify any problems with the evidence or interpretations presented, and they contradict what you have already learned, then it is probably time to re-examine your views and perhaps to modify them. Don't be afraid to let new evidence challenge you to discard your own preconceptions and to throw out old dogmas.

The last point is crucial. I have emphasized hypothesis testing as a means of obtaining information on which to base policy decisions. But, this method has another advantage. Because it provides a framework within which we can critically re-examine our own views, testing them against evidence and modifying or discarding them when appropriate, the scientific method can, if we are willing to let it do so, teach us a process we can use for our personal as well as professional growth. Although the scientific method has its limitations, it does have built-in safeguards against its own excesses. Barbara McClintock's biographer Evelyn Fox Keller put it this way: "However severely communication between science and nature may be impeded by the preconceptions of a particular time, some channels always remain open; and, through them, nature finds ways of reasserting itself" (Keller 1983:197).

References

Caughley, G. (1985). Harvesting of wildlife: past, present, and future. In *Game Harvest Management*, ed. S. L. Beasom and S. F. Roberson, pp. 3–14. Kingsville, TX: Caesar Kleberg Wildlife Research Institute.

Gould, S. J. (1989). *Wonderful Life: The Burgess Shale and the Nature of History*. New York, NY: W. W. Norton.

Keller, E. F. (1983). *A Feeling for the Organism: The Life and Work of Barbara McClintock*. New York, NY: W. H. Freeman.

Luoma, J. R. (1987). Black duck decline: an acid rain link. *Audubon* **89**(3):18–24.

Norris, S. (2000). Reading between the lines. *BioScience* **50**:389–394.

Swetnam, T. W., C. D. Allen, and J. L. Betancourt (1999). Applied historical ecology: using the past to manage for the future. *Ecological Applications* **9**:1189–1206.

U.S. Army Corps of Engineers (2001). Ecosystem Management and Restoration Information System CD. Vicksburg, MS: U.S. Army Corps of Engineers.

The figure opposite illustrates some of the features of landscapes that have been managed with a strictly utilitarian approach to conservation. The landscape has been greatly simplified by dispersed clearcuts, a tree plantation with a single-aged stand of one species of trees, and monolithic croplands. Remnants of forest are present only in small, isolated fragments with lots of edge. The woodlot in the foreground has been clearcut. The stream channel has been straightened, and there are no trees or shrubs along its banks. These actions have created a relatively homogenous landscape.

PART ONE

Management to maximize production of featured species – a utilitarian approach to conservation

1

Historical context – the commodification of resources and the foundations of utilitarian resource management

When we conserve biological, that is living, natural resources such as wild plants and animals, we make decisions about their use, management, or protection in order to prevent their depletion and insure that they will continue to be around in the future. In this chapter we will look at the historical conditions that created a need for a formal approach to natural resource conservation in the western world and its colonies. We will see how the disciplines of wildlife management, forestry, and range management arose in response to threats to living natural resources that followed the commodification of resources in Europe, the Americas, Africa, Asia, and the Australian region. These disciplines are utilitarian in their approach; they focus on the exploitation of economically valuable species. The word "exploitation" has several connotations. To exploit a resource is to use it, but the term often carries an implication of excessive use, unfair use, or use without appropriate compensation. In this book, the term exploitation is meant to be synonymous with utilization, without a connotation of exorbitant or inappropriate use. We will, however, see many cases of unregulated or excessive exploitation that resulted in resource depletion or environmental degradation. All societies, of course, use living natural resources, and thus they are utilitarian in their approach to those resources (although this is not necessarily the only way that they relate to the natural world). Chapters 2 and 3 present some of the central concepts of the utilitarian approach, and Chapters 4 to 6 examine the principal techniques for accomplishing its goals.

1.1   Historical background

All human societies use living natural resources. Throughout history, people have altered their environments by hunting, fishing, foraging, cultivating, raising livestock, burning, and moving water (Thomas 1956; Pyne 1982). Many rulers of ancient and medieval societies issued decrees regulating the use of wild plants and animals (Leopold 1933). Sometimes these are considered the earliest "conservation" measures (Alison 1981), but that designation assumes that societies based on hunting and gathering, shifting agriculture, or nomadic pastoralism were incapable of regulating resource use. It ignores arrangements regulating the use of communally owned resources, as well as taboos or other traditional restrictions on what, when, where, and how wild plants and animals can be used (see Chapter 12). The restrictions on resource use that were promulgated by kings, emperors, and czars, on the other hand, were often aimed primarily at preventing peasants from using resources that belonged to the ruling classes, rather than at conservation *per se*.

The expansion of western Europe into other parts of the world resulted in profound changes in ecological and economic conditions, and these changes created a need for conservation measures beyond those that already existed in the colonized societies. Beginning in the sixteenth century, western Europeans colonized North and South America, Africa, South Asia, and the Australian region. This expansion was followed by marked ecological changes associated with increased rates of resource exploitation (particularly minerals, timber, meat, furs, and in Africa ivory), widespread changes in land use, and the arrival of numerous exotic species of microorganisms, plants, and animals.

Europeans encountered many different cultures in the lands that they colonized, and these cultures had diverse forms of social and economic organization. Some made their living by hunting and gathering, others were farmers or pastoralists or practiced a combination of these modes of resource utilization. Because it is risky to generalize about such variety, we will look at one example where the changes following colonization have been documented in some detail.

Historian William Cronon argues that the changes that took place after Europeans arrived in New England resulted from differences in Native American and colonial concepts of land ownership (Cronon 1983). The sixteenth century in England had witnessed the rise of a country gentry, a class that regarded land as a capital investment. To obtain a return on that investment, it was necessary to improve the land, and the most important way of doing that was by enclosing common lands. This trend led to a growing privatization and commercialization of the English landscape (Ingrouille 1995). Colonists emigrating from England in the seventeenth century brought with them this conception of land ownership. They believed that development of the land's resources legitimized ownership. Land owned in this way could be defined by abstract boundaries and traded as a commodity.

In contrast, the Indians of southern New England exploited plants and animals by moving around to follow patterns of seasonal abundance. They planted crops and fished spawning runs in spring; gathered seafood in summer; and fished, hunted, and harvested crops in autumn. Their concepts of property rights differed radically from the colonists' ideas on the subject. Indian villages had collective sovereignty to the territory they used throughout the year. In addition, families owned the crops they produced as well as the products they gathered from the land, and they also had the right to set snares or traps in specific hunting territories (Cronon 1983). Although the Indians recognized these different forms of property rights to land, their concepts of ownership did not allow land to be treated as a commodity.

These indigenous forms of land ownership were unfamiliar to the colonists. In fact, except for agricultural land, the colonists did not recognize Indian forms of ownership as legitimate. Their position was based on the view that land ownership entailed an obligation to "improve" lands. In the colonists' eyes, lands from which wild products were harvested was unimproved and therefore the Indians did not own such lands. (Europeans brought the same perspective to other lands they colonized. For example, colonial administrators in Africa did not recognize Maasai claims to their land because by European standards nomadic pastoralism did not improve the land (Collett 1987).)

Cronon suggests that the concept of land as a commodity played a crucial role in the ecological changes that occurred in the wake of European settlement:

To the abstraction of legal boundaries was added the abstraction of price, a measurement of property's value assessed on a unitary scale. More than anything else, it was the treatment of land and property as commodities traded at market that distinguished English conceptions of ownership from Indian ones. . . . It was the attachment of property in land to a marketplace, and the accumulation of its value in a society with institutionalized ways of recognizing abstract wealth . . . that committed the English in New England to an expanding economy that was ecologically transformative. (Cronon 1983:75,79)

1.2   The result: Habitat alteration, declines, and extinctions

European institutions came to dominate the New World and other colonized lands, and the European conception of land as a commodity replaced Native American views. Intensified resource use and the myriad ecological changes that followed stemmed in part from this difference in how property was treated (Cronon 1983). Not only did colonists exploit resources more intensively than the Indians had, but natural resources were also shipped from the colonized regions to distant markets in Europe. The harvest of commercially valuable species such as furbearers, fish, and trees had marked effects on populations of the exploited plants and animals. Although living natural resources are capable of replenishing themselves, the high levels of harvest that followed colonization often exceeded that capacity.

In addition, habitats were transformed as a result of the European attitude that a landowner could and should increase the value of his land by subduing it and making it productive, that is by using it to produce crops or livestock. In New England, the clearing and fencing of fields and pastures led to a host of ecological changes. Land cover was altered on an unprecedented scale, resulting in changes in habitat structure and in the distribution and abundance of wild plants and animals. Although data are scarce, it is likely that these changes were accompanied by changes in microclimate (local climatic conditions), nutrient cycles, and hydrology. Dramatic ecological changes were also caused by the exotic organisms that the colonists brought with them, deliberately (such as crops and livestock) or unintentionally (such as disease-causing microorganisms and weeds) (Crosby 1972, 1986; Cronon 1983) (see Chapter 7).

Changes in the intensity of resource exploitation, in habitat structure, and in community composition followed each new wave of settlers. Just as settlers in the northeastern United States altered the ecology of that region, other parts of the New World were transformed by the economic, social, and biological alterations that accompanied Euroamerican settlement. White (1980) documented such changes in the extreme northwestern corner of the continental U.S.A. Likewise, in the Southwest and in South and Central America, major ecological changes followed Spanish and Portuguese settlement. Changes in systems of land ownership and production, along with the introduction of cattle and the reintroduction of the horse (which had gone extinct in the New World about 10 000 years before) led to fundamental changes in the culture and ecology of those regions. When Spaniards settled El Salvador, for example, they used precolonial agricultural techniques to produce cocoa for export. This did not involve a major change in land use, since the Indians had raised cocoa before Europeans arrived, but it did increase the intensity of production. They also marketed balsam (which was valued for medicinal purposes and as an ingredient in perfumes), harvested from the sap of the balsamo tree, and indigo, a perennial shrub grown on plantations. Both of these developments contributed to deforestation. Another round of deforestation was initiated in the nineteenth century, with the development of coffee plantations in El Salvador's volcanic highlands (Browning 1971). As these fertile lands were converted from producing subsistence crops to producing coffee for export, the displaced indigenous population moved onto marginal lands, and these, too, were cleared for subsistence farming, creating additional ecological pressures on the land (Durham 1979). (We will return to this point in Chapter 12.)

In the colonies, land was abundant and "available" (if one did not recognize the property rights of the indigenous inhabitants). The colonists' actions were based on the assumption of a limitless supply of land and its products – trees, wildlife, and minerals. This assumption led to repeated cycles of depletion and expansion, which are described in more detail below. In North America, this process came to an end only when westward expansion was stopped by the Pacific Ocean.

European colonization of many other parts of the world led to similar changes in land tenure. Communally owned resources were often converted to privately owned, marketable land (Browning 1971; Grove 1990; Gadgil and Guha 1992, 1995). Dramatic and widespread changes in levels of resource exploitation were associated with this change, and as a result many habitats were altered and populations of numerous species declined. In Australia, changes in land use after European convicts arrived in the nineteenth century converted the native semiarid vegetation to pioneer grass and later to pine scrub, habitats that were less favorable to native mammals than the original native plant communities. These problems were exacerbated by introduced predators, such as foxes and cats, and competitors, such as rabbits and sheep. Many native Australian mammal species declined as a result of these changes, and some became extinct (Caughley and Gunn 1996). Dutch settlers deforested parts of Mauritius, an island in the Indian Ocean (Grove 1992), and large areas of communally managed forest were expropriated and cleared by the British in India (Agarwal 1992; Gadgil and Guha 1992, 1995). The wood from these forests was used to build ships and railroads, and mixed forests were converted to single-species stands of commercially valuable trees. Populations of big-game animals were depleted by British hunters in India and by European and American hunters in Africa (MacKenzie 1987; Grove 1992).

For example, in the 1870s the big-game hunter W. H. Drummond praised southeastern Africa as "the finest game country in the world," but lamented the fact that "day by day, almost hour by hour, and with ever increasing rapidity, the game is being exterminated or driven back." He especially feared that the "wanton and wasteful wholesale destruction" of elephants for ivory could not "last much longer" (Drummond 1972:viii,220,221).

Recall from the Preface that coincidence in timing does not prove causality. The ecological changes that followed European settlement were not necessarily caused by the newcomers. Some were undoubtedly part of long-term trends or were caused by a combination of European influences and other factors. Although we may not be able to prove that a specific environmental effect would not have occurred in the absence of European immigration, a general pattern is clear: colonial expansion was followed by widespread and relatively rapid ecological change characterized by increased levels of resource exploitation, changes in land use, and introductions of exotic species. As a result of these changes many species declined in abundance and others became extinct.

In North America, beaver was the first resource to be intensively exploited for European markets. Thousands of animals were trapped and shipped to Europe to supply the market in men's hats. Because they were easy to find and had fairly low reproductive rates, beavers were vulnerable to exploitation. It is likely that the species would have gone extinct if fashions had not changed in the 1840s causing the market for beaver pelts to collapse (Udall 1974).

When settlers arrived in eastern North America, forests were cleared for farming and to provide fuel and timber. Clearing the forest was considered a prerequisite to taming the wilderness. Americans consumed far more firewood in the New World than they had in Europe, where wood was scarce (Cronon 1983). When the supply of wood was exhausted locally, the circle of influence expanded. Forests were first depleted in the East and then in the Great Lakes region. As the supply of timber dwindled, forests even further west were cut. Federal and state governments sold forested land at bargain prices. Labor and land were cheap, and the "cut-and-take" mentality prevailed, which emphasized making a quick profit and then moving on. After a site was logged, it was abandoned, leaving large amounts of slash (branches, limbs, damaged trees, and debris) behind. When these dried they were highly flammable, and when they burned they did so with unprecedented intensity. In some areas forests still have not recovered from these fires (Alverson *et al.* 1994). There were no incentives for conservation during this period. The average life of a sawmill was 20 years, and the towns that had grown up around the mills often folded when the mills moved on (Udall 1974).

Likewise, minerals and oil were extracted in the nineteenth century using methods that involved wasting resources and making profits quickly. In California in the 1870s, hydraulic mining of gold washed tons of soil and gravel downslope, causing major problems for people living in the valleys and irreparable ecological impacts upslope (Udall 1974). In mining as in timber harvest, big profits could be reaped by those who extracted the resources before anyone else did, regardless of the waste or ecological damage they caused.

An estimated 50 million "buffalo" or bison roamed the plains of North America when Europeans arrived on the continent. Because of their numbers and biomass, these animals played a pivotal role in the midwestern steppes (grasslands dominated by perennial grasses). They provided a prey base for predators; influenced nutrient cycling through their grazing, defecation, and urination; and modified the physical structure of the vegetation by trampling and wallowing.

But because of their great numbers, bison were also easy to find and to kill. Hunting for commercial markets played a major role in their decline. (See Chapter 4 for more discussion of the different kinds of hunts.) Much of the carcass was usually wasted; often only the tongues and hides were taken (Hornaday 1971). The elimination of bison as a major ecological force in the American plains was motivated by political factors as well. The persistence of massive herds of bison in the Midwest was not consistent with the prevailing vision of how the frontier should develop (Figure 1.1). Some government officials wanted to reduce bison populations in order to subjugate the tribes of the plains, which were dependent upon the herds. The connection between eliminating bison and Indians was identified by Representative Garfield of Ohio in 1874 when he commented on a bill that would have limited the killing of bison:

I have heard it said . . . by a gentleman who is high in authority in the Government, the best thing which could happen for the betterment of our Indian question . . . would be that the last remaining buffalo should perish, and he gave . . . as his reason . . . that so long as the Indian can hope to subsist by hunting buffalo, so long will he resist all efforts to put him forward in the work of civilization. . . . The Secretary of the Interior said that he would rejoice, so far as the Indian question was concerned, when the last buffalo was gone. (Quoted in Allen 1962:15.)

Representative Conger of Michigan, commenting on the same bill, explicitly stated that the bison herds were incompatible with settlement. He argued that the bill granted a "privilege"

to the wild, savage Indian that is not given to the poor civilized settler. [The buffalo] eat the grass. They trample upon the plains upon which our settlers desire to herd their

Figure 1.1. *American Progress*, by John Gast, 1872, oil on canvas. This painting reflects nineteenth-century attitudes about settlement of the American frontier. As settlers moved west, bringing "civilization," Native Americans and wildlife were driven further west, and farming transformed the landscape. For a more detailed discussion of the mentality captured in this painting see Merchant (1996). (With permission of the Autry Museum of Western Heritage, Los Angeles, CA.)

cattle and their sheep. . . . They range over the very pastures where the settlers keep their herds of cattle and sheep to-day. They destroy that pasture. They are as uncivilized as the Indian. (Quoted in Allen 1962:15–16.)

This statement clearly shows that the enormous herds of bison were considered an impediment to settlement of the frontier. Congress did pass the bill, but President Grant failed to sign it. By 1890, there were fewer than 1000 bison in North America. Most of these were in Canada.

As the settlers proceeded westward, many habitats – including forests, steppes, and wetlands (areas of land that support vegetation adapted to life in saturated soil) – declined because they were converted to croplands or used for grazing. As livestock replaced native herbivores, problems from overgrazing developed. Livestock and bison have quite different ecological effects. Bison herds grazed an area intensively and then moved on, allowing the area to recover, but livestock grazing is more prolonged. The settlers were unfamiliar with the dry climates of the Midwest and failed to appreciate the potential impacts of overgrazing and loss of plant cover in such a setting. Furthermore, because the impacts of grazing are gradual, most people did not recognize what was happening until impacts on the native vegetation were dramatic, soil erosion had become severe, and alien weeds had become entrenched.

In open habitats of the Midwest, colonies of prairie dogs dug extensive underground burrow systems and lived in towns covering as much as 100 ha. Because these rodents fed on grasses and forbs (broad-leaved, herbaceous plants), ranchers viewed them as potential competitors of livestock. By the early 1900s, extensive public and private poisoning programs were directed at prairie dogs (Nowak and Paradiso 1983). As a result, they declined in geographic range and abundance, as did their principal predator, the black-footed ferret, which is highly endangered today.

In the last half of the nineteenth century, women's hats containing ornate feathers (and even whole, mounted birds) were fashionable in North America and Europe. As a result, a thriving trade in feathers, known as the plume trade, developed. In both Europe and North America, large numbers of many species that nested colonially, such as herons, grebes, terns, ibises, and egrets, were killed for their plumage. Like the bison, their tendency to concentrate in groups made these birds vulnerable to exploitation. Populations of colonial nesting birds declined dramatically as a result of exploitation. (We will see below that the plume trade was a major impetus for the establishment of the first national wildlife refuges in the U.S.A.)

Marine mammals provide other examples of overexploitation followed by population declines. Prior to the eighteenth century, coastal peoples exploited seals in the North Pacific, the North Atlantic, and the Mediterranean for their products – meat, blubber, and skins. Seals were important in many of these cultures. In Scottish and Irish folklore, for example, "silkies," creatures that are seals in the sea and transform into people on land, figure prominently. There is no evidence that this level of use caused seal populations to decrease, except perhaps for the monk seal in the Mediterranean (Hewer 1974).

In the late eighteenth century, however, a thriving trade in seal pelts developed. By the end of the nineteenth century, many commercially hunted species of seals had declined markedly in both the northern and the southern hemispheres, and entire rookeries (breeding colonies) had been eliminated from the Antarctic, at which point two factors came into play that probably saved several seal species from extinction. First, as it became harder and harder to find seals, commercial exploitation dwindled. In addition, the demand for seal oil shrank as alternative sources of fuel were developed.

Along the coast of California, elephant seals and sea otters were hunted

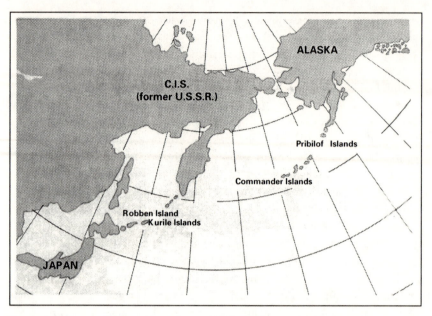

Figure 1.2. Range of the northern fur seal. (After Baker *et al.* 1970.)

almost to extinction in the nineteenth century. Sea otters were killed for their pelts, and elephant seals for their oil (Carroll 1982). We shall see in Chapter 8 that the genetic consequences of that population reduction remain to this day.

The process of intensive commercial exploitation followed by depletion is clearly illustrated by the fate of the northern fur seal. This species breeds on a chain of islands stretching across the North Pacific Ocean, including Robben Island, the Kurile Islands, the Commander Islands, and the Pribilof Islands (Figure 1.2). After the Russian explorer Gerassim Pribilof found the Pribilof rookeries in 1786, many small companies began to kill fur seals for their pelts. The number of animals they took is not known, but between 1786 and about 1820 the fur seal population declined precipitously (Gentry 1998). (Efforts to reverse this trend are discussed below and in Chapter 4.)

The story of whale exploitation is similar. Whaling has a long history, probably dating back to the Neolithic period. Inuit (Eskimos) along the North Pacific coasts of Asia and North America as well as Native American tribes of the northwest coast pursued whales long before the arrival of Europeans, and native whalers of northern Japan, Kamchatka, and the Aleutian and Kurile Islands hunted whales as well. In Europe, Basques on the French and Spanish coasts took whales from boats as early as the twelfth century. When Europeans arrived in New England, however, they established a prosperous

(A)

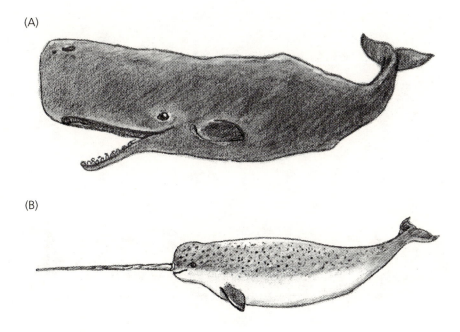

(B)

Figure 1.3. Toothed whales. (A) Sperm whale; (B) narwhal. (Drawn by M. Rockwood.)

whaling industry on an unprecedented scale. Whales provided many products before the age of modern chemistry. Whale oil was a valuable source of light until petroleum became available in the nineteenth century. Ambergris, a substance formed in the intestines of sperm whales, was used as a fixative for perfumes. It brought high prices in Paris and other centers of the perfume industry. Spermaceti, a waxy substance found in the large reservoir at the front of the sperm whale's head, was valued as an industrial lubricant. Whale ivory was obtained from the teeth and jaw bones of toothed whales (Figure 1.3). Baleen (fringed plates that hang from the upper jaws of blue, fin, right, humpback, gray, bowhead, and minke whales and form a massive strainer used to separate small marine organisms from seawater) provided a flexible, springy, strong material that was used for many items made today from steel or plastic, such as hairbrush handles, umbrella ribs, and buggy whip handles (Figures 1.4 and 1.5). When shredded into fibers, baleen became the "horse hair" used to stuff chairs and sofas, or was used for the plumes on soldiers' hats. When made into thin strips and woven, it was used for chair seats (similar to cane-bottomed chairs). Also known as whalebone, baleen formed the stays of women's "whalebone" corsets.

(A)

(B)

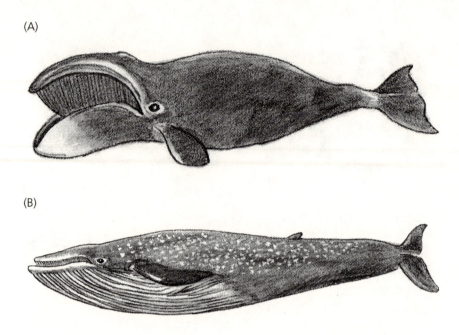

Figure 1.4. Baleen whales. (A) Bowhead whale; (B) blue whale. Note the manner in which the baleen plates are suspended from the upper jaw of the bowhead whale, and the blue whale's throat grooves, which expand when water is taken into the mouth. (Drawn by M. Rockwood.)

In the eighteenth and nineteenth centuries whaling was a dangerous business. Crews pursued whales in small boats launched from the side of the mother ship. The whale was struck with a harpoon, and the whalers simply hung on until the whale died or the crew drowned. Sperm, bowhead, and right whales were taken using this method. These species can be handled from small boats because their carcasses float. In 1864 a Norwegian captain invented a cannon-powered harpoon that exploded after it entered a whale. This "Foyn gun" was much more efficient at killing whales than the hand-held harpoon. Thus, the whaling industry became more efficient at finding, killing, and processing whales (Deason 1946). As a result, populations of all the large whale species were reduced to low levels. (Chapter 10 covers efforts to reverse this trend.) As each species became harder to find, whalers moved on to harvest other kinds of whales, much as settlers had moved west to exploit new frontiers when timber and game were depleted.

Declining habitats and species were not limited to regions that Europeans colonized. From the sixteenth through the nineteenth centuries, many of the

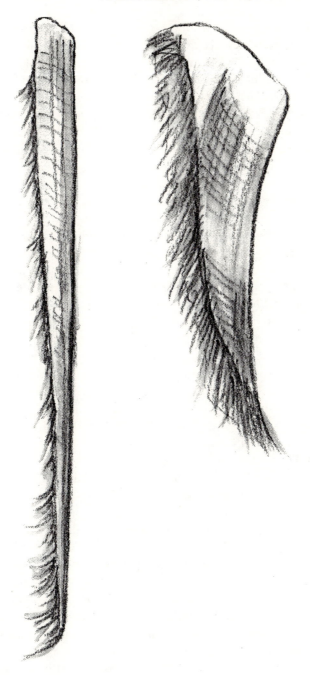

Figure 1.5. Plates of baleen from a right whale (left) and blue whale (right). (Drawn by M. Rockwood.)

same problems occurred in Europe as well. The combination of population growth and agricultural and industrial development led to widespread and relatively rapid changes in habitat. At the beginning of the nineteenth century, the British agricultural landscape was a diverse mosaic of fields, woodlands, wetlands, heaths, and meadows. But as the human population grew, the "influence of the marketplace on the countryside" increased, wetlands were drained, fertile uplands were cultivated, and heaths were limed (Ingrouille 1995:273). These changes resulted in a more homogeneous landscape.

By the nineteenth century, native plant diversity was dwindling in Britain, as populations of many native vascular plants declined or disappeared and the ranges of numerous alien plant species increased (Perring 1974). Habitat changes, combined with increasing exploitation and attempts to reduce populations of undesirable species, contributed to the declines of numerous European animals. The aurochs, the ancestor of domestic cattle, disappeared from western and central Europe in the Middle Ages, probably because of hunting and habitat alteration. The last remaining population persisted in Poland until 1627 (Nowak and Paradiso 1983). The wild horse became extinct in Europe by the nineteenth century, so wild horses remained only in Mongolia and China. (See Chapter 9 for information on captive breeding of these horses.) Like its American relatives, the European bison, or wisent, barely escaped extinction. Hunting combined with conversion of forested habitat to cropland contributed to this species' decline. By the early twentieth century, wisent survived only in the Caucasus Mountains and the Bialowieza Forest on the Polish–Russian border. By 1919, the species was extinct in the wild, although a small number of individuals survived in zoos (Nowak and Paradiso 1983).

Many species of predators were systematically killed in an effort to minimize conflicts with livestock and people. The gray wolf, which formerly had the largest geographic range of any modern terrestrial mammal other than *Homo sapiens* and occurred in all habitats in the northern hemisphere except deserts and tropical forests, probably became extinct in England around 1500 but survived in Scotland and Ireland until the eighteenth century (Corbet 1974). By the early twentieth century, it had virtually disappeared from most of western Europe as well as Japan (Nowak and Paradiso 1983). Other carnivorous mammals declined as well. Bears were eradicated from Germany by the end of the seventeenth century, although they persisted in Scandinavia. In the nineteenth century, however, the governments of Sweden and Norway attempted to eradicate bears because of predation on livestock, especially sheep. Sweden paid bounties for bears until 1893; in Norway, the national government paid bounties on bears until 1930, and local bounties continued

Table 1.1. *Examples of British or Irish bird species that reportedly declined between 1800 and 1849 because of exploitation or control programs*

Species	Contributing factor
Great-crested grebe	Birds killed to obtain feathers for the plume trade
Gannet	Birds collected for food
Black-headed gull	Eggs collected for food
Red-backed shrike	Eggs collected by egg collectors
Golden eagle	Birds killed to protect game animals
White-tailed eagle	Birds killed for taxidermy
Woodpigeon	Birds killed to protect crops
Goldfinch	Live birds trapped for pets

Source: Based on data in Sharrock (1974).

until 1972. In the mid 1800s, there were about 3100 bears in Norway and 1650 in Sweden, but over 7000 bounties were paid between 1856 and 1893. By the second decade of the twentieth century, Scandinavian bear populations had dwindled and become fragmented. Most of the remaining populations subsequently became extinct, except for those that persisted in the mountains of western Sweden (Figure 1.6). It is estimated that by 1930, there were only about 130 bears left in Norway and Sweden (Swenson *et al.* 1995).

The lynx and wild cat were eradicated from Germany by the end of the eighteenth century (Leopold 1936), and in Britain the geographic ranges of wild cat, polecat, and pine marten were sharply reduced in the nineteenth century (Corbet 1974). Exploitation, pest control, and habitat alteration also reduced populations of many European birds. Between 1800 and 1849, decreases occurred in populations of 71% of the British or Irish bird species that were exploited or considered pests (Table 1.1) and 57% of bird species associated with wetland habitats (Sharrock 1974).

Some species were not as lucky as bison, beaver, and fur seals. Several examples of species or subspecies that became extinct after AD 1600 are discussed in Boxes 1.1 to 1.5. Certain characteristics, such as limited geographic distribution and defense strategies that worked against nonhuman predators but not with people, made these species especially vulnerable to extinction. Market hunting, efforts to control predators and pests, habitat alteration, and the arrival of non-native species were usually the circumstances that actually precipitated their declines, however. Note that regulated sport hunting (as distinct from unregulated hunting for profit) did *not* play a role in any of these extinctions. The causes of extinctions are discussed in more detail in Chapter 8.

Figure 1.6. Distribution of the brown bear in Norway and Sweden in the period from 1910 to 1920. Only the populations in shaded areas persisted. (After Swenson *et al.* 1995.)

Box 1.1   The extinction of the dodo on the Mascarene Islands

The dodo was a member of a family of flightless birds that is believed to have descended from pigeons. It occurred only in the Mascarene Islands in the Indian Ocean east of Madagascar, where it probably fed on seeds and fruits. A dodo was about the size of a turkey, with bluish or brownish-gray plumage, a curly tail, and a heavy, hooked bill (Figure 1.7).

The extinction of the dodo is a well-known story. The Mascarene Islands were visited by Portuguese sailors early in the sixteenth century and occupied by the Dutch in 1598. Sailors killed the birds for their meat, and some were captured alive and shipped as curiosities to Europe and Asia. In addition, the sailors introduced pigs, monkeys, rats, and cats, which probably preyed on the dodo's eggs and chicks and competed with them for food. By 1640 the birds were extinct on the island of Mauritius, although they lingered on some of the smaller islands of the Mascarene group a little longer. By the 1660s the species was extinct (Cheke 1985).

The dodo is usually portrayed as slow, stupid, and easily killed, driven to extinction because it stood around "allowing itself to be clubbed to death" (Caughley and Gunn 1996:74). It is likely, however, that competition and predation from pigs and rats contributed to the dodo's decline.

Figure 1.7. The dodo. (Drawn by M. Rockwood.)

Box 1.2   The extinction of South Africa's quagga

The quagga was a zebra that inhabited the steppe, or veld, of southern South Africa. Named for its unusual, shrill cry, this member of the horse family had a brown body, white legs and tail, striped head and neck, and short mane (Figure 1.8). In 1652, the Dutch East India Company established a settlement at the southern tip of Africa. The Dutch settlers and their descendants, known as Boers or Afrikaners, found quagga herds easy to locate, and the animals easy to kill. The meat was used to feed servants, and the hides were used locally or exported. Some were taken alive as well. These were shipped to England, where the animals were popular in zoos (but generally too high-strung to breed in captivity) or used by local farmers to guard their livestock (Day 1961).

The last wild quagga died in 1878, and the last captive died in the Amsterdam Zoo in 1883. Little is known about the specific causes of the quagga's extinction; it is likely that its limited distribution and tendency to congregate in large groups made it vulnerable, and that exploitation reduced the population. It is not known whether habitat alteration or competition from domestic animals played a role in its demise.

Figure 1.8. The quagga. (Drawn by M. Rockwood.)

Box 1.3   The extinction of North America's passenger pigeon

In the middle of the nineteenth century, the passenger pigeon (Figure 1.9) may have been the most abundant bird on earth. This species was once present throughout much of eastern North America, in such great numbers that it was difficult for people to imagine its disappearance. The birds traveled in huge groups, accounts of which are so amazing that they might seem exaggerated were they not so well documented. The ground underneath roosting flocks appeared to be covered with snow because of a thick layer of droppings, and it was common for tree trunks and limbs to break under the birds' weight. In 1806 the ornithologist Alexander Wilson estimated that a flock he saw in Kentucky contained over 2 billion birds. In 1813 the naturalist John James Audubon reported seeing a flock on the Ohio River that obscured the sun at midday (Trefethen 1975).

By forming such large, nomadic groups, it is likely that the passenger pigeon was able to overwhelm animals that preyed on it. Although predators were attracted to the pigeons' roosting and breeding aggregations, there were so many birds that predators became satiated, and large numbers of individuals always escaped being eaten. The flocks never stayed in one place long enough for predator populations to build up to levels that would exert sustained pressure on the birds (Blockstein and Tordoff 1985).

Even though the enormous flocks were an effective adaptation for coping with animal predators, they made the birds especially vulnerable to hunters. The passenger pigeon was a market hunter's dream. It took little effort to find and to kill large numbers of birds, and the railroads made it possible to ship birds to markets in distant cities. Professional market hunters trapped them in baited nets capable of killing hundreds of birds at a time. In addition, local hunters shot nesting birds and took young from the nest, sometimes killing virtually all the young from a nesting colony. In 1878, hunters in Michigan took over a million birds from the last large nesting colony. Commercial killing stopped when the birds became too scarce to hunt profitably, but unfortunately by that time it was too late for the species to recover. The last individual died in the Cincinnati Zoo in 1914.

The decline of the passenger pigeon was remarkably abrupt. Hundreds of millions of birds persisted into the 1870s, but by the 1890s they were very rare, and two decades later they were extinct. How was it possible for a species that had been so abundant to go extinct so quickly?

Figure 1.9. The passenger pigeon. (Drawn by M. Rockwood.)

To answer this question, it is necessary to understand how the ecology of the passenger pigeon made it vulnerable in spite of the enormous size of its populations. This species was endemic to (that is, it occurred only in) the deciduous forests of northeastern North America. It fed on acorns, beechnuts, and chestnuts. Oaks, beeches, and chestnuts are mast trees, which means they produce fruits erratically. Every few years, the mast trees throughout a region produced abundant crops of fruits synchronously. This is thought to be adaptive for the tree species that produce mast because predators cannot consume all the fruits that are produced in a mast season, so some seeds survive and germinate. Because availability of their food was unpredictable, passenger pigeons had to search a large geographic area to find food. The large size of their flocks may have been an adaptation that allowed them to scan large expanses of the landscape. When some of the birds detected food, they would call to the other flock members. It has been suggested that when passenger pigeon populations were reduced by a combination of hunting and habitat loss, their ability to locate food was compromised, even though the flocks were still large by conventional standards. Thus it seems likely that their dependence on erratically available food supplies that could only be found by enormous aggregations of birds moving through the eastern deciduous forests made passenger pigeons uniquely vulnerable to the combined effects of exploitation and deforestation (Bucher 1992).

Like the bison, the passenger pigeon might have been intolerable to settlers. It is doubtful whether the landscapes of the East and Midwest could have supported passenger pigeon flocks along with intensive agriculture and urbanization.

Box 1.4   The extinction of the heath hen, mid-Atlantic subspecies of prairie-chicken

The heath hen of New England (Figure 1.10) was one of three subspecies of the greater prairie-chicken, a grouse of open or brushy habitats. It was once abundant in woodland clearings of New England and the middle Atlantic states. The heath hen was "so common . . . that laboring people or servants stipulated with their employers not to have *Heath-Hen* brought to table oftener than a few times in the week" (Nuttall 1832:662). Market hunting and habitat destruction reduced heath hen populations, and by the middle of the nineteenth century only a single population – on the island of Martha's Vineyard, off the coast of Massachusetts – remained. The plight of the bird was recognized at that point, and efforts were made to protect it. In 1860 state regulations outlawed hunting of the heath hen, and in 1902 the Commission of Fisheries and Game released three greater prairie-chickens in an effort to bolster the Martha's Vineyard population. This may have done more harm than good, however. The released birds were adapted to midwestern prairies, rather than to the more humid environment of the east coast; furthermore, they could have introduced diseases or parasites, and if they had succeeded in interbreeding with heath hens, the latter's unique genetic adaptations would have been diluted by being mixed with a different subspecies (see Chapter 8 for more on this topic).

In 1907 a reserve for heath hens was established on Martha's Vineyard, and intensive efforts to suppress fires and to control predators began. These measures seemed to be beneficial, and the population increased to at least 800 birds. But a catastrophic fire swept the island in 1916, followed by

Figure 1.10. The heath hen. (Drawn by M. Rockwood.)

an unusually large influx of predatory goshawks in the winter of 1917. Fire and predation killed a disproportionate number of females, which made it difficult for the surviving males to find mates. Losses from dogs and cats abandoned by tourists and from a disease introduced by domestic turkeys probably also took their toll on this small population. In 1930 only a single male was observed, and in 1933 none were found (Trefethen 1975).

The fate of the heath hen illustrates the vulnerability of small populations to chance events. The state of Massachusetts spent more than $56 000 in efforts to save the heath hen, but ultimately these efforts failed. The remnant population was driven to extinction by a combination of unfavorable circumstances, including (1) an environmental catastrophe (fire), (2) interactions with other species (predation), and (3) fluctuation in the makeup of the population (the loss of most females). A large population might have been able to withstand predation and catastrophe and would have been less likely to have an unbalanced sex ratio. For a small population, however, these misfortunes were fatal. We shall return to this topic in Chapter 8.

Box 1.5   The extinction of the marsupial Tasmanian wolf

The Tasmanian wolf was a carnivorous marsupial of the Australian region (Figure 1.11). Although it is not closely related to dogs and true wolves, the Tasmanian wolf's teeth, build, and feet are remarkably doglike. (Its scientific name means "pouched dog with a wolf head.") This species was widespread in Australia and New Guinea until about 3000 years ago, but subsequently it became extinct everywhere except in Tasmania. It is likely that its disappearance was due to competition with dingos, which were introduced by aboriginal hunters about 10000 years ago (Archer 1974). These dogs spread throughout the Australian region but did not become established in Tasmania.

When Europeans arrived in Tasmania and began raising sheep, predator control programs were initiated. Tasmanian wolves were shot, trapped, and poisoned. Between 1888 and 1909, the government paid bounties for over 2000 individuals, and others were killed for private bounties or for their pelts. Habitat loss, competition with dogs, and disease may have contributed to their decline (Caughley and Gunn 1996). By 1905 the Tasmanian wolf population had declined markedly, and by the 1930s the species was extinct (Nowak and Paradiso 1983).

Figure 1.11. The Tasmanian wolf. (Drawn by M. Rockwood.)

The preceding discussion is not meant to imply that precolonial societies had only benign effects on their environment or that all environmental degradation stems from colonial expansion or from resource commodification. There are examples of subsistence hunts that have depleted resources and of commercial harvests that have been carried out sustainably. It was not the existence of markets themselves that increased the level of exploitation in the colonized regions; rather, it was the size of the new markets. By linking resource use in the colonies to distant markets, European expansion increased the intensity of resource consumption and exploitation, and this had far-reaching effects. There is a second reason why I have focused on the effects of resource commodification. The ecological results of this process had consequences that were serious enough to stimulate significant developments in resource conservation.

1.3   Diagnosing the problem

By the end of the nineteenth century it was becoming clear in many parts of the western world and its colonies that resources were being used at a rate that could not be sustained for very long. The immediate causes of the problem seemed fairly straightforward: high rates of use, waste, and lack of regard for restoration. More careful management of living natural resources was obviously necessary to stem the tide of declining populations and habitats. Some examples of the responses that followed are discussed below.

1.4   The response to the problem

1.4.1   Regulation and protection

Restrictions on harvest
As the ecological impacts of exploitation, habitat loss, and species introductions mounted, governments enacted laws to protect what was left. For example, in 1769 the French government in Mauritius adopted a forest protection ordinance, forest reserves were established on the Caribbean island of Tobago in 1864, and measures to conserve forests were undertaken in India in the latter half of the nineteenth century (Grove 1992). Between 1860 and 1892, the British Parliament passed eight statutes relating to game licenses, the protection of wild birds, and small mammal conservation (Alison 1981).

Measures to limit exploitation were rarely initiated by private commercial interests, but the Russian American Company, which held the concessions for sealing on islands where northern fur seals bred, is an exception. In about 1834, the company prohibited the killing of females on the Pribilof Islands, and about a decade later it placed the same restriction on the Commander Islands harvest. Under this early management plan, the herd increased from about 300 000 to over 2 million in less than 50 years.

Although limitations on the seal kill worked for a while, changing market conditions prevented a prolonged recovery. In the 1860s, in response to a rise in the price of pelts, pelagic (open ocean) sealing was undertaken. This type of harvest was extremely wasteful, because many animals that were killed sank and were not retrieved, so the number of animals that were killed far exceeded the usable take. A second serious decline in northern fur seal populations followed, and by 1910 the species was down to about 10% of its 1867 level. In 1911 the International North Pacific Fur Seal Treaty between Britain (for Canada), the U.S.A., Japan, and Imperial Russia was adopted. Under the provisions of this treaty, the U.S.A. and Russia, which contained breeding islands, shared the seal take with Canada and Japan, which agreed to stop pelagic sealing. After the treaty was signed, fur seal populations again rebounded, reaching a high in 1941 (Baker *et al.* 1970; Gentry 1998).

This example illustrates how complicated management can become with mobile wild animals that do not respect political boundaries. Like seals, migratory birds move between national jurisdictions, and this complicates efforts to manage them. By the late 1890s, North American waterfowl populations had dropped to very low levels because of a combination of drought and market hunting. But, since waterfowl migrate long distances between their breeding and their wintering grounds, state governments were reluctant to pass protective legislation limiting hunting within their own borders. People reasoned that if they did not shoot the migrants, someone in another state or province would.

An attempt to solve this problem in the U.S.A. with federal legislation was challenged in the courts as a states' rights issue, but before the Supreme Court decided the case, the problem was circumvented by the Migratory Bird Treaty between the U.S.A. and Canada. (The U.S.A. subsequently signed similar treaties with Mexico, Japan, and the Soviet Union.) The original treaty was signed in 1916, and with the passage of the Migratory Bird Treaty Act shortly thereafter, it became law. The treaty and act outlawed the killing of nongame birds and established restrictions on the taking of migratory game birds such as waterfowl. States were permitted to pass more restrictive laws if they wished to. The act set a precedent for considering migratory waterfowl a federal resource.

Jurisdictional problems also confounded efforts to manage wildlife within national borders. In the U.S.A., if a plume hunter killed birds in violation of a state's laws and transported the feathers to another state, neither the state nor the federal government had any jurisdiction. This situation changed with the passage of the Lacey Act in 1900, which outlawed interstate shipment of any wild birds or mammals or their parts or products (including feathers and eggs) that had been taken in violation of state laws.

Designation of public lands for regulated resource use

There are basically two types of reserves, those that prohibit or severely curtail resource extraction within their borders, and those in which such use is allowed. The first category includes national parks or other areas set aside to protect their scenic values or the habitats and populations within their borders. This type of preserve is discussed in Chapter 10. The second category includes lands set aside for the express purpose of regulating resource use on those lands. Beginning in the 1890s, substantial areas of public land in the U.S.A. were set aside for this reason. The resources of concern included timber (national forests), wildlife (national wildlife refuges), and livestock forage (lands managed by the Bureau of Land Management). The first American national forests were set aside in 1891, when a rider to a public lands bill authorized President Harrison to create forest reserves by executive order. Before leaving office, Harrison set aside 13 million acres in the West. These lands and additional forest reserves subsequently added to the system are now termed national forests and are administered by the U.S. Forest Service within the Department of Agriculture. In 1898 Gifford Pinchot took office as the Department of Agriculture's chief forester. As a result of Pinchot's advocacy, the concept of regulated use was adopted as the guiding principle of national forest management. Under this philosophy, forests were managed to maximize the amount of wood that could be produced, but timber harvests were regulated, with the goal of insuring a future supply.

In 1903 Pelican Island, off the coast of Florida, became the first national wildlife refuge in the U.S.A. It was set aside by President Theodore Roosevelt to protect pelicans, egrets, herons, and spoonbills that were hunted by market hunters for the plume trade. Additional land was rapidly added to the system. By the end of 1904, Roosevelt had set aside 51 refuges. National wildlife refuges are administered by the U.S. Fish and Wildlife Service within the Department of the Interior.

The word "refuge" implies a safe place or sanctuary. People are often surprised to learn that hunting and fishing are allowed on many national wildlife refuges. The earliest national wildlife refuges in the U.S.A. were indeed intended to serve as sanctuaries. Many of the refuges that were subsequently created, however, were established for game species such as waterfowl or ungulates (hoofed mammals), with the explicit purpose of providing hunters with a supply of game.

The Bureau of Land Management, within the Department of the Interior, is another federal agency in the U.S.A. that administers large areas of federal land on which extractive activities, including grazing, timber harvest, and mining, are permitted. This agency was established in 1946 to take over the administration of rangelands (lands that are unsuited for cultivation but instead produce forage for livestock or wildlife) from the General Land Office. The Bureau of Land Management manages more than half the federal lands of the U.S.A. Most of the land administered by this agency is rangeland located in the western states.

1.4.2   The disciplines of natural resource management

In the 1930s, forestry and game management emerged as formal areas of study in North America. American forestry drew upon but modified an older tradition of European forestry. The discipline of game management was launched when Aldo Leopold, a professor of forestry at the University of Wisconsin, published a text on *Game Management* in 1933. (*"Game management"* eventually came to be called wildlife management, and the term *"wildlife"* itself expanded over the decades. At first it was used to refer to game species; later it came to mean terrestrial vertebrates. In current usage, the term wildlife often denotes all forms of wild organisms, including animals, plants, and microorganisms.) Range management did not emerge as a formal area of study until several decades later. (The first issue of the *Journal of Range Management* was published in 1948, 45 years after the first issue of the *Journal of Forestry* and 11 years after the birth of the *Journal of Wildlife Management*.) Range management differs from the management of forests or game because it manages domestic organisms produced from wild forage, whereas foresters and game managers are concerned with harvesting wild products directly.

These emerging disciplines emphasized regulated use of resources and manipulation of the environment to maximize the productivity of desired species and minimize undesirable species. They drew on European traditions of resource management but were fundamentally different in approach. European forestry, game management, and animal husbandry involved intensive management of scarce, privately owned resources, and efficient utilization with minimal waste. For example, in nineteenth-century Germany stands of spruce were planted and thinned, resulting in monocultures (holdings dominated by a single species) of small, symmetrical, clean-limbed trees of the same species and age (Brown 1935; Leopold 1936).

American management of living natural resources was primarily directed at publicly owned resources, and although it strove to eliminate the prodigious waste of the previous centuries, managers recognized that the European style of intensive management and efficient use would not be accepted by the American public. A forestry text published in 1935 stated that "until wood becomes more scarce and more valuable, we will be able to practice only an extensive rather than an intensive system of forestry in this country" (Brown 1935:15).

Although American resource managers rejected the European approach as too unnatural, they still saw themselves as analogous to farmers. Leopold defined game management as "the art of making land produce sustained annual crops of wild game for recreational use" (Leopold 1933:3). The comparison to agriculture was explicit: "Like the other agricultural arts, game management produces a crop by controlling the environmental factors which hold down natural increase, or productivity of the seed stock" (Leopold 1933:3). In this view, agriculture and natural resource management differ only in the degree of domestication of the product. Farmers raise domesticated crops; game managers raise wild organisms. Both manipulate critical factors to enhance production. We will see later (Chapter 6), however, that although Leopold was a utilitarian manager concerned with game species, he was ahead of his time in realizing that even predators play important ecological roles.

Similarly, early foresters and range managers emphasized maximizing the productivity of wild crops. A U.S. Department of Agriculture Bulletin published in 1926 stated that "more and better forage, as well as the maximum production of beef, wool, and mutton, is a primary object of grazing management" (Sampson 1926:1). Forestry was defined as "the art and science of managing our forests and converting the product to best serve mankind," and its central theme was "continuity of production and use" (Brown 1935:1).

This was to be done in a regulated fashion that would insure a continuous flow of products. Managers recognized that America's living natural resources could not sustain the levels of harvest they had been subjected to:

Heretofore, our forests, during the colonial and expansion periods of westward development, have been viewed largely as a resource to be exploited. Little thought was given to the future of our natural resources. Forests served as a deterrent rather than as an aid to civilization during the early generations. Now, sustained yield management, treating the forest as a source of periodic crops, has replaced the former practice of exploitation and wastage. (Brown 1935:1)

Thus, the depletion of natural resources set the stage for natural resource management aimed at regulating the harvest of valuable species of wild plants and animals, so as to guarantee future supplies. The next two chapters describe the scientific concepts upon which this utilitarian approach to management is grounded.

References

Agarwal, A. (1992). Sociological and political constraints to biodiversity conservation: a case study from India. In *Conservation of Biodiversity for Sustainable Development,* ed. O. T. Sunderlund, K. Hindar, and A. H. D. Brown, pp. 293–302. Oslo: Scandinavian University Press.

Alison, R. M. (1981). The earliest traces of a conservation conscience. *Natural History* **90**(5):72–77.

Allen, D. L. (1962). *Our Wildlife Legacy,* 2nd edn. New York, NY: Funk and Wagnalls.

Alverson, W. S., W. Kuhlman, and D. M. Waller (1994). *Wild Forests: Conservation and Biology.* Washington, DC: Island Press.

Archer, M. (1974). New information about the quaternary distribution of the thylacine (Marsupialia, Thylacinidae) in Australia. *Journal of the Royal Society of Western Australia* **57**, Part 2:43–50.

Baker, R. C., F. Wilke, and C. H. Baltzo (1970). *The Northern Fur Seal.* U.S. Department of the Interior, U.S. Fish and Wildlife Service, Bureau of Commercial Fisheries, Circular no. 336, Washington, DC.

Blockstein, D. E. and H. B. Tordoff (1985). Gone forever: a contemporary look at the extinction of the passenger pigeon. *American Birds* **39**:845–851.

Brown, N. C. (1935). *A General Introduction to Forestry in the United States.* New York, NY: John Wiley.

Browning, D. (1971). *El Salvador: Landscape and Society.* Oxford: Clarendon Press.

Bucher, E. H. (1992). The causes of extinction of the passenger pigeon. *Current Ornithology* **9**:1–36.

Carroll, G. F. (1982). Born again seal. *Natural History* **91**(7):40–47.

Caughley, G. and A. Gunn (1996). *Conservation Biology in Theory and Practice.* Cambridge, MA: Blackwell Science.

Cheke, A. S. (1985). Dodo. In *A Dictionary of Birds,* ed. B. Campbell and E. Lack, p. 152. Calton, Staffordshire: T. and A. D. Poyser. (Published by the British Ornithologists' Union.)

Collett, D. (1987). Pastoralists and wildlife: the Maasai. In *Conservation in Africa: People, Policies, and Practice,* ed. D. Anderson and R. Grove, pp. 129–148. Cambridge: Cambridge University Press.

Corbet, G. B. (1974). The distribution of mammals in historic times. In *The Changing Flora and Fauna of Britain,* ed. D. L. Hawksworth, pp. 179–202. London: Academic Press.

Cronon, W. (1983). *Changes in the Land: Indians, Colonists, and the Ecology of New England.* New York, NY: Hill and Wang.

Crosby, A. W. (1972). *The Great Columbian Exchange: Biological and Cultural Consequences of 1492.* Westport, CT: Greenwood Press.

Crosby, A. W. (1986). *Ecological Imperialism: The Biological Expansion of Europe, 900–1900.* Cambridge: Cambridge University Press.

Day, D. (1961). *The Doomsday Book of Animals.* New York, NY: Viking Press.

Deason, H. J. (1946). Conservation of whales: a world-wide project. *Transactions of the North American Wildlife Conference* **11**:260–273.

Drummond, W. H. (1972). *The Large Game and Natural History of South and South-east Africa,* Heritage Series, vol. 4. Facsimile reprint of the 1875 edition. Salisbury, Rhodesia: The Pioneer Head.

Durham, W. H. (1979). *Scarcity and Survival in Central America: Ecological Origins of the Soccer War.* Stanford, CA: Stanford University Press.

Gadgil, M. and R. Guha (1992). *This Fissured Land: An Ecological History of India.* Berkeley, CA: University of California Press.

Gadgil, M. and R. Guha (1995). *Ecology and Equity: The Use and Abuse of Nature in Contemporary India.* New Delhi: Penguin Books India.

Gentry, R. L. (1998). *Behavior and Ecology of the Northern Fur Seal.* Princeton, NJ: Princeton University Press.

Grove, R. H. (1990). Colonial conservation, ecological hegemony, and popular resistance: towards a global synthesis. In *Imperialism and the Natural World,* ed. J. M. McKenzie, pp. 15–50. Manchester: Manchester University Press.

Grove, R. H. (1992). Origins of western environmentalism. *Scientific American* **267**(July):42–47.

Hewer, H. R. (1974). *British Seals.* New York, NY: Taplinger.

Hornaday, W. T. (1971). *The Extermination of the American Bison, With a Sketch of its Discovery and Life History.* Reprinted from Report of the National Museum, 1887. Seattle, WA: Shorey Book Store.

Ingrouille, M. (1995). *Historical Ecology of the British Flora.* London: Chapman and Hall.

Leopold, A. (1933). *Game Management.* New York, NY: Charles Scribner's Sons.

Leopold, A. (1936). Deer and Dauerwald in Germany. I. History. *Journal of Forestry* **34**:366–375.

MacKenzie, J. M. (1987). Chivalry, social Darwinism, and ritualised killing: the hunting ethos in Central Africa up to 1914. In *Conservation in Africa: People, Policies, and Practice,* ed. D. Anderson and R. Grove, pp. 41–81. Cambridge: Cambridge University Press.

Merchant, C. (1996). Reinventing Eden: western culture as a recovery narrative. In *Uncommon Ground: Rethinking the Human Place in Nature,* 2nd edn, ed. W. Cronon, pp. 132–159. New York, NY: W. W. Norton.

Nowak, R. M. and J. L. Paradiso (1983). *Walker's Mammals of the World,* vol. 2. Baltimore, MD: Johns Hopkins University Press.

Nuttall, T. (1832). *Manual of the Ornithology of the United States and of Canada.* Cambridge, MA: Hilliard and Brown.

Perring, F. H. (1974). Changes in our native vascular plant flora. In *The Changing Flora and Fauna of Britain*, ed. D. L. Hawksworth, pp. 7–25. London: Academic Press.

Pyne, S. (1982). *Fire in America*. Princeton, NJ: Princeton University Press.

Sampson, A. W. (1926). *Grazing Periods and Forage Production on the National Forests*. U.S. Department of Agriculture, Department Bulletin no. 1405, Washington, DC.

Sharrock, J. T. R. (1974). The changing status of breeding birds in Britain and Ireland. In *The Changing Flora and Fauna of Britain*, ed. D. L. Hawksworth, pp. 203–220. London: Academic Press.

Swenson, J. E., P. Wabakken, F. Sandegren, A. Bjärvall, R. Franzén, and A. Söderberg (1995). The near extinction and recovery of brown bears in Scandinavia in relation to the bear management policies of Norway and Sweden. *Wildlife Biology* 1:11–25.

Thomas, W. L., Jr. (ed.) (1956). *Man's Role in Changing the Face of the Earth*, 2 vols. Chicago, IL: University of Chicago Press.

Trefethen, J. B. (1975). *An American Crusade for Wildlife*. New York, NY: Winchester Press and the Boone and Crockett Club.

Udall, S. L. (1974). *The Quiet Crisis*, 12th edn. New York, NY: Holt, Rinehart, and Winston.

White, R. (1980). *Land Use, Environment, and Social Change: The Shaping of Island County, Washington*. Seattle, WA: University of Washington Press.

2

Central concepts – population growth and interactions between populations

In the previous chapter we saw how overexploitation of living renewable resources created a need for more rational management of wild plant and animal populations. In response to this situation, conservationists in the young disciplines of wildlife management, forestry, and range management adopted a utilitarian approach, which emphasized the economic values of species. Utilitarian managers seek to regulate the exploitation of economically valued plant and animal species and to minimize populations of species that are considered weeds or pests.

In this and the following chapter we will consider the central concepts underlying utilitarian management. This type of management focuses on certain phenomena in nature, including population regulation in resource-limited systems, predation, the specific requirements of organisms of interest, and changes in community composition over time. The last three chapters in this section illustrate how this understanding of the natural world is applied by utilitarian managers.

2.1 Adding to and subtracting from populations

A population can be defined as a group of organisms of the same species occupying a defined area during a specific time. For example, we may want to refer to the population of people in Germany in 1910 or in 1999, the population of aphids on a rose bush, spruce trees in a forest, fish in a lake, or pronghorns in Wyoming.

The rate at which a population grows depends upon how many individuals are added to and removed from it during a given period of time. Members can be added to a population by birth or by immigration (permanent movement into a population), and members can be removed either by death or by emigration (permanent movement out of a population). If birth and immigration into a population exceed death and emigration from it, the population will grow, and if more individuals die or leave than are born or join, it will decline.

In practice, it is easier to get information about death and birth than about emigration and immigration, so demographers (scientists who study changes in population size) usually focus on mortality and reproduction when analyzing population processes. The percentage mortality over a given time period plus percentage survival during the same period always total 100% because every individual is either alive or dead at the end of that period.

The potential rate at which a species can increase, under ideal environmental conditions, is called its biotic potential. The biotic potential of a species is a genetically determined characteristic of that species. It depends on such things as the number of eggs laid (clutch size) or young born (litter size), the frequency of reproduction, the age at which individuals first reproduce, and the age at which reproduction ends. The age at which reproduction begins has a particularly strong influence on the rate of population growth; populations grow much more rapidly when reproduction begins at a young age.

Biotic potential is high for some organisms, such as bacteria and dandelions, and low for others, such as people and elephants. We will see in Chapter 8 that biotic potential is one factor that influences the likelihood that a species will become threatened or endangered.

2.2   Limits to population size

2.2.1   Mortality

Individual organisms die from many causes. Mortality occurs when individuals are killed outright by enemies that eat them, uproot them, or parasitize them. Mortality is also caused by accidents, or it may result if competition between members of the same species or with members of other species prevents individuals from obtaining resources they need for survival. Finally, organisms die if they are both unable to tolerate environmental conditions (for example, if it is too cold, too hot, or too dry for them) and unable to move elsewhere. (See Chapter 3 for a discussion of the requirements of plants and animals.) An otherwise suitable habitat may become intolerable because of unusual short-term fluctuations in weather (such as droughts or blizzards), long-term environmental changes (for example, an ice age or global warming), high levels of physical disturbance (such as wave action or winds), or catastrophes (such as earthquakes, fires, volcanic eruptions, storms, or floods). When such a change occurs, a species will become extinct if mortality is high and its members are unable to emigrate from the unsuitable location to a place where conditions can be tolerated.

2.2.2   Reproduction

The reproductive rate of a population under a specific set of conditions depends upon a species' biotic potential as well as on the survival of its members. Thus, the factors that affect mortality directly influence reproduction indirectly. In addition, reproductive rate is also affected by physiological state. Nutritional status, disease, body size, and parasite loads can influence reproductive rate. Individuals that are in poor-quality habitat or that are prevented from obtaining resources by competitors may survive but fail to reproduce or may produce few young or young that are unlikely to survive.

2.3   Types of population growth

2.3.1   Exponential growth

Under ideal conditions (where there is no crowding; resources such as light, water, nutrients, cover, and nesting sites are abundant; and no substances that inhibit growth or reproduction are present), organisms are capable of increasing to extremely high numbers. Some bacteria can divide to produce two cells every 20 minutes if conditions are favorable. Starting with a single bacterial cell, over 250 000 cells can be produced in just six hours. In the example in Table 2.1, the population doubles each generation, and a generation lasts 20 minutes. Growth in which a population is multiplied by a constant in a given time interval is said to be exponential or geometric (as opposed to arithmetic).

When the size of an exponentially growing population is graphed as a function of time (with the number of individuals on the y axis and time on the x axis), a characteristic curve is produced (Figure 2.1A). The population builds up slowly at first, but later it increases extremely rapidly. Exponential or geo-

Table 2.1. *Exponential population growth in bacteria*

Time (minutes)	Population size	Log_{10} of population size	Population size expressed as an exponential function of 2
0	1	0.00	2^0
20	2	0.30	2^1
40	4	0.60	2^2
60	8	0.90	2^3
80	16	1.20	2^4
100	32	1.51	2^5
120	64	1.81	2^6
140	128	2.11	2^7
160	256	2.41	2^8
180	512	2.71	2^9
200	1024	3.01	2^{10}
220	2048	3.31	2^{11}
240	4096	3.61	2^{12}
260	8192	3.91	2^{13}
280	16384	4.21	2^{14}
300	32768	4.52	2^{15}
320	65536	4.82	2^{16}
340	131072	5.12	2^{17}
360	262144	5.42	2^{18}

metric growth is sometimes termed logarithmic growth, because when it is graphed using a logarithmic scale on the *y* axis (that is, a scale on which each number is a multiple of the preceding number, for instance, progressive intervals of 1, 10, 100, 1000, etc., instead of 1, 2, 3, 4, etc.) the resulting graph is a straight line (Figure 2.1B).

Bacteria have a very short generation time and a high reproductive rate. Exponential growth can also take place in organisms that reproduce more slowly, but it takes longer to reach a large population size. Charles Darwin pointed out that even elephants, which are notoriously slow breeders, have the potential for explosive population growth. Thus, although different organisms are capable of increasing at different rates, any type of organism can increase exponentially under ideal conditions. The maximum possible reproductive rate of a species is represented by *r*, its intrinsic rate of increase.

Obviously, ideal conditions are rarely encountered in nature except, perhaps, when a species is introduced into an environment from which it was

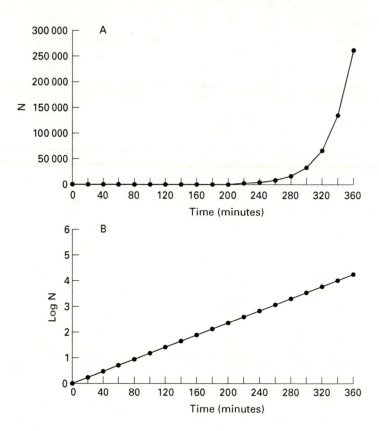

Figure 2.1. Exponential population growth. (A) A characteristic exponential growth curve is produced when the size (*N*) of a population is multiplied by a constant (in this case, 2) in each successive time interval. Note that the population grows slowly at first, but eventually it increases very rapidly. (B) The growth of the same population, with population growth shown on a logarithmic scale. Note that the shape of the logarithmic population growth curve is a straight line.

previously absent. In this situation, there may be a temporary period when resources are abundant and there are no predators, parasites, diseases, or competitors. Although unlimited population growth is rare, exponential growth can also occur when conditions are not ideal. Even a population with a relatively low reproductive rate can periodically increase by a constant factor (Figure 2.2).

An important characteristic of exponential growth is the way in which the number of organisms increases slowly at first but then increases very rapidly. The following legend from ancient Persia illustrates this:

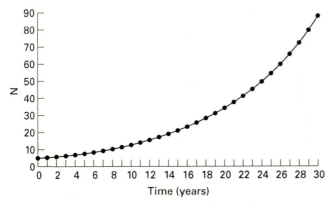

Figure 2.2. Exponential growth of a population that starts with five individuals and increases by 20% annually. Although the population is small initially and is multiplied by only 1.2 each year, within 30 years $N = 87$, an increase of over 1700%.

A clever courtier . . . presented a beautiful chessboard to his king and requested that the king give him in return 1 grain of rice for the first square on the board, 2 grains for the second square, 4 grains for the third, and so forth. The king readily agreed and ordered rice to be brought from his stores. The fourth square of the chessboard required 8 grains, the tenth square 512 grains, the fifteenth required 16,384, and the twenty-first square gave the courtier more than a million grains of rice. By the fortieth square a million million rice grains had to be brought from the storerooms. The king's entire rice supply was exhausted long before he reached the sixty-fourth square. Exponential increase is deceptive because it generates immense numbers very quickly. (Meadows *et al.* 1974: 36–37.)

In 1798 the Reverend Thomas Malthus published *An Essay on the Principle of Population,* in which he noted that the human population has a tendency to increase geometrically and suggested that unchecked population inevitably leads to resource scarcity, conflict, and mortality (Malthus 1986). His ideas influenced Darwin, who saw population pressure as the driving force for evolution. (See Chapters 7 and 12 for a discussion of contemporary neo-Malthusian views about the role of population growth in generating current environmental problems.)

Malthus was correct in noting that populations cannot continue growing indefinitely. The exponential growth curve is not a realistic model of how populations behave for very long. An alternative model is the logistic growth curve, proposed by the Belgian mathematician Pierre-François Verhulst in 1849. It describes the behavior of populations that stabilize at a level determined by limiting resources.

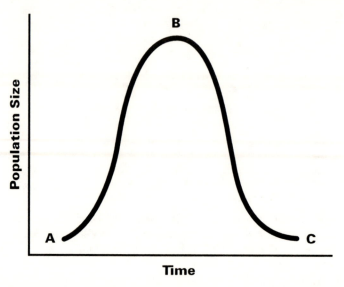

Figure 2.3. Population growth in a closed system. (A) Initial phase of exponential growth; (B) plateau phase; (C) decline phase.

2.3.2   Logistic growth and density-dependent population regulation

To return to the example involving bacteria, if organisms are in a closed flask and no new resources are added, a population will undergo an initial phase of exponential growth (Figure 2.3A), but as resources are used up, the population will level off and reach a plateau phase (Figure 2.3B). Finally, when resources are used up and waste products begin to accumulate, the population will enter a decline phase (Figure 2.3C).

In this example, the organisms are in a closed system in which resources are limiting and are eventually used up. If a fresh supply of resources is added from time to time, the population growth curve may remain at the plateau phase instead of dying out (Figure 2.4). This is called logistic growth, and the characteristic graph of such growth is termed a sigmoid (S-shaped) growth curve. The number of individuals of a given species that can be maintained in a particular environment is sometimes termed the carrying capacity (K) of that environment (Leopold 1933; Edwards and Fowle 1955). If a population drastically exceeds the carrying capacity of its environment, environmental damage may result, and, if the damage is severe enough, carrying capacity can be permanently diminished. When a population stabilizes at carrying capac-

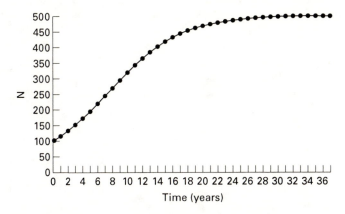

Figure 2.4. Density-dependent population growth. The solid line represents the growth of a hypothetical population that is limited by resources and consequently experiences logistic growth. In this example, the initial population size (*N* at time 0) is 100, carrying capacity (*K*) is 500, and the intrinsic rate of increase is 0.20.

ity, additions to the population are balanced by losses. In other words, the population is at equilibrium (Figure 2.5).

Sigmoid population growth is regulated by density-dependent processes. In a density-dependent system, as population density increases, population growth slows because of increased mortality and/or declining reproduction. When a population is at the carrying capacity of its environment, additions to the population are balanced by losses, and the population is stable. This density-dependent slowing of population growth can result from intraspecific competition, that is, competition between members of the same species. Plants that are closely spaced will die if they fail to obtain enough water, food, light, or nutrients to survive. In animal populations, high population density can lead to increased fighting, or it can result in increased stress that physiologically weakens individuals and leaves them susceptible to disease.

Disease often operates in a density-dependent fashion, because contagious diseases spread from one individual to another more rapidly in a dense population than where individuals are widely spaced. Crowding may cause individuals to be susceptible to other mortality factors as well. For example, predators may be able to find and kill more prey where prey populations are dense than where they are sparse.

Density-dependent processes can affect reproduction as well as mortality. If individuals do not die from the effects of high population density, they may nevertheless fail to reproduce, or they may produce fewer healthy offspring or

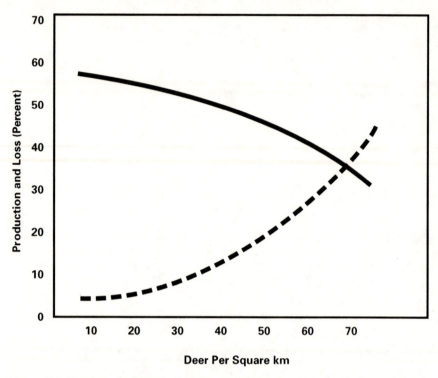

Figure 2.5. Diagram of a hypothetical deer population in which density has stabilized at carrying capacity. As population density (*x* axis) increases, reproductive rate decreases (solid line) and mortality increases (dashed line), until they equal each other and the population is at equilibrium. (After Leopold 1955.)

reproduce less frequently because of stress or resource shortages associated with high population density.

In addition to limiting population growth by decreasing survival and reproduction at high population densities, density-dependent processes can also promote rapid population growth at low densities. If mortality and reproduction are regulated by density-dependent processes, then mortality will be low and survival will be high when population density is low. This means that a population that is subject to density-dependent regulation will increase relatively rapidly at low densities.

It is easy to understand why utilitarian resource managers pay particular attention to density-dependent population processes. If survival, reproduction, and physical condition tend to decline at high population densities, then removing some individuals to thin, that is, decrease the density of, a population is a logical basis for harvest (see Chapter 4). On the other hand, density-dependent responses undermine efforts to control overabundant species (Box 2.1). This density-dependent response makes control efforts inefficient and presents obvious practical difficulties for managers interested in keeping down the densities of pest species (see Chapter 6).

Box 2.1   Density-dependent reproduction in feral horses

Two herds of feral (undomesticated) horses inhabit Assateague Island, a barrier island off the coast of Maryland and Virginia. Both herds occur on lands managed by federal agencies, but the herds are managed quite differently. This difference in management strategies creates a natural experiment that allows us to test the hypothesis that reproduction is density-dependent in these two herds.

The Virginia herd occurs on Chincoteague National Wildlife Refuge and consists of about 200 horses. For more than three decades, the U.S. Fish and Wildlife Service has intensively managed this herd by removing over 80% of the foals each spring, in an effort to prevent overpopulation. The approximately 150 horses in the Maryland herd, on the other hand, inhabit Assateague Island National Seashore, which is managed by the National Park Service. In keeping with this agency's policy of nonintervention, managers do not attempt to control population growth in the Assateague herd.

To find out whether population control measures were associated with a compensatory increase in reproduction, and if so, what caused that increase, Kirkpatrick and Turner (1991) followed the fates of 88 sexually mature mares with home ranges confined to either the national wildlife refuge or the national seashore. Specifically, they hypothesized that the proportion of surviving fetuses would be higher in the population that was subjected to population control than in the unmanaged herd. All of the sample mares could be individually identified from unique markings, which were sketched or photographed. To determine pregnancy rates, urine or feces or both were collected from each individual in the fall of 1989 and analyzed in a laboratory, using techniques previously tested in domestic or feral horses. During the following summer and fall, the pregnant mares were observed to determine whether they were nursing foals.

Two-thirds of the 48 sample mares from the refuge herd, which was subject to population control measures, tested positive for pregnancy,

Table 2.2. *Reproductive characteristics of feral mares at Assateague Island National Seashore and Chincoteague National Wildlife Refuge*

	Assateague	Chincoteague	Difference
Total mares tested	40	48	
Diagnosed pregnant	35.0%	66.6%	significant
Foaling rate	32.5%	66.6%	significant
Fetal loss rate	7.1%	6.2%	not significant

whereas the pregnancy rate for the 40 mares from the unmanaged national seashore herd was only 35%. Not surprisingly, the refuge herd also produced more foals (32) than the national seashore herd (13). These differences were statistically significant. On the other hand, the herds did not differ significantly in fetal loss rates (Table 2.2).

Thus, the general hypothesis that reproductive rate responds to population control in a density-dependent fashion was supported by Kirkpatrick and Turner's findings, but the specific hypothesis about what caused that response was not supported. The Assateague Island refuge herd did respond to population control measures with a compensatory increase in reproduction, but the mechanism responsible for that increase came as a surprise. The researchers had anticipated similar pregnancy rates in the two herds but higher fetal survival in the herd from which foals had been removed. Instead, they found that it was higher rates of pregnancy, not higher rates of fetal survival, that were responsible for the compensatory increase in reproduction in the managed herd.

Populations are not always limited by resources, however. In spite of the popularity of the concept of density dependence with utilitarian resource managers, biologists who focus on insect populations have long argued that too much attention is paid to density-dependent regulation and, consequently, to resource limitation and competition. For example, in the 1950s Andrewartha and Birch (1954) suggested that unfavorable weather, physical disturbances, or environmental catastrophes typically operate in a density-independent fashion to keep populations well below the level at which resources become limiting. (For another example of density-independent mortality, see Box 9.1.) This point of view, however, is not the dominant one among utilitarian resource managers.

2.4   Interactions between populations: Predation

Populations do not exist in isolation. Every species interacts with other species that prey on it, parasitize it, compete with it, or provide it with necessary components of its habitat. An assemblage of such interacting populations is termed a community. Within a community, every species plays a unique functional role; this is termed its niche. (A description of a species' niche includes the resources it utilizes, the type of habitat it lives in, specific features of its habitat, the seasons when it is active, and so on.) Scientists and managers working within a utilitarian perspective pay particular attention to interactions in which one species of animal eats another.

2.4.1   Predation as a mechanism of limiting prey populations

Since predatory mammals and birds kill and eat many of the same animals that people feed upon, such as livestock, waterfowl, chickens, and deer, utilitarian managers tend to view predators as economically detrimental. The concept underlying this perspective is that predators limit the size of prey populations. The idea that more prey would be available if fewer prey were eaten by predators seems obvious. Unfortunately, however, for many years hunters, ranchers, and resource managers simply accepted that this must be the case on the basis of intuition or anecdotal evidence, a position which both reflected and contributed to a rather simplistic perspective in which species were viewed as either "good" or "bad" on the basis of their utilitarian values. In their enthusiasm for this view, utilitarian managers have sometimes focused on predation as a limiting factor to the exclusion of other variables.

This was the case with events on the Kaibab Plateau in Arizona in the early part of the twentieth century. Parts of the plateau were designated as a forest reserve in 1893, and other parts were set aside as a game reserve in 1906. Several changes in land use followed. The number of domestic sheep grazing in the area was reduced, deer hunting was prohibited, fires were suppressed, and government agents were employed to kill predators in an effort to increase populations of game species such as mule deer. After the reserve was created, the deer population of the plateau initially grew, but subsequently it declined.

In 1941 Irvin Rasmussen published an ecological monograph on the Kaibab Plateau. He presented estimates of trends in the deer population from three sources: the forest supervisor, park visitors, and himself. The forest supervisor's estimates, which were probably the most reliable since he was most familiar with the situation, were much lower than those of Rasmussen and the other observers. Nevertheless, Rasmussen used the most extreme data points in his analysis, particularly the estimate that the herd had reached 100 000 in 1924. By connecting these data points, he came up with a curve that depicted a population explosion followed by a dramatic crash (Rasmussen 1941) (Figure 2.6A).

Managers accepted Rasmussen's scenario and assumed that predator removal had allowed the deer population to build up to the point where it exceeded the plateau's carrying capacity. In a publication that appeared in 1943, wildlife biologist Aldo Leopold recounted how predator removal on the Kaibab Plateau led to a dramatic irruption (a steady rise in population followed by a dramatic decline, also termed eruption) of the deer herd, followed by range degradation and a population crash (Leopold 1943). This version of events was subsequently repeated in text books by Allee et al. (1949), Davis and Golley (1963), and others (Figure 2.6B).

Finally in 1970, population ecologist Graeme Caughley re-examined the Kaibab deer story and concluded that "little can be gleaned from the original records beyond the suggestion that the population began a decline sometime in the period 1924–1930, and that the decline was probably preceded by a period of increase." He pointed out that it is not possible to determine in retrospect whether the increase in deer was caused by the decline in predators, because predation was not the only factor that changed, so that "the factors that may have resulted in an upsurge of deer are hopelessly confounded" (Caughley 1970:56).

No one has suggested that Rasmussen deliberately misconstrued the Kaibab deer data, but this classic example from the early days of the profession of wildlife management illustrates the shortcomings of before-and-after comparisons conducted without adequate controls and with subjectively chosen data sets. Fire suppression and decreased grazing might have resulted in an increase in vegetation available to the deer and contributed to the rise in their population. Although the Kaibab deer data suggest that predators may regulate ungulate populations, this example certainly cannot be considered conclusive evidence, because more than one variable changed simultaneously.

To determine whether predation limits the size of prey populations, it is necessary to find out if populations that are preyed upon are smaller than populations that are not preyed upon. Natural and deliberate introductions of herbivores into predator-free environments are a source of data on this question. When prey are introduced into novel environments that lack predators, their populations often increase steadily at first. This has been documented for ungulates, rodents, rabbits, and insects. Population growth cannot increase

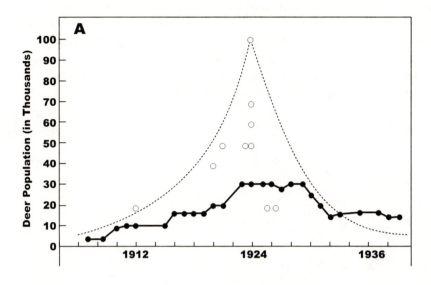

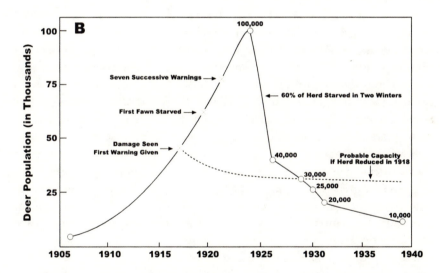

Figure 2.6. Estimated changes in the population of deer on the Kaibab Plateau from 1906 to 1939. (A) Trends in the Kaibab deer herd from 1906 to 1939 according to Rasmussen (1941). Three estimates of the deer population are shown. The solid line connecting solid circles denotes the forest supervisor's estimate; open circles are the estimates of other visitors; and the dashed line is Rasmussen's estimate. (B) Trends in the Kaibab Plateau deer population as portrayed in a textbook on animal ecology by Allee *et al.* (1949). (After Allee *et al.* 1949 and Caughley 1970.)

indefinitely, however. If resources are overgrazed, the population may crash, like the hypothetical population in Figure 2.3. This occurred after 29 reindeer were introduced on St. Matthew Island in the Bering Sea (Klein 1968). Nineteen years later, the reindeer population numbered about 6000, but by the following winter all but 42 animals had died. Similar irruptions occurred after moose dispersed to Isle Royale in Lake Superior, and when red deer were introduced to New Zealand. Similarly, a population of Himalayan thar, a goat-like member of the cow family, irrupted after being released on New Zealand (Caughley 1970).

These examples suggest that predation does indeed play a role in regulating populations of some herbivorous animals, particularly ungulates. Note, however, that we are still dealing with before-and-after studies. In the case of introductions into predator-free environments, a nearby population of the same species subjected to predation could serve as a control. If the habitat and other features of the environment at the two sites were similar, and the control population did not increase simultaneously with the introduced population, that would lend support to the conclusion that release from predation allowed the introduced population to expand in its new environment.

A variety of other approaches have been used to understand the effects of predation on prey populations, including mathematical models and laboratory experiments. Perhaps the best way to determine whether predators limit prey populations in natural environments, however, is with controlled field experiments. For example, in a six-year study of the effects of predation on nesting ducks at Agassiz National Wildlife Refuge in Minnesota, predatory mammals and birds were removed annually from half of the study area for three years (Balser *et al.* 1968). At the end of that time predators were removed only from the other half of the study area. Researchers recorded the number of nests and ducklings on the treated and untreated halves. They also assessed predation by placing chicken eggs in simulated nests and recording whether or not the eggs were attacked. On the areas from which predators had been removed, 7571 ducklings were recorded, compared to 4858 ducklings on the sites with predators, a statistically significant difference. In addition, predators attacked fewer artificial nests in the predator-removal areas than in the untreated areas. These findings suggest that predation limited the reproduction of duck populations on the refuge. However, although the study demonstrated higher duckling production on the predator-free areas, it did not show that the greater productivity caused an increase in the duck populations. If increased duckling production were followed by higher duckling mortality, duck populations would still not be greater on the predator-free areas.

2.4.2   Factors that compensate for predation

In the hypothetical scenario suggested above, where an increase in productivity is accompanied by a rise in death rate, mortality can be said to have compensated for reduced predation. The idea that predation and other mortality factors substitute (or compensate) for each other was suggested by wildlife biologist Paul Errington in the 1940s. In his field studies of muskrats, Errington found that populations were regulated by social behavior rather than by predation. Population size was determined by the number of individuals that older muskrats would tolerate in a given habitat. Mortality was very high among young animals driven from their lodges by older muskrats. As they dispersed into unfamiliar country, juveniles died from intraspecific strife, motor traffic, predation, and a variety of other causes. Errington argued that among muskrats and many other species, predation does not limit the size of prey populations, because approximately the same number of individuals will die regardless of whether predation occurs (Errington 1946). This phenomenon is termed compensatory mortality. When mortality from predation is compensatory, the same amount of mortality will occur with or without predation.

2.4.3   Management implications

From this brief excursion into the topic of predator–prey interactions, we can conclude that the effects of predation on prey populations are complex and depend on a variety of factors including habitat and behavior of the prey species. Predators sometimes exert a limiting effect on prey populations, but this is by no means always the case because sometimes compensatory population processes counteract losses due to predators.

Either viewpoint – that predators limit prey populations or that mortality from predation merely substitutes for other forms of mortality – is compatible with the objectives of utilitarian resource managers. Both perspectives suggest justifications for harvesting economically valued species: hunters harvesting deer kill individuals that would otherwise be taken by predators; trappers remove muskrats that would die regardless of whether they were killed by predators. Similarly, foresters thin forests to remove competitors and allow commercially valuable trees to grow larger. In each case, people act as surrogates for natural processes. This topic will be covered in more detail in Chapter 4.

Ecological studies undertaken in the context of utilitarian resource management have contributed to our understanding of population growth and regulation and of competitive and predatory interactions between species. This body of work is especially useful in elucidating how populations behave when resources are limiting. Utilitarian resource managers are also interested in the habitat requirements of species and communities of interest. This topic is explored in the next chapter.

References

Allee, W. C., A. E. Emerson, O. Park, T. Park, and K. P. Schmidt (1949). *Principles of Animal Ecology*. Philadelphia, PA: W. B. Saunders.

Andrewartha, H. G. and L. C. Birch (1954). *The Distribution and Abundance of Animals*. Chicago, IL: University of Chicago Press.

Balser, D. S., H. H. Dill, and H. K. Nelson (1968). Effect of predator reduction on waterfowl nesting success. *Journal of Wildlife Management* **32**:669–682.

Caughley, G. (1970). Eruption of ungulate populations, with emphasis on Himalayan thar in New Zealand. *Ecology* **51**:53–72.

Davis, D. E. and F. B. Golley (1963). *Principles in Mammalogy*. New York, NY: Van Nostrand Reinhold.

Edwards, R. Y. and C. D. Fowle (1955). The concept of carrying capacity. *Transactions of the North American Wildlife Conference* **20**:589–598.

Errington, P. L. (1946). Predation and vertebrate populations. *Quarterly Review of Biology* **21**:144–177, 221–245.

Kirkpatrick, J. F. and J. W. Turner (1991). Compensatory reproduction in feral horses. *Journal of Wildlife Management* **55**:649–652.

Klein, D. R. (1968). The introduction, increase, and crash of reindeer on St. Matthew Island. *Journal of Wildlife Management* **32**:350–367.

Leopold, A. (1933). *Game Management*. New York, NY: Charles Scribner's Sons.

Leopold, A. (1943). Deer irruptions. *Transactions of the Wisconsin Academy of Science, Arts and Letters* **35**:351–366.

Leopold, A. S. (1955). Too many deer. *Scientific American* **193**(110):101–108.

Malthus, T. R. (1986). *An Essay on the Principle of Population*, 7th edn. Fairfield, NJ: Augustus M. Kelley.

Meadows, D. H., D. L. Meadows, J. Randers, and W. W. Behrens III (1974). *The Limits to Growth*, 2nd edn. New York, NY: Signet.

Rasmussen, D. I. (1941). Biotic communities of Kaibab Plateau, Arizona. *Ecological Monographs* **11**:229–275.

3

Central concepts – habitats

I have no doubt about it, the time has arrived when we must manage specifically for anything we want from the land. . . . Our renewable resources will be renewed only if we understand their requirements and plan it that way.

(Allen 1962:22)

We noted in Chapter 1 that when the disciplines of forestry, wildlife management, and range management developed, their practitioners saw themselves as analogous to farmers, manipulating the environment to produce crops of desirable organisms. Their perspective led them to focus on the environmental requirements of species of interest, and they reasoned that by providing those requirements they could maximize production.

In seasonal environments, plants and animals are likely to have different requirements at different times of the year. In addition, the requirements of animals depend upon sex, age, and breeding condition. As managers came to understand this, they realized that in order to provide for a species of interest, it was necessary to understand that species' requirements throughout its entire life cycle (King 1938).

Although these developing disciplines emphasized economically valuable products, resource managers came to understand that meeting the habitat requirements of species of interest meant managing for the plant communities on which those species depended. This insight was a significant contribution. Managers realized that even the most stringent restrictions on grazing, hunting, and logging would not conserve future stocks in the face of severe habitat degradation. This led them to ask important questions about how habitats provide the resources that organisms need; how soil, water, and vegetation change over time; and how those changes can be molded to favor species of interest. This chapter examines these ideas in a general way. Chapter 5 illustrates how an understanding of these concepts is applied by resource managers.

3.1   Ecosystems

3.1.1   The ecosystem concept

Living organisms depend upon and modify their physical environment. In the nineteenth century, several European and American scientists published studies of interactions between organisms and their environment, but it was not until 1935 that the term "ecosystem" was used to denote a system formed by the interaction of a community of organisms with each other and with their physical environment (Tansley 1935). The British plant ecologist A. G. Tansley argued that the fundamental concept for ecological study should be

the whole *system* (in the sense of physics), including not only the organism-complex, but also the whole complex of physical factors. . . . Though the organisms may claim our primary interest, when we are trying to think fundamentally we cannot separate them from their special environment, with which they form one physical system.

It is the systems so formed which, from the point of view of the ecologist, are the basic units of nature on the face of the earth. . . . These *ecosystems*, as we may call them, are of the most various kinds and sizes. (Tansley 1935:299; emphasis in original)

As the last sentence indicates, defining the boundaries of an ecosystem is somewhat subjective, because biophysical interactions occur at a variety of spatial scales and because ecosystems are open and interconnected across a landscape (see Chapter 10). An ecosystem might be thought of as a pond, a marsh, a forest, a rotting log, or the entire biosphere (the portion of the earth that contains life).

3.1.2   Ecosystem components

Ecosystems are composed of living (biotic) and nonliving (abiotic) components. Solar radiation, oxygen, rocks, and water are examples of abiotic components of ecosystems. The biotic components of an ecosystem are divided into autotrophs and heterotrophs. Autotrophs (*auto*, self; *troph*, feeding) are organisms that do not have to feed on organic matter derived from the tissues of other organisms because they are able to manufacture their own food. They do this in one of two ways. Green plants and some one-celled organisms contain the pigment chlorophyll, which traps energy from sunlight and uses it to synthesize organic compounds in the process of photosynthesis. In the chemical reactions of photosynthesis, carbon dioxide is incorporated into organic (carbon-containing) compounds, such as carbohydrates, that can be used by other organisms. Some bacteria and cyanobacteria (blue-green algae) employ another method of manufacturing organic matter; they use the energy from chemical bonds to make organic compounds in the process of chemosynthesis. In most ecosystems, photosynthesis is more important than chemosynthesis; but Box 13.7 describes an ecosystem in which the autotrophs are chemosynthetic bacteria.

Heterotrophs (*hetero*, different; *troph*, feeding) are organisms that cannot manufacture their own food; they obtain energy by consuming autotrophs. Animals, fungi, and some microorganisms are heterotrophs. Heterotrophs have evolved diverse means of meeting their nutritional requirements. Herbivores, such as butterflies, aphids, tadpoles, geese, grouse, hummingbirds, elk, horses, and rats, feed primarily upon plant material. Carnivores, on the other hand, feed primarily on animal tissue. Some plants, such as pitcher plants and Venus'-flytraps, are able to trap and digest small animals. These adaptations allow them to function as heterotrophs.

The biotic components of an ecosystem can also be classified into a series of functional levels termed trophic levels. Autotrophs comprise the first trophic level of an ecosystem, the producers. Herbivores and carnivores are consumers; they obtain their energy by ingesting producers. Herbivores are primary consumers because they feed directly on the first trophic level, the producers. Carnivores are secondary consumers if they feed on herbivores, or tertiary consumers if they feed on other carnivores. Heterotrophs that chemically break down dead organisms from any of the other trophic levels are known as decomposers. Fungi, bacteria, and earthworms are examples of decomposers. They convert the complex molecules of plant and animal tissues into substances that can be used by plants, thus allowing nutrients to be recycled.

The relationships between trophic levels can also be represented by a pyramid of biomass (Figure 3.1). Biomass is the total dry weight (usually expressed as g/m^2) of the organisms at any trophic level. In most cases, the biomass of each trophic level is greater than the biomass of the next trophic level. In other words, biomass pyramids are usually right-side-up (Figure 3.1A). Occasionally, where the producers are small, highly efficient organisms with rapid turnover, the biomass of the producers at a given point in time can actually be lower than that of the consumers. When this happens, the biomass pyramid is inverted. This is the case for some open-water communities where the producers are algae (Figure 3.1B). The fact that biomass decreases as we

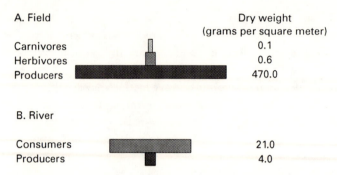

Figure 3.1. Pyramid of biomass for a field (A) and a river (B). Because of high turnover among the producers in the aquatic ecosystem, the biomass of producers can actually be lower than the biomass of the consumers that feed on them. Bar widths indicate relative biomass. (After Woodwell 1967.)

ascend the food chain means that persistent toxins become more concentrated at each trophic level (see Box 7.1).

3.2   How habitats provide the resources needed by organisms

The needs of organisms are termed resources. The basic requirements of plants are light, water, air, minerals, and a medium to grow in (usually soil or water). Animals require food (as a source of energy and of specific nutritional requirements such as minerals), oxygen, water, and space. In addition, animals usually require certain structural features in the spaces they inhabit, which are generally referred to as cover. These components of animal and plant habitats are described briefly below.

A habitat consists of the physical attributes of an environment that make it habitable. Thus habitat is the "suite of resources . . . and environmental conditions . . . that determine the presence, survival, and reproduction of a population" (Caughley and Sinclair 1994:4). To understand the value of a habitat, it is necessary to look at the resources it provides.

3.2.1   Resources

Soil
Soil is a link between organic and inorganic parts of the environment, formed when rock weathers and the decomposed tissues and products of organisms

mix with the resulting particles. The development of soil is influenced by geology; climate; topography; hydrology; and the plants, animals, and microorganisms present on a site.

Because soil influences the distribution and abundance of organisms, it is of interest to managers seeking to produce trees, game, cattle, or other products. Soil fertility, texture, chemistry, water-holding capacity, and depth influence plant growth. These properties of soil also affect the physical condition of animals, both indirectly, through their influence on plant condition and distribution, and directly. In turn, the plants and animals on a site also influence soil properties. Dead plant parts form a layer of litter on the soil surface. Minerals that leach out of this litter are added to soil along with the decomposed tissues of dead plants and animals. Animals also influence soil through activities such as burrowing, wallowing, trampling, and deposition of feces and urine. Microorganisms modify soil fertility and properties such as texture and chemistry.

The density and height of aboveground vegetation is not necessarily a good indicator of soil fertility. Temperate grasslands have less aboveground vegetation than temperate forests, yet in general grasslands have more fertile soils. Vegetation in the moist tropics is lush, but it sits atop soils that are typically low in fertility because of high rates of leaching in humid climates. Soil fertility is linked to vulnerability to disturbance, and the infertility of soils in moist tropical regions is one reason why these areas are slow to recover after vegetation is removed. Failure to recognize the limited fertility of these soils has had dire consequences (see Chapter 13).

Nutrients and oxygen

Green plants do not require food in the ordinary sense, but they do require nutrients. (For an example of a situation in which nitrogen was a limiting factor for habitat restoration efforts, see Chapter 10.) Most plants, except for those that live in saturated soils or in water, are able to obtain sufficient oxygen easily. (See Boxes 13.7, 13.8, and 13.9 for more information about wetlands and aquatic habitats.)

It is obvious that all animals must be able to obtain food. Utilitarian wildlife managers often focus on the food habits of the species being managed. In the early issues of the *Journal of Wildlife Management* (which initiated publication in 1937), a great many studies addressed the food habitats of game species.

In animals, requirements for nutrients and oxygen are intimately related, because oxygen is used in the breakdown of high-energy molecules, such as glucose, in the process of cellular respiration. The food requirements of warm-blooded animals depend in part upon the temperature of the surrounding environment. In cold environments, the energy requirements of birds and mammals increase. Consequently, their intake of food and/or mobilization of fat stores must increase if a relatively high body temperature is to be maintained in cold conditions. This is especially true if unusual circumstances occur that reduce the effectiveness of the body's insulation, for example when fur or feathers are coated with oil.

The amount of oxygen dissolved in water depends upon the water's temperature and salinity, on whether the water is moving, and on how much oxygen is consumed by aquatic organisms. Cold, unpolluted, fast-moving waters of mountain streams are rich in oxygen; warm, stagnant waters and waters with high levels of bacteria or algae are oxygen-poor. Aquatic organisms that have high oxygen requirements may have difficulty meeting their oxygen needs in such water.

In seasonal environments, food is more abundant at some times of year than at others. Many animals store food in preparation for seasonal shortages or for times of exceptional energy drain. External reserves of food are termed caches; internally, food can be stored as fat. An animal's food requirements change seasonally and also depend upon its age, sex, health, and reproductive state.

During any period when an organism has special requirements, it becomes particularly vulnerable if these requirements are not met. Many birds that are primarily herbivorous feed on insects or other invertebrates when they are young. The advantage of this strategy stems from the fact that young birds can grow rapidly on a high-protein diet of animal matter. These critical invertebrate foods may be depleted in areas affected by acid rain (Chapter 7), which can have serious consequences for breeding birds. For birds that undertake long-distance migrations, the greatest energy drain occurs during migration. Failure to lay down sufficient fat stores in preparation for migration or to replenish fat at migratory stopover points can lead to mortality.

Water

Natural resource management and water management are intimately related, especially in the semiarid and arid regions in parts of Africa, Australia, South America, central Asia, and western North America. For healthy populations of plants and animals to be maintained, adequate supplies of unpolluted water must be available.

Managers of wildlife, forests, and rangelands may become involved with water management for several reasons. First, populations of plants and animals can sometimes be enhanced simply by providing additional sources of water. For instance, structures that trap and retain water may benefit deer in desert habits, and irrigation can coax plants to grow in regions where they would otherwise be absent. (This approach is not without its problems, however. The concentration of herbivores near water sources can lead to overgrazing and may attract predators, and irrigation can cause salts to build up in the soils of arid regions.) Second, resource managers must assess the effects of water diversion. Organisms are often negatively impacted when water levels are manipulated or water is moved from one place to another. This occurs when marshes are drained to create cropland, water is used for irrigation, river channels are straightened to minimize flooding, or rivers are dammed. (The impacts of these activities are discussed in Chapter 7.) Third, managers of living natural resources also become involved in water management issues when water supplies become polluted. Acid rain and oil spills are two familiar examples of types of water pollution that can cause substantial ecological damage (see Chapter 7).

Cover

Cover refers to structural features of an animal's habitat that fulfill certain functions such as providing visual obstruction or shelter. Different types of cover – such as hiding cover, nesting cover, thermal cover, or resting cover – allow animals to perform different functions. Plants often provide cover, but soil, rocks, snow, water, and artificial structures can serve as cover too.

For prey animals, visual obstruction from potential predators is often a critical element of cover. Thermal cover aids animals in the regulation of body temperature. In winter, thermal cover allows warm-blooded animals such as birds and mammals to conserve body heat. Vegetation that intercepts wind and precipitation and reduces the amount of heat a body radiates to the night sky contributes to winter thermal cover. In summer, thermal cover provides shade and allows air movement, thereby allowing animals to minimize heat gain and maximize heat loss (Figure 3.2).

Special habitat features

Some species require special microhabitats (small habitats characterized by conditions that differ from those of the surrounding area) or specific structural habitat features. Standing dead trees (snags), woody debris (in stream beds or on forest floors), surface water, cliffs, and caves are examples of habitat structures that are important to some species of wildlife (Thomas 1979). When in short supply, these features may limit population size. For instance, populations of cavity-nesting birds – such as woodpeckers, swallows, swifts, wood ducks, chickadees, nuthatches, tits, bluebirds, and small owls – are

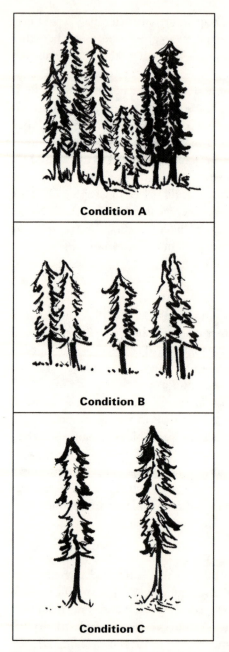

Figure 3.2. Thermal cover. Vegetation structure influences heat loss and heat gain and, therefore, the value of a stand in providing thermal cover. (A) provides better thermal cover than (B), and (B) provides better cover than (C). (After Thomas *et al.* 1979*b*.)

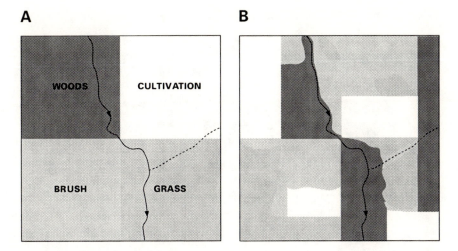

Figure 3.3. Edge effect. The amount of each type of habitat is the same in (A) and (B), but (B) has more edge than (A). (After Leopold 1933.)

frequently limited by the availability of suitable cavities in standing dead trees (Bruns 1960; Thomas *et al.* 1979*a*).

3.2.2   Juxtaposition of habitat patches

In his text on game management, Aldo Leopold (see Chapter 1) noted the importance of habitat juxtaposition for game animals:

game is a phenomenon of *edges*. It occurs where the types of food and cover which it needs come together, *i.e.*, where their edges meet. Every grouse hunter knows this when he selects the edge of a woods, with its grape-tangles, haw-bushes, and little grassy bays as the place to look for birds. The quail hunter follows the common *edge* between the brushy draw and the weedy corn, the snipe hunter the *edge* between the marsh and the pasture, the deer hunter the *edge* between the oaks of the south slope and the pine thicket of the north slope, the rabbit hunter the grassy *edge* of the thicket. Even the duck hunter sets his stool on the edge between the tules and the celery beds. . . . In those cases where we can guess the reason [for these effects], it usually harks back either to the desirability of *simultaneous access* to more than one environmental type, or the *greater richness* of border vegetation, or both. (Leopold 1933:131; emphasis in original)

The positive influence of the amount of patch border on the number and abundance of organisms in an area is termed the edge effect (Figure 3.3). The

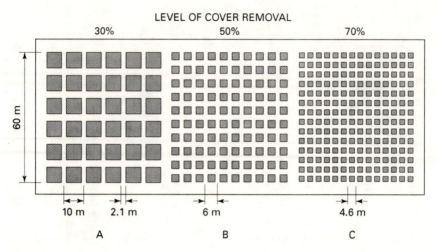

Figure 3.4. Arrangement of test plots used to evaluate the effect of habitat interspersion on densities of dabbling ducks. Stippled areas represent patches of cattails; clear areas represent open water. (A) 30% open water:70% vegetation; (B) 50% open water:50% vegetation; (C) 70% open water:30% vegetation. These patterns were created by removing vegetation from a homogeneous stand of cattails. Initially, increasing the number of channels increases the amount of edge between surface water and vegetation (compare (A) and (B)). Eventually, however, a point is reached where further division divides the cattails so finely that the resulting pattern begins to appear uniform instead of patchy (compare (B) and (C)). (After Murkin *et al.* 1982.)

area where two habitats come together is termed an ecotone. The boundary zone between a patch of forest that burned 20 years ago and a stand of 200-year-old trees is an ecotone, as is the transition zone at timberline, where the forest borders tundra. Ecotones may be abrupt or gradual. Abrupt boundaries between two plant communities can result from a change in the physical environment, such as a change in soil type, from competitive interactions between plant species, or from land management. Leopold noted that ecotones are more important for animals of relatively low mobility than for highly mobile species.

The edge effect is a testable hypothesis. It predicts that the number of species and the density of individuals of a given species will be higher where two habitats meet than in the interior of either habitat. Studies of birds and game species lend some support to this hypothesis. For example, Lay (1938) censused birds in the interior of forest patches in Texas and along the margins of clearings in the forest. He reported that the margins averaged 16.6 birds of 6.5 species, whereas the woodland interiors averaged 8.5 birds of 4.6 species.

The results of these studies need to be interpreted with caution, however, because the methods used to census bird populations may produce biased results. Typically such studies are done by counting singing males, but males of many species select perches near clearings, even if their territories extend far into the forest interior. Counting these males as edge inhabitants may artificially inflate the estimated edge population and underestimate the forest population.

A more rigorous way to test the hypothesis that edges are beneficial to wildlife involves experimental habitat manipulations. Murkin *et al.* (1982) used this approach at Delta Marsh in Manitoba. By cutting a series of perpendicular channels (Figure 3.4), they created openings in cattails that resulted in blocks with 30%, 50%, and 70% open water. The greatest densities of dabbling ducks (mallard, blue-winged teal, pintail, gadwall, northern shoveler, and green-winged teal) occurred on plots with a 50:50 ratio of cattails to open water. This treatment (Figure 3.4B) provided more edge than either of the other two treatments. The treatment with 30% water and 70% vegetation (Figure 3.4A) had less interspersion between water and edge because the blocks of vegetation were larger. On the other hand, in the treatment with 70% open water and 30% vegetation, the cattail patches were so finely divided that they apparently did not provide adequate cover (Figure 3.4C).

We shall see later (Chapter 7) that in their enthusiasm for edges as a way to increase game populations, wildlife managers failed for a long time to appreciate the fact that although creating small patches of habitat with a high proportion of edge will benefit some species, this practice is detrimental to species that require forest interiors.

3.2.3   Range of tolerance

Every organism is adapted to exist within a specific range of environmental conditions, termed its range of tolerance. For example, all organisms are adapted to function within a specific range of temperatures; physiological activity ceases and death results if temperatures rise above or drop below that range for very long. For some plants and animals, the range of conditions that can be tolerated may be quite broad; these organisms can live in a variety of environments. Others may require conditions within a narrow range; they need a rather specific environment. (These specific requirements allow us to use organisms that have narrow ranges of tolerance as indicators of environmental conditions; see box on pp. 20–1).

Within a population, different individuals will differ slightly in their ability to tolerate environmental conditions. If the environment of a species changes, those individuals that are able to tolerate the new conditions will be favored by natural selection (see Chapter 8). If the environmental change exceeds a species' range of tolerance and the species is unable to adapt to the change, it will become extinct unless it is able to move to an area where environmental conditions are tolerable.

It is important for managers to understand the tolerance ranges of organisms they are interested in for several reasons. Clearly, organisms can persist only in habitats that allow them to meet their requirements. If a population is limited by a requirement in short supply, it may be possible to increase that population by providing more of the limiting resource. Similarly, it may be possible to keep down populations of pest species by manipulating environmental factors. Finally, when managers seek to move populations of plants or animals around, it helps if they know whether the new site meets the requirements of the organisms they are trying to translocate. (This might seem obvious, but recall from Chapter 1 that when an attempt was made to augment the last remaining population of heath hens, this was done with little regard to the requirements of the transplanted birds, which came from a very different environment.)

3.2.4   Seasonal variations in resource availability

Organisms need to cope with seasonal fluctuations in the availability of resources. In temperate and arctic regions, seasonal fluctuations in resource ability result from changes in temperature and day length, whereas in the tropics, annual cycles are typically marked by wet and dry seasons.

Some animals are only able to meet all the requirements of their annual cycle by undertaking seasonal migrations (round-trip movements). These movements allow them to take advantage of resources in different locations. Migratory behavior is found in many animal groups, including insects (the monarch butterfly), fishes (salmon and eels), amphibians (some salamanders), reptiles (sea turtles and rattlesnakes), birds (some birds of prey, shorebirds, waterfowl, and songbirds), and mammals (some bats, whales, seals, and ungulates). Anadromous fishes are hatched in fresh water, migrate to salt water, and return to fresh water to breed. Salmon, for example, hatch in streams, migrate to the ocean as juveniles, and return to their natal stream to breed. In contrast, eels and some populations of alewives (a type of herring) have a catadromous life cycle, which is the reverse of the salmon's: they migrate from fresh water to the ocean to spawn.

The best-known migrations are those of birds that winter in the tropics and breed in the northern hemisphere. Millions of waterfowl and shorebirds migrate thousands of kilometers from their southern wintering grounds to breeding grounds in the arctic tundra. Similarly, many songbirds that breed in the northern hemisphere, including most flycatchers, warblers, and vireos, winter in the tropics. Elk, moose, and deer often participate in altitudinal migrations, which allow them to winter at low elevations where snow cover is likely to be sparse and migrate to high-elevation breeding grounds in spring when green forage is available. In arid regions, seasonal migrations occur in response to changes in moisture rather than temperature. Insects, birds, and grazing mammals inhabiting the plains of sub-Saharan Africa migrate to follow the seasonal rains.

An understanding of migrations is important to managers for two reasons. First, the requirements of migrants are geographically separated, sometimes by very long distances. Making sure that all the appropriate links (breeding grounds, wintering grounds, and stopover points) in the chain of required habitats are protected requires administrative and political coordination. Recall from Chapter 1 that the Fur Seal Treaty and the Migratory Bird Treaty were early attempts to address this problem. Second, as noted above, long-distance migrations require a great deal of energy. In order to complete their long-distance movements, migrants must accumulate significant stores of fat prior to migration; in addition, they may need to replenish their fat stores at stopover points along their migratory route. If migrants are unable to store up sufficient fat reserves before or during migration because critical habitat has been altered, mortality may increase.

3.3   Changes in communities over time: Succession

3.3.1   The concept

Habitats often change in predictable ways. Biologists have been aware of this phenomenon for centuries. As early as 1685, W. King described habitat changes in Ireland that led to the formation of bogs (Clements 1916). In 1895, the German scientist E. Warming attempted to state some general principles about transitions in plant communities through time, and in 1899, the American ecologist H. C. Cowles published the results of his studies of sand dunes on the shores of Lake Michigan, in which he described a complete sequence of communities beginning with the plants that colonize windblown sand and ending with the development of a forest (Cowles 1899) (see below).

Two years later, Cowles suggested classifying "plant societies" according to their relationship to a sequence of stages (Cowles 1901).

The ecologist F. E. Clements (1916), and others such as A. G. Tansley (1920, 1935), elaborated upon these ideas to develop a comprehensive theory of changes in community composition. They pointed out that when a bare area is created by a disturbance (something that removes vegetation, such as a volcano, flood, landslide, receding glacier, or fire), it is initially colonized by organisms that are able to tolerate conditions on the unvegetated surface. These organisms are termed pioneers. They possess adaptations that allow them to survive and reproduce in the harsh conditions present on bare sites. As colonization proceeds, pioneers gradually change their environment, making it less favorable for some living things and more favorable for others. The pioneers modify wind speed, temperature, light availability, and moisture in a host of ways, and when they die their decomposed tissues add organic matter to the soil. These changes provide conditions that allow other species to become established. The species in this new group again alter the site, allowing still other species to colonize it. This process may be repeated several times, so that different plant communities follow each other in a fairly predictable series of stages. The unidirectional change in community composition, in which plant communities succeed one another in a predictable sequence, is termed succession. Ultimately a stage is reached at which the species on a site are capable of reproducing under the conditions they create; this is termed a climatic climax, and the climax plant community is known as an association.

Clements and his colleagues recognized that most of the earth's vegetation was not actually in a climax state, but they argued that these deviations from equilibrium were due to the activities of people: "Man alone can destroy the stability of the climax during the long period of control by its climate" (Weaver and Clements 1938:80–81). According to Clementsian ecologists, the preponderance of early successional communities was due to disturbances such as cultivation and deliberate burning; natural disturbances played only a minor role.

The communities preceding the climax are termed seral stages (see Box 3.1). The plant community that ultimately develops on a particular site depends upon climate and soil conditions. Unusual conditions, such as special soils or topography, result in the development of plant communities that differ from a region's typical climatic climax. Soils that are very shallow, excessively stony or sandy, much wetter or drier than normal, or unusually low in fertility support alternative sequences of communities. A climax community that develops as a result of such special soil conditions is termed an edaphic climax. (Edaphic means pertaining to soil.) Bogs are examples of edaphic climaxes.

Box 3.1   Major plant community types occurring within the western hemlock association of western Washington and Oregon

The western hemlock zone is the most extensive type of vegetation in the area of Washington and Oregon west of the Cascade Mountains and Coast Ranges. This region experiences a maritime climate, and the typical undisturbed vegetation there is moist temperate forest. This type of forest supports large, long-lived trees that produce valuable timber. Franklin and Dyrness (1988) have described the vegetation of the western hemlock zone in terms of the stages that lead up to a climax community. The material below is condensed from their discussion.

Western hemlock is able to reproduce beneath the canopy of a mature forest, whereas Douglas-fir is not. Thus, the potential climax association is dominated by an overstory of western hemlock. It can take hundreds of years for western hemlock to replace Douglas-fir, however. Even stands of old growth that are 400 to 600 years old retain substantial numbers of Douglas-fir trees. This, combined with the fact that much of the region has been logged within the past 200 years, means that true climax stands of hemlock are rather rare, and large areas are dominated by seral forests in which Douglas-fir is the dominant tree.

On sites that are clearcut and then burned to remove slash, a sequence of several short-lived seral stages has been documented. Sometimes mosses and liverworts are dominant for a year, and sometimes this phase is skipped and the first-year community is characterized by residual species and invading forbs. In the following year, wood groundsel, an invading annual forb, makes up a high proportion of the vegetation. This species has very high nutrient requirements, which are satisfied on recently burned sites. By the fourth or fifth year, perennial herbaceous species such as bracken fern, common thistle, and fireweed invade. This "weed stage" is gradually supplanted by a shrub stage. The vegetation of the weed stage includes residual members of the pre-fire community – such as vine maple, Pacific blackberry, Cascade Oregongrape, salal, and western rhododendron – as well as invaders such as willows and ceanothus. Eventually trees, especially Douglas-fir and in some areas big-leaf maple, become dominant, and if the stand does not experience a major disturbance for several hundred years, western hemlock may finally replace the Douglas-fir.

Sites that are unusually dry or wet or that occur on special substrates support different climaxes. On very dry sites, western hemlock never becomes dominant, and so Douglas-fir retains its dominance indefinitely. On wet sites, western redcedar is an important constituent of the climax association. South of Puget Sound, unusual "prairies" occur on gravelly soils derived from glacial outwash materials. These sites support plants such as Idaho fescue, sedges, and camas (a blue-flowered member of the lily family whose bulbs were an important item in the Native American diet), which are able to tolerate wet soils in spring.

Several unusual associations grow on talus slopes (slopes where rock debris has accumulated). These communities are often dominated by the shrub vine maple, accompanied by a variety of ferns, grasses, and forbs. Another group of special climaxes is found on intrusive igneous rock in the Oregon Coast ranges. These areas support a type of meadow vegetation termed "grassy balds."

In temperate and polar regions, microclimate is strongly influenced by aspect, the direction of a slope. Slopes that face the poles (north-facing slopes in the northern hemisphere and south-facing slopes in the southern hemisphere) receive less solar radiation than slopes that face the equator. Consequently, in the northern hemisphere, south-facing slopes are typically warmer and drier than north-facing slopes. Microclimate is also influenced by a site's position on a slope. The tops of hills and ridges may experience unusual conditions because of excessive drainage or exposure to drying winds; valley bottoms are wetter and at night they are also cooler because cold air, which is denser than warm air, flows downslope and settles in low places. Thus topography, like soil conditions, influences the development of vegetation. A community that develops as a result of topographic conditions and differs from the climatic climax of a zone is termed a topographic climax.

A plant community that can be maintained only by repeated burning is termed a fire climax. In parts of the southeastern United States, periodic fires have favored fire-tolerant longleaf pines in areas that would develop forests dominated by broad-leaved trees if fire were absent. Similarly, in parts of east Africa, savanna vegetation, consisting of grasses and other low-growing plants accompanied by widely scattered trees and shrubs, is maintained by frequent burning, even though the environment is actually capable of supporting forests (see Chapter 13).

Finally, some plant communities, termed zootic climaxes, are maintained only by the interactions between animals and vegetation. Zootic climaxes occur where grazing mammals have an overriding influence on the structure and composition of vegetation or where dense aggregations of nesting birds impact plant communities through trampling and the deposition of feces.

Clements and Tansley envisioned the climax association that develops at the end of a successional sequence as a stable community, analogous to an organism. In this view, although a climax community might contain some local variation, it remains in a "permanent or apparently permanent" state of equilibrium (Clements 1916; Tansley 1920, 1935:293).

Not all plant ecologists of the early twentieth century believed that communities are tightly integrated associations of species. European ecologists Leonid Ramenski and Josef Paczoski and the American Henry Gleason considered communities to be fairly loose collections of species (Gleason 1917; Maycock 1967; Rabotnov 1978; Barbour 1996). Gleason argued that plant associations result from assemblages of organisms responding to the environment individualistically, according to their own ranges of tolerance (Gleason 1917, 1926; McIntosh 1975). In this view, a community is a collection of individual plants with similar environmental requirements, not a highly integrated unit. Clements's views tended to prevail in North America until fairly recently, however.

3.3.2   Examples

Primary succession

Succession occurring on a previously unvegetated substrate such as rocks or sand is termed primary succession. Bare substrate can result from geologic activity, the retreat of glaciers, or erosion. In primary succession, colonization takes place on relatively unweathered rock that was not previously vegetated, so the development of soils is gradual. For this reason, primary succession proceeds slowly.

In his work on primary succession on sand dunes, Cowles described the rigors of the environment facing the first dune pioneers. They must be able to tolerate conditions of low moisture availability and to withstand exposure to wind and to extreme temperatures. They also need to be able to cope with a shifting substrate. Colonizers of windblown sand must be able to spread rapidly, both horizontally and vertically. Their roots need to be capable of withstanding exposure when the sand covering them shifts, and their shoots must tolerate burial. "In short, a successful dune-binder must be able at any moment to adapt its stem to a root environment or its root to a stem environment" (Cowles 1899:177).

Only a few species of plants can thrive in this demanding environment.

These include the sand reed – a dune-binding grass – as well as a few other species of grasses and some willows. Once these pioneers become established, their presence reduces wind speed, traps moisture, provides shade and nutrients, and retards the downslope movement of sand. Other, less hardy, plants can then colonize the dune. Like other types of primary succession, the development of vegetation on dunes is slow.

The colonization of bare rock is likewise a slow process. Initially bare, flat rock surfaces can be colonized only by certain organisms that have special adaptations allowing them to eke out a living in such a demanding setting. Usually the only organisms capable of exploiting this habitat are lichens, mosses, and clubmosses. (Succession in rock crevices proceeds a bit more rapidly, because soil accumulates in these cracks more rapidly than on flat rock surfaces.) These organisms have numerous adaptations that allow them to meet their requirements on rocks. Because most rocks are impermeable to water, periods in which water stands on the surface alternate with dry intervals. Lichens are able to endure prolonged periods of desiccation and are able to carry on photosynthesis at low temperatures. This allows them to be active in winter, when moisture is most likely to be available in depressions in the rocky substrate (Daubenmire 1968).

These pioneers alter a rock outcrop by accelerating the disintegration of the substrate. Rootlike structures penetrate the rock and break it into small fragments; this process is aided by the carbonic acid they produce. As the rock weathers, a thin layer of soil forms. The lichens themselves trap wind-borne debris and dust, and they also contribute organic matter when their tissues die.

As a rock outcrop is modified by these pioneers, the site becomes hospitable to other organisms. In the next stage, a few kinds of grasses or forbs colonize the rock surface. These modify the environment further. Because they have substantially taller aboveground parts, these plants are able to intercept still more windblown material as well as rocks and leaves that fall downslope. This accelerates soil development further.

Secondary succession

Succession also occurs when an existing community is disturbed by the removal of vegetation. This type of succession, in which an area that has been cleared of pre-existing plants is recolonized, is termed secondary succession. Many types of disturbances, both natural and man-made, lead to secondary succession. Fires, insect outbreaks, storms, cultivation, logging, and trampling are examples of disturbances that remove vegetation and initiate secondary succession.

This type of succession proceeds far more rapidly than primary succession. I have already mentioned one reason why this is the case: soils on previously vegetated sites are more developed. In addition, the soils of previously vegetated areas contain microorganisms, seeds, and spores that greatly facilitate revegetation. The soil on such sites is likely to contain burrows made by earthworms and other animals and the root channels of predisturbance plants. These passageways aerate the soil and make it more permeable to water. Finally, disturbances of vegetated sites rarely remove all pre-existing vegetation. The plants that remain moderate the environment locally, creating microhabitats with more shade and higher moisture, and if they are alive and mature these plants may provide seeds. Consequently, the communities that develop after a disturbance are often characterized by a mixture of pioneer species and residual species from prior communities.

Almost all terrestrial plant communities experience repeated fires, except those in which vegetative cover is too sparse to allow fires to spread. In most cases fire damages or kills some plants, but the effects of this destruction on community composition depend on a fire's size and intensity. Cool fires typically kill only some plants, whereas hot fires may burn everything above ground as well as the organic matter in the upper layer of soil.

Paradoxically, fires often create improved soil conditions but are accompanied by a harsher microclimate at the soil's surface. Burning returns nutrients to the soil, so that soil fertility is temporarily enhanced after a fire. Furthermore, the decayed roots of fire-killed plants add organic matter, and the increase in sunlight following fire promotes nitrogen fixation by soil bacteria. Because burning decreases the amount of vegetation on a site, the total amount of water lost through the leaves of plants decreases, and soil moisture increases. On the other hand, the decrease in shade after a fire occurs results in higher ground temperatures.

The pioneers that invade after a fire find an environment with relatively few competitors for space, light, water, and nutrients, but to take advantage of these resources they must be able to tolerate the high temperatures and bright light on burned sites. Many also possess adaptations that allow them to regenerate from pre-existing plant parts or to reproduce from seeds after a fire. Some fire-adapted species regenerate by sprouting from roots. Most grasses, many shrubs, and a number of seral tree species including quaking aspen, paper birch, and redwood, have this capability. (Redwood also has thick bark which allows it to survive most fires.) Other species have adaptations that promote reproduction by means of seeds after a fire. For instance, lodgepole pine, jack pine, and black spruce produce special cones that open only after they are heated. After a fire, the seeds of these cones fall on soil that has been cleared by burning (see Box 9.1). Similarly, some shrubs, particularly those that occur in chaparral vegetation (a type of community dominated by thick-leaved, highly flammable, evergreen shrubs in Mediterranean climates), produce seeds that require high temperatures in order to germinate (Biswell 1974) (see Box 13.4).

3.3.3   Implications

The concept of succession leading to a final climax allows managers to think about the potential natural vegetation of a site as well as the vegetation that happens to be there at a particular moment in time. This is a useful approach because it provides insights about how the characteristics of a species change in different successional contexts. For example, Douglas-fir grows more rapidly when it is a member of a seral community than where it is the climax dominant. Such information has obvious practical implications for utilitarian managers interested in maximizing wood production.

The concept of succession has other practical applications as well. Some wildlife species utilize late successional habitats, such as mature and old-growth forests, whereas others utilize early successional habitats, such as clearings and groves of saplings (Thomas *et al.* 1975; Thomas 1979). By understanding these relationships, wildlife managers can manipulate habitats to favor species of particular interest. This topic is treated in more detail in Chapter 5.

The next three chapters will examine how utilitarian resource managers put the concepts discussed in this and the preceding chapter into practice.

References

Allen, D. L. (1962). *Our Wildlife Legacy*, 2nd edn. New York, NY: Funk and Wagnalls.

Barbour, M. G. (1996). Ecological fragmentation in the fifties. In *Uncommon Ground: Rethinking the Human Place in Nature*, 2nd edn, ed. W. Cronon, pp. 233–255. New York, NY: W. W. Norton.

Biswell, H. H. (1974). Effects of fire on chaparral. In *Fire and Ecosystems*, ed. T. Kozlowski, pp. 321–364. New York, NY: Academic Press.

Bruns, H. (1960). The economic importance of birds in forests. *Bird Study* 4:193–208.

Caughley, G. and A. R. E. Sinclair (1994). *Wildlife Ecology and Management*. Cambridge, MA: Blackwell Science.

Clements, F. E. (1916). *Plant Succession*. Washington, DC: Carnegie Institution of Washington.

Cowles, H. C. (1899). The ecological relations of the vegetation on the sand dunes of Lake Michigan. *Botanical Gazette* **27**:95–117, 167–202, 281–308, 361–391.

Cowles, H. C. (1901). The physiographic ecology of Chicago and vicinity: a study of the origin, development, and classification of plant societies. *Botanical Gazette* **31**:73–108, 145–182.

Daubenmire, R. (1968). *Plant Communities*. New York, NY: Harper and Row.

Franklin, J. F. and C. T. Dyrness (1988). *Natural Vegetation of Oregon and Washington*. Corvallis, OR: Oregon State University Press.

Gleason, H. A. (1917). The structure and development of the plant association. *Bulletin of the Torrey Botanical Club* **44**:463–481.

Gleason, H. A. (1926). The individualistic concept of the plant association. *Bulletin of the Torrey Botanical Club* **53**:7–26.

King, R. T. (1938). The essentials of a wildlife range. *Journal of Forestry* **36**:457–464.

Lay, D. W. (1938). How valuable are woodland clearings to birdlife? *Wilson Bulletin* **50**:254–257.

Leopold, A. (1933). *Game Management*. New York, NY: Charles Scribner's Sons.

Maycock, P. F. (1967). Josef Paczoski: founder of the science of phytosociology. *Ecology* **48**:1031–1034.

McIntosh, R. P. (1975). H. A. Gleason – "individualistic ecologist" 1882–1975: his contributions to ecological theory. *Bulletin of the Torrey Botanical Club* **102**:253–273.

Murkin, H. R., R. M. Kaminski, and R. D. Titman (1982). Responses by dabbling ducks and aquatic invertebrates to an experimentally manipulated cattail marsh. *Canadian Journal of Zoology* **60**:2324–2332.

Rabotnov, T. A. (1978). Concepts of ecological individuality of plant species and of the continuum of the plant cover in the works of L. H. Ramenskii. *Soviet Journal of Ecology* **9**:417–422.

Tansley, A. G. (1920). The classification of vegetation and the concept of development. *Journal of Ecology* **8**:118–149.

Tansley, A. G. (1935). The use and abuse of vegetational concepts and terms. *Ecology* **16**:284–307.

Thomas, J. W. (tech. ed.) (1979). *Wildlife Habitats in Managed Forests: The Blue Mountains of Oregon and Washington*. U.S. Department of Agriculture Forest Service, Agriculture Handbook No. 553, Washington, DC.

Thomas, J. W., G. L. Crouch, R. S. Bumstead, and L. D. Bryant (1975). Silvicultural options and habitat values in coniferous forests. In *Proceedings of the Symposium on Management of Forest and Range Habitats for Nongame Birds*, pp. 272–287. U.S. Department of Agriculture Forest Service, General Technical Report GTR-WO-1, Washington, DC.

Thomas, J. W., R. G. Anderson, C. Maser, and E. L. Bull (1979a). Snags. In *Wildlife Habitats in Managed Forests: The Blue Mountains of Oregon and Washington*, tech. ed. J. W. Thomas, pp. 60–77. U.S. Department of Agriculture Forest Service, Agriculture Handbook no. 553, Washington, DC.

Thomas, J. W., H. Black, Jr., R. J. Scherzinger, and R. J. Pedersen (1979*b*). Deer and elk. In *Wildlife Habitats in Managed Forests: The Blue Mountains of Oregon and Washington*, tech. ed. J. W. Thomas, pp. 104–127. U.S. Department of Agriculture Forest Service, Agriculture Handbook no. 553, Washington, DC.

Weaver, J. E. and F. E. Clements (1938). *Plant Ecology*, 2nd edn. New York, NY: McGraw-Hill.

Woodwell, G. M. (1967). Toxic substances and ecological cycles. *Scientific American* **213**(3):24–31.

4

Techniques – harvest management

4.1 Classifying species on the basis of utilitarian values

The disciplines of wildlife management, range management, and forestry in North America developed in large part as a reaction to the excessive exploitation of birds and mammals by market hunters and to overgrazing and excessive timber cutting (see Chapter 1). Hence, it is not surprising that these disciplines sought to regulate exploitation by managing harvests of economically valuable species.

This approach to management embodies a utilitarian view of species. In 1885 the U.S. Department of Economic Ornithology (which was at that time a part of the Department of Agriculture) began studies of the economic value of birds. These studies classified species as good or bad, according to whether they were deemed beneficial or harmful to agriculture. Harmful species were those that consumed crops; beneficial ones ate the harmful species. These studies promoted the cause of conservation through their claim that many nongame birds played an important role in the control of insects and weeds. The underlying message, however, was that some species were better than others and that as many pests as possible should be killed.

Species that were considered useful were managed to enhance their populations. Under the new resource management policies, these species were protected (by the Lacey Act, for example) or harvested in a controlled fashion. On the other hand, species that were considered pests were managed to reduce their populations. This chapter looks at the regulated harvest of wild plants and animals. Chapter 5 covers the management of habitats to benefit featured species, and Chapter 6 covers the control of unwanted species.

4.2   Kinds of harvest

4.2.1   Commercial harvests

Harvests can be classified according to whether their main objective is to provide recreation, profit, or subsistence and also whether or not they are legal. In commercial harvests, the objective is to sell the harvested product at a profit. Unregulated market hunting, which depleted or exterminated numerous species of wildlife before conservation laws were enacted, is one form of commercial harvest (see Chapter 1). As a result of the excesses of market hunting, the commercial harvest of wild game is now illegal in the United States. Commercial fish harvests occur in the oceans or large bodies of fresh water and involve either marine fishes, anadromous fishes, or freshwater fishes. Commercial fisheries also harvest invertebrate food resources, such as abalone, oysters, crabs, shrimp, and lobsters.

Commercial timber harvests take either hardwoods or softwoods. Hardwoods are broad-leafed, deciduous tree species, such as maple, hickory, or oak. In North America, these species grow mainly in the eastern United States. Their high quality wood is used for furniture and flooring. Softwoods are coniferous trees, which are usually evergreens. Douglas-fir, pine, cedar, and spruce are among the principal commercially valuable softwoods. They are typically used for paper, lumber, or pulpwood.

A variety of nontimber products are commercially harvested from forests, including, in temperate forests, mushrooms, decorative greenery for the floral industry, and pharmaceutical products. The best-known of these is the anti-cancer drug taxol, which can be obtained from the bark of the Pacific yew. The harvest of nontimber products from tropical forests is discussed in Chapter 14.

4.2.2   Recreational harvests

Recreational harvest refers to the legally regulated, noncommercial harvest of wildlife or plants. Flowers, herbs, mushrooms, berries, mosses, and lichens are sometimes harvested for recreational purposes, but usually the term applies to hunting, trapping, and sport fishing. Trophy hunting is a type of sport hunting in which the primary goal is the killing of an animal that is rare or dangerous or is an unusual physical specimen because of its large size, antlers, or other characteristics.

Most terrestrial vertebrates that are hunted for sport are either large mammals or birds. Hunted mammals include ungulates, such as deer, elk, and moose, or predators, such as black bear and cougar. Game birds include waterfowl (ducks, geese, and swans), a few other game birds associated with water (such as the sandhill crane), and upland game birds (hunted birds that are not associated with wetland habitats). The latter group includes the wild relatives of domestic poultry (turkeys, grouse, pheasant, partridge, and chukar) as well as mourning doves, snipe, and woodcock.

In recreational harvests, recreation is part of the product obtained from the resource. Managers must therefore take into account the quality of the recreational experience; sometimes this is more important to participants than the take. The definition of what constitutes a high-quality recreational experience depends on cultural values and history (see Box 4.1). Arguments for and against recreational harvest of game and fish are summarized in Box 4.2.

Box 4.1   The hunting experience in the United States and Germany

In America, hunting is regarded as an egalitarian activity that anyone can participate in. These attitudes reflect the idea that settlement of the frontier was a romantic challenge which can be recapitulated by hunting, fishing, and trapping. Leopold (1966) termed this ability of hunting to remind people of their history the "split-rail value." In *A Sand County Almanac*, first published in 1949, he wrote that the boy scout who "has tanned a coonskin cap and goes Daniel Booneing in the willow thicket" is "reenacting American history," and the farmer boy who "arrives in the schoolroom reeking of muskrat [because] he has tended his traps before breakfast" is "reenacting the romance of the fur trade" (Leopold 1966:211–212). Leopold identified two other values of hunting: it reminds people of their connection to nature and it encourages the exercise of ethical restraints (although he acknowledged that these values can be undermined by excessive reliance on gadgetry). Thus, in the U.S.A., hunters expect game to be available in habitats that are perceived as natural. Although hunting is controversial in the U.S.A., Kellert (1993) reported that most Americans he surveyed approved of hunting if the meat was used.

German hunting contrasts markedly with the American situation. In Germany, hunting has a different history and different meanings. Not surprisingly, Germans also value different aspects of the hunting experience.

For centuries, hunting in Europe was confined to the nobility and perhaps the clergy. German hunting traditions reflect these roots. Even today, hunting is considered a status symbol, and only a small proportion of the population (less than 1%) hunts. The requirements to qualify as a hunter are stringent, and the activity of hunting itself embodies a formality and a concern with tradition that is foreign to American hunting. Hunters wear ceremonial garb, use a formalized vocabulary, and observe special hunting ceremonies, including blowing the hunting horn. In this context, trophies confer considerable status, and management is geared toward the production of individuals with large antlers and other highly valued attributes (Webb 1960; Gottschalk 1972). Paradoxically, although hunting is considered a status symbol in Germany, 85% of the Germans surveyed by Kellert disapproved of hunting (Kellert 1993).

Box 4.2   Arguments for and against recreational harvest of fish and game

Hunting, trapping, and fishing are emotionally charged issues, with passionate convictions held on both sides. These controversies revolve around issues of ethics, emotions, ecology, management, and economics. Those who oppose hunting and trapping emphasize the suffering and death of *individual* animals, whereas the pro-hunting and trapping camp tends to stress the effects of these activities on animal *populations*. Some of the most common arguments for and against these activities are briefly summarized below.

For

- **Death by hunting, fishing, or trapping is faster and more humane than a slow, lingering death from a natural cause.** It is often asserted that the deaths of animals killed by hunters or trappers substitute for painful, prolonged deaths that would otherwise occur as a result of disease, starvation, or predation. This argument applies only to situations where hunting and trapping cause compensatory mortality and only to types of harvest that kill quickly.
- **Recreational harvests can prevent the harvested species from exceeding the carrying capacity of the environment.** This argument

is related to the previous one and applies mainly to populations of herbivores, especially where native predators have been removed.

- **Recreational harvests provide useful products.** The products obtained from hunting and trapping include meat, leather, and furs.
- **Pest individuals can be removed from areas where they cause economic damage.** Hunted and trapped species are responsible for several kinds of economic damage. Deer and elk damage haystacks, crop fields, pastures, and orchards. Furbearers such as beaver, muskrat, and nutria damage dikes and plug culverts, causing local flooding. Large predators sometimes threaten people, and smaller predators such as otter, mink, fox, and coyote kill poultry and, in the case of the coyote, sheep. (The control of wildlife damage is considered in more detail in Chapter 6.) Proponents of hunting and trapping point out that these activities can reduce damage while providing recreational and financial benefits.
- **Recreational harvests provide valuable information for managers.** Data from hunting and trapping on the age, sex, and physical condition of the harvested individuals would often be difficult and expensive to obtain in any other way. This information provides the basis for managing harvested populations. Without this data it would be difficult to assess population trends and to implement appropriate management policies.
- **Hunting, fishing, and trapping provide recreation and can promote appreciation of nature, an understanding of ecological relationships, and the development of a sense of sportsmanship and ethics.**
- **In the U.S.A., income from the sale of hunting and fishing licenses and from taxes on guns, ammunition, and fishing gear provides funding for conservation.**

Against

- **Hunting and trapping are inhumane because of the trauma and suffering they produce.** Many people argue that people should not be responsible for inflicting suffering on wild animals.
- **The products obtained by recreational hunting and trapping are not necessities.** In developed countries it is not necessary to kill wild animals in order to obtain food and clothing. The killing of furbearing animals to provide expensive furs, a luxury item, is considered particularly offensive.

- **Hunters and trappers kill or wound many unintended victims.** This includes individuals that are not killed outright as well as mortality among nontarget species. People are also killed and wounded in hunting accidents.
- **It is unethical to kill for recreation.** Although the arguments for and against recreational hunting and trapping touch on numerous issues, the crux of the controversy seems to be this ethical issue. It is on this point that people on opposite sides of the question seem unable to understand each other's point of view.

Many people who are opposed to recreational hunting and trapping find some forms of these activities more objectionable than others. For instance, most anti-hunters find hunting less objectionable if the hunters make use of meat or other products from the animals that are harvested, and many express the highest degree of disapproval for forms of trophy hunting in which the harvested individuals are not utilized.

4.2.3   Subsistence harvests

Another type of harvest is for the purpose of subsistence. In this case, a harvester's family or community is somewhat dependent upon the products that are obtained. In many parts of the world, gathering wood for fuel and hunting are significant forms of subsistence harvest.

4.2.4   Illegal harvests

Regardless of whether the objective is commercial gain, recreation, or subsistence, harvests may be legal or illegal. A "poacher" may be a subsistence hunter who is prohibited from meeting his or her needs in traditional ways, a recreational hunter who hunts an abundant game species but does so without a license or takes more than the legal bag limit or kills an animal out of season, or a professional who illegally kills game or fish for profit (see Chapter 7).

The intensity of illegal harvest depends on a variety of factors, including economic, social, and cultural factors such as the profits to be made, the likelihood of being caught, the severity of the penalty, whether the regulations against hunting are regarded as legitimate and fair, and the standard of living of the people governed by the regulations. Clearly, poaching is likely to be prevalent where enormous profits can be made and the likelihood of being caught is low or penalties are small compared to potential profits. Also, officials involved in regulating trade in harvested products are more likely to be corrupted in situations where big proceeds can be obtained (Parker and Amin 1983). This is one reason why the contemporary illegal traffic in wild plants and animals is a serious problem.

Where hunting regulations are made by one class of people to govern others who have little or no say in the matter, compliance is likely to be low. In Europe, game management has a long association with protecting the property of the wealthy classes. In this setting, poaching was often seen as a right of the common man, a way to assert oneself against the aristocracy, and therefore people were generally more sympathetic to poachers than to enforcers. This attitude is reflected in stories of the cultural hero Robin Hood killing the king's game and giving it to landless peasants and in children's books such as Roald Dahl's *Danny, The Champion of the World*. The situation became even more complicated when colonial authorities denied people access to resources that had cultural significance and that they had utilized and been dependent upon for generations. We shall return to this topic in more detail in Chapter 12.

These categories are not mutually exclusive. Recreational hunters often use the meat they obtain, although they would not usually be classified as subsistence hunters. Similarly, many fur trappers seek both recreation and financial gain. Some species are taken by more than one type of harvest. For instance, salmon are sought for recreation, commerce, and subsistence. This chapter covers the regulation of recreational and commercial harvests.

4.3   Managing for sustained yield

4.3.1   In theory

Maximum sustained yield

Resource managers often have the job of deciding whether a species should be harvested and, if so, how many individuals should be removed from a population. To do this, they obtain data on population dynamics, use it to set harvest levels, and enforce regulations. In this way, they attempt to prevent both overpopulation and overharvesting of the harvested species. The basic question posed by utilitarian resource managers is: How many individuals can be removed from a population without compromising its ability to recover by replacing the harvested individuals? Or, stated another way, if we want to harvest a population without causing it to decline permanently how much can we take?

If we take only an amount that is equivalent to what the population produces (and if the harvest itself does not cause reproduction to decline), then we should be able to harvest repeatedly, without triggering a population decline. This harvest could be considered sustainable, that is, capable of being continued indefinitely.

Suppose that we want to know the maximum amount we can take. This would be the population's maximum sustained yield (the largest number of organisms that can be removed repeatedly under existing conditions without causing the population to decline). To find out how we can do that, we need to consider the effects of harvesting at different population densities.

Assume that we are dealing with a population that shows logistic growth (see Chapter 2). In other words, our hypothetical population is limited by resources; the curve showing population growth through time has a sigmoid shape (see Figure 2.4). Now imagine the population consists of N_1 individuals at time t_1 and grows to N_2 individuals by time t_2. At time t_2, this population has almost reached K, the carrying capacity of its environment (Figure 4.1C), and we decide to remove enough animals to return the population to the level it had previously attained at time t_1. (In this example the interval between successive ts is one year. We return the population to the level it had the previous year by reducing it from its level at t_2 to its level at t_1.) To do this, we remove $(N_2 - N_1)$ individuals. The population continues to grow, but now it grows with the growth rate characteristic of a population consisting of N_1 individuals. That is, it displays the growth rate that is associated with the population size at t_1.

Now imagine that we follow the procedure described in the preceding paragraph, but we begin our harvest when the population is very small (Figure 4.1A) or when the population has achieved moderate size (Figure 4.1B). Any of these harvests should be sustainable indefinitely, if the population is regulated by density-dependent processes, if we do not take more than the yearly increment with each year's harvest, and if carrying capacity does not change.

But if we want to maximize the sustainable yield, we should harvest when the population has reached about one-half its carrying capacity, because that is the density at which the population growth rate will be at its maximum (Figure 4.1B). Therefore, if we are harvesting a population that displays logistic growth, we would expect the maximum sustained yield to be obtained if the harvest takes place when $N = K/2$. If the population we wish to harvest is below $K/2$, then we shouldn't harvest it at all; if it is above $K/2$, then in principle we will obtain maximum sustained yield if we temporarily harvest at high enough levels to drive the population down until it is at half the carrying capacity. The implication is that to maximize harvest, populations should be

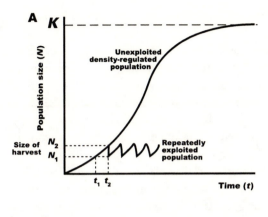

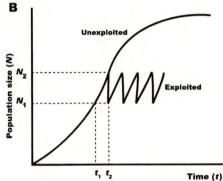

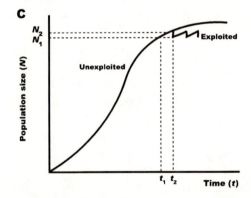

Figure 4.1. The effects of harvest on a population that exhibits density-dependent growth. (A) Effect of harvesting when population density is well below the carrying capacity of the environment. (B) Effect of harvesting when the population is at one-half of carrying capacity. (C) Effect of harvesting when the population approaches carrying capacity. Notice that the rate of population growth following the removal of animals is greatest when the medium-density population is harvested, that is, when N equals $K/2$. (After Begon et al. 1996.)

reduced well below the carrying capacity that their habitat is capable of maintaining (Gross 1969).

Notice, however, that this conclusion applies only to populations whose growth is described by a sigmoid growth curve, that is, populations in which density-dependent effects related to limiting resources occur as K is approached. Furthermore, we have assumed that removing individuals from a population does not interfere with subsequent population growth. That is, we expect that each time we reduce the population, it rebounds with the growth rate predicted by the logistic growth equation.

If our assumptions that the population is regulated by density-dependent processes and that harvest does not negatively impact reproduction are correct, and if we have correctly gauged the carrying capacity of the habitat and the population's reproductive rate, then the method described above should work quite well to identify the harvest level that will give maximum sustained yield. If we overestimate carrying capacity, then our estimate of $K/2$ will actually be greater than the true value for $K/2$. In that case, we will have underestimated the maximum sustained yield. The same is true if we underestimate the annual reproduction. We won't harvest as much as we could, but we will not deplete the population either. But if we err in the other direction, each time we take more than the maximum sustained yield we will drive the population lower and lower, sending it into a downward spiral. For this reason, it is important to be conservative and set harvest levels well above the estimated maximum sustained yield.

There is one other assumption that is usually overlooked but which must be met if a harvest is to be sustainable. That is the assumption that the harvest itself does not disturb the population or otherwise change ecological conditions in ways that affect the population growth rate or the habitat's carrying capacity. If this assumption is not met, then losses caused indirectly by hunting must be added to the direct losses when considering the effects of harvest. Game species are usually fairly tolerant of people (this is one reason why they are game species), and so in many cases disturbance from hunters is not a serious problem for the harvested species (although other species that are incidentally disturbed may be more sensitive). But even with game species, harvesting may influence population productivity in subtle ways. After reviewing studies of the impacts of hunting on European waterbirds (such as waterfowl, shorebirds, and wading birds), Madsen and Fox (1995) concluded that because hunting reduces the amount of time birds spend feeding, hunting has the potential to reduce reproductive output the following spring. This winter-spring link occurs because reproduction in many species of waterbirds is limited by the amount of fat reserves that are accumulated prior to the breeding season. In addition, by-products of managing for harvested species have the potential to cause substantial ecological impacts that are often ignored. Roads built to facilitate timber harvests and fences built to manage livestock are examples of such by-products. These indirect impacts should be considered when the effects of hunting are evaluated.

Additive and compensatory mortality

In order to assess the effects of harvest on the size of a harvested population, it is helpful if a resource manager knows whether the mortality caused by harvest substitutes for mortality from other causes. In other words, is mortality compensatory? (See Chapter 2.) If hunting mortality merely substitutes for other causes of death, and "the harvesting of 1 animal saves the life of another" (Caughley 1985:5), then the total number of individuals that die should not increase as mortality from hunting increases. When hunters kill 50 animals in a population that would normally lose 75 animals to predators or other natural causes, then natural mortality should drop by 50 animals to compensate for the 50 animals taken by hunters, if hunting mortality is truly compensatory. We would expect that a graph of the relationship of total mortality to hunting mortality would not show any effect of hunting on the total number of deaths. This is the pattern represented in Figure 4.2A.

If hunting mortality is not compensatory, the deaths caused by hunters add to other causes of death, and deaths rise with an increase in hunting. This is termed additive mortality. If 50 animals are killed by hunters, then the total mortality should go up by 50 deaths. This is the pattern shown in Figure 4.2B. If the situation is somewhere in between these two extremes, there is partial compensation for hunting mortality, and hunting causes total mortality to increase some but not as much as the total number of harvested individuals (Figure 4.2C).

In principle harvest mortality can be compensatory in any organism that is regulated by density-dependent effects. For instance, when a large number of fruits are crowded beneath a parent tree, seedling mortality is likely to be high and density-dependent. In this situation, it may be possible to harvest many fruits without having any impact on the tree's reproductive rate (see Chapter 14).

If mortality is truly compensatory, then harvest levels can be increased without increasing a population's losses, that is, there is a shootable surplus. This is often a desirable situation from a manager's perspective. The other side of this coin, however, is that if a population that experiences compensatory hunting mortality begins to decline, protection from hunting pressure would not be expected to be helpful, because natural mortality would substitute for hunting losses.

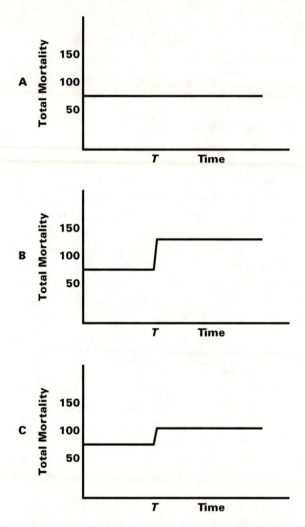

Figure 4.2. The effects of hunting on mortality in a hunted population. Assume that before hunting begins, 75 animals die from other causes and that hunting, which begins at time *T*, kills a total of 50 animals. (A) Hunting mortality is compensatory. If hunters kill 50 animals, then only 75 − 50 or 25 animals will die from natural causes, and total mortality with hunting will equal 50 deaths from hunting plus 25 deaths from natural causes or 75 individuals. The total number of deaths does not increase as hunting mortality increases. (B) Additive mortality. Hunting causes an additional 50 deaths; total mortality with hunting rises from 75 deaths to 125 deaths. (C) Hunting mortality is partially additive and partially compensatory. When hunters kill 50 animals, natural mortality drops, in this case by 25 animals, so that 75 − 25 or 50 individuals die from natural causes. Total mortality with hunting equals 50 deaths from hunting plus 50 deaths from natural causes, or 100 deaths.

In reality, mortality often operates in a manner that is somewhere between these two extremes, that is, there is partial compensation. For decades, the red grouse of the British Isles was considered a classic example of a species with compensatory hunting mortality, and territorial behavior was thought to be the mechanism that produced this compensation (see Box 4.3), but several lines of evidence point to the conclusion that the situation is actually more complex. Although hunting mortality in grouse may compensate for natural deaths during the breeding season, losses that occur during winter appear to be added to hunting losses (Bergerud 1985, 1988).

Box 4.3   Is hunting mortality compensatory or additive in red grouse populations?

The red grouse is an upland game bird of Scottish and Irish moors (open habitats dominated by heather and similar vegetation). This species is popular with Scottish hunters. The birds are taken in late summer or early fall. Since the turn of the century, their habitat has been intensively managed to benefit grouse populations and thus provide more game (see Chapter 5).

Red grouse are relatively easy to observe because these large, diurnal (daytime-active) birds inhabit open country. For more than four decades, a team of scientists has been studying red grouse populations in northeast Scotland. This long-term investigation provided an excellent opportunity to evaluate the mechanisms controlling population and the effects of hunting on this species. The studies involved behavioral observations of individually marked birds, as well as experimental manipulations to test hypotheses about how population density is regulated.

The researchers found that red grouse males are territorial – that is, they attempt to defend defined areas – throughout most of the year and that winter grouse populations consist of males on territories, hens paired with them, and nonterritorial individuals. They also found evidence that birds without territories move into marginal habitats. In these habitats, which are presumably of poorer quality, they suffer heavier losses than territorial birds. The grouse that survive in marginal habitats move onto vacated territories when territory holders die, however, and once they obtain a territory, their chances of surviving to breed improve. Mortality rates for banded adult grouse in hunted populations suggested that fall shooting had no effect on population density the following spring because hunting eliminated mainly individuals without territories. In other words, hunting mortality was compensatory. On the basis of these findings, the researchers concluded that social behavior in the form of male territoriality regulates population density in red grouse and that the nonterritorial males are expendable "surplus non-breeders" (Jenkins *et al.* 1963:373).

To test the hypothesis that nonterritorial breeders were prevented from breeding because they lacked territories, researchers experimentally removed grouse from territories and determined whether nonterritorial birds took over the vacated territories and bred the following spring. As predicted, nonterritorial birds did move onto unoccupied territories and breed, whereas on the control areas there were few changes in occupancy (Watson and Jenkins 1968).

These findings suggested that hunters could harvest grouse without depressing population size. The researchers concluded that a failure to harvest this "over-production" amounted to "under-exploiting a genuine population surplus" (Jenkins *et al.* 1963:356, 373).

Detailed studies of the population dynamics of some large herbivorous mammals such as elk suggest that they exhibit partial compensation for hunting mortality (Caughley 1985). With ungulates, hunting mortality often partially substitutes for death from predation or, if predators have been removed, from starvation. Deer populations that are not kept in check by predation or hunting have the potential to deplete their food supplies, especially in winter. When this happens, starvation and economic damage to crops or ornamental plantings usually follows (see Chapter 6).

Like game managers, forest managers seek to maximize harvest, but the harvested product is wood. They do this by creating conditions favorable to the growth rates of individual trees of economically valuable species. (Techniques for manipulating the composition of forest communities to maximize wood production will be explored in Chapter 5.) When the trees in a stand are spaced far enough apart to allow for maximum growth, but not so far apart that space will be wasted, a stand is said to be fully stocked (Wiley 1959).

In managing rangeland for sustained yield of livestock, the goal is to avoid overharvesting forage plants. Livestock are the products of this enterprise, but wild plant communities are the resource that managers seek to use sustainably. In the words of one textbook, "grass is [ranchers'] product and livestock their manner of harvesting and marketing it" (Highsmith *et al.* 1969). In principle, this means regulating the levels of grazing so that livestock do not remove plants at a rate that exceeds their rate of recovery.

4.3.2   In practice

Managing harvests is a three-step process. Data must be obtained, limits on the harvest (also termed the "take" or "offtake") must be set, and those limits must be enforced.

Getting data

Only occasionally can population growth rate be measured directly. If it is possible to count all the individuals in a population at several successive time intervals, then managers can determine whether a population is increasing or decreasing and at what rate. They can use this information as a basis for adjusting harvest levels so that fewer individuals are taken from declining populations or more individuals are removed from dense ones. In most cases, it is not possible or practical to count all the individuals in a population, however. Instead population growth rates must be estimated from a sample of the population of interest.

Managers also make use of information about the physical condition of harvested individuals. If animals typically lack fat stores or show signs of disease or parasite infestations, or if trees have slow growth rates, then managers look for the causes of these conditions. Malnutrition, disease, and slow growth rates can be signs of high population density, a situation that can sometimes be alleviated by increasing the level of harvest.

In exploited populations, the harvest itself can be used to obtain information about the composition of the harvested population. At hunter check stations, for instance, wildlife biologists obtain information about the age, sex, and physical condition of harvested animals. Similarly, fisheries biologists aboard commercial fishing and whaling vessels obtain data on harvested stocks. For example, the Soviet take of southern right whales between 1951 and 1971 and the Japanese catch of minke whales since 1987 have provided substantial data on the distribution, diet, movements, reproduction, and population structure of the harvested species (Tormosov et al. 1998; Normile 2000). These research expeditions were highly controversial, however, because in both cases whales were hunted without the approval of the International Whaling Commission, the international body that regulates commercial whaling (see Chapter 9).

Setting harvest limits

Fish and game managers have several methods of limiting harvest, including bag limits or quotas (limits on the number of animals that can be killed by an individual hunter, a business, or a nation), season limits (the dates during which hunting may take place), limits on the sex or age (or, for fish, size) of the animals that may be harvested, restrictions on the methods or equipment that can be used, and restrictions on the areas where hunting is allowed.

Foresters managing for sustained yield usually focus on tree maturation instead of reproduction. Rather than estimating how many individuals are produced and can be harvested in a given period of time, they attempt to determine how much wood is produced during a given interval and then adjust harvest so that no more than that increment is removed during an equivalent interval. The maximum volume of timber that may be harvested sustainably is termed the "allowable cut." The crucial management decision is not how many trees can be cut, but how often a stand can be cut, or rotation length. To achieve sustained yield, a forester must not allow a forest to be cut until it has reached the same stage of development it had before it was cut. If parts of a large forested area are cut sequentially, with each stand being allowed to grow back to the stage of development it was at before it was cut, and stands are not cut again until what was removed has been replenished, this procedure should be sustainable. Note that this approach involves a shift in perspective from individuals to the arrangement of trees across a landscape. Timber harvest regulations may stipulate the techniques that can be used (for example clearcutting or selective cutting), the type and intensity of harvest permitted under certain ecologically sensitive circumstances (such as where endangered species are involved or on highly erodible slopes or near streams), and requirements for postharvest reforestation.

The main tool that range managers use to regulate yield is manipulation of the intensity and timing of grazing (see Chapter 5). Grazing pressure can be manipulated by regulating the stocking level, that is, the number of grazing animals, as well as the timing of grazing. A number of different grazing systems have been developed that aim to minimize the negative impacts of livestock grazing on rangelands. Many of these replace continuous, year-round grazing with a short period of intense grazing followed by a rest period when plants can recover.

Although there is a firm theoretical framework for harvesting on a sustained yield basis, in practice setting harvest limits usually involves a fair amount of trial and error (see Boxes 4.4 and 4.5). Most of the time, carrying capacity and reproductive rate are not known precisely, and the mechanisms of population regulation are not fully understood. Instead of determining carrying capacity and then using that as a basis for setting harvest limits, managers make an educated guess about the level of harvest that a population can sustain, monitor the response of the population, and adjust harvest levels accordingly if necessary.

Box 4.4   Setting harvest limits for greywing francolin populations in South Africa

The greywing francolin is a popular South African upland game bird. In the Stormberg Plateau, eastern Cape Province, hunting parties pay landowners for the privilege of hunting on their farms. In the traditional method of hunting, once a year two to seven hunters walk in a line behind several pointing dogs, and when a covey is found, the hunter who is nearest to the birds flushes them. The coveys are then shot at until a preset bag limit is reached.

To assess the impact of harvesting on hunted greywing populations, Little and Crowe (1993) compared the effects of two harvest levels during a four-year period. Hunting parties were either told to limit their shooting after 50% of a covey was shot or allowed to bag an unlimited number of birds. Data were obtained for a total of 123 hunts. The number of birds found per minute of searching along predetermined routes was used as an index of abundance. A "population" was defined as the birds on a given farm. (Note that this definition is based on administrative considerations, and that a population designated in this way may or may not coincide with a biologically discrete population.)

When the researchers looked at the relationship of harvest levels to abundance the following year, they found that harvest levels below 50% were not associated with a decrease in estimated abundance, but populations from which an unlimited number of birds were harvested were significantly lower a year later. On the basis of these findings, Little and Crowe recommended that greywing francolins should be harvested no more than once a year and that the take should be limited to not more than 50% of a covey. They predicted that this level of harvest would be ecologically sustainable.

In addition, the researchers looked at factors that influence the economic and social success of greywing hunts. Group size affects the economic viability of the hunt, but there are tradeoffs. Landowners prefer large hunting parties, because larger groups pay higher fees. On the other hand, hunter satisfaction must also be maintained if hunting is to continue. To evaluate the social dimensions of the hunt, Crowe and Little tallied the number of shots fired and birds bagged per hunter. These measures were used as indicators of hunter satisfaction. Smaller groups were associated with significantly greater hunter satisfaction. (This was true regardless of which measure — shots fired or birds killed — was used.) This analysis led Little and Crowe to recommend that group size be maintained at four to seven hunters, a size which they felt would be small enough to maintain hunter satisfaction while providing adequate remuneration to landowners.

Box 4.5   Managing polar bear harvests in the Canadian arctic

Bears have a number of characteristics that make them vulnerable to overharvesting. It takes several years to reach sexual maturity, litter sizes are small, and the intervals between births are long. For these reasons, bear populations do not recover rapidly after they decline. This makes it especially important to understand the impacts of harvest on these large carnivores.

Canada's western Hudson Bay population of polar bears is utilized in several ways. Prior to 1960, bears were harvested for York Factory, a fur trading post to the south. In addition, adult females and cubs were harvested by Inuit hunters for hides and dog food, and military personnel stationed at Churchill, Manitoba killed an unknown number of bears. Except for subsistence hunts and control of problem individuals, harvest in the vicinity of Churchill ended by the middle of the 1960s. In addition, viewing bears (primarily adult males, which predominate near the coast) has become a popular tourist activity along the western coast of Hudson Bay. Managing polar bears for these multiple uses and values (both positive and negative) requires accurate information on population size and trends, but polar bear populations are not easy to study. They are dangerous and occur in inaccessible terrain under rigorous environmental conditions. For these reasons, it is generally difficult to obtain large enough sample sizes of polar bear populations to arrive at accurate estimates of population size. At Hudson Bay, however, the bear population concentrates in a restricted area during the four months of the year when sea ice melts, and this facilitates data collection.

Derocher and Stirling (1995) obtained data on the western Hudson Bay polar bear population during a 16-year period. From 1977 through 1992, they trapped bears, immobilized them with drugs, recorded their sex and reproductive status, extracted a vestigial premolar (which was subsequently used to estimate the individual's age), and marked the bears with a tattoo on each side of the upper lip and a plastic tag in each ear. After the initial trapping effort, some of the trapped bears were ones that had been previously captured and marked. A number of models have been developed that generate population estimates from this kind of mark-and-recapture data. The basic idea underlying these models is that the proportion of recaptures in the trapped population can be used to derive an estimate of population size. If many of the marked animals disappear, then survival is apparently low. If a high proportion of the marked individuals are subsequently recaptured, this suggests that survival is high. The models involve certain assumptions, however. For example, it is necessary to assume there is no emigration. If some individuals permanently leave the population, researchers know only that these individuals are missing from the population; it is not possible to tell if they died or moved away. Previous research on the study population had shown that the bears did not often settle in adjoining populations, so the researchers felt that emigration probably did not introduce any serious errors into their analysis.

The average size of the autumn polar bear population from 1978 to 1992 was 1000 individuals, and an average of 191 cubs was recruited into the population each year. Since 1980 the population neither increased nor decreased, presumably because the population stabilized after the level of harvest was reduced. Because the population size was high and stable and recruitment was fairly high, the researchers concluded that the current level of consumptive uses (subsistence harvest and control for the purpose of protecting people) and nonconsumptive uses (tourism) is sustainable; however, they cautioned that if harvesting reduces the population of adult males, opportunities for tourism could decline.

Enforcing regulations

Even the best harvest regulations work only if they are followed. Compliance depends upon a variety of factors, including whether or not the system for setting regulations is respected, whether harvesters believe that their competitors (either sport hunters or commercial operations) will comply, whether they expect to get caught if they violate regulations, and, finally, the ratio of expected gains to expected penalties for noncompliance. As noted above, enforcement is difficult if profits are high or the people affected by antipoaching regulations do not support those regulations. Enforcement is particularly difficult when a resource is harvested by several nations (Chapter 1).

4.3.3   Social, economic, and political considerations

Biological considerations are not necessarily the major force determining harvest levels. A host of nonbiological issues also enter into decisions about harvest levels (see Box 4.4). This has not always been bad for the harvested species. As noted above, hunters like to have high populations of game animals. For this reason, game harvests are usually set at levels that are much lower than what the harvested populations could probably sustain. Tradition plays a big role as well. People resist change, so existing practices have a certain inertia. Thus, although our knowledge of the dynamics of hunted populations has become more sophisticated in recent decades, this knowledge has not necessarily been translated into practice. One biologist suggests that the major difference between wildlife management at the turn of the century and today is that there has been "a relaxation of strictures against hunting on Sundays" (Caughley 1985:4). These conservative tendencies have tended to protect game species from overexploitation.

On the other hand, when recommended harvest levels are exceeded, serious problems can result. Any time a commercially valuable resource is regulated, particularly when the potential short-term economic gains of exploitation are large, there is a danger that scientists' recommendations will be ignored. Managers can be subtly or openly pressured into acquiescing to harvest levels that are not sustainable, especially where large profits are at stake, regardless of whether the resource is fish, timber, ivory, whales, or forage.

4.3.4   How successful has harvest management been?

In many cases, managed game harvests have been quite successful for long periods of time, for two reasons. First, the risks of managing for maximum sustained yield are fairly low with populations that have high reproductive rates and are regulated by density-dependent processes, because they rebound when their density declines. Many game species have these characteristics; if they didn't they might well have been overharvested to extinction long ago (Caughley and Sinclair 1994). Second, as noted above, wildlife management typically aims to keep game populations far above $K/2$, because hunters and nonhunters alike want to have dense populations of game animals, and they resist attempts to reduce population levels to half of carrying capacity, even though such reductions might result in greater yields.

On the other hand, attempts to regulate commercial harvests of whales and fish have been less successful. The harvest of ocean resources such as fish and marine mammals poses special problems for both biological and economic reasons. One problem is that marine environments sometimes vary in ways that are complex and unclear (see Box 4.6). A second difficulty in managing marine resources is that fish populations, or stocks, mingle at sea. This makes it difficult for managers to obtain the kinds of data they need to set sustainable harvest levels. Managers cannot reliably predict the level of harvest that a population can sustain if they have inadequate data on that population. As a result of these problems, managed fishery resources have often been overexploited. In 1992, 59% of 78 European stocks and approximately 45% of 156 U.S. populations for which assessments of resource status were available were classified as overutilized (Rosenberg *et al.* 1993).

Box 4.6   The harvest of northern fur seals in the Pribilof Islands

The breeding behavior of fur seals has several characteristics that make it fairly easy for managers to get data on sex and age structure, and to gauge the effects of harvest. Fur seals are sexually dimorphic (*di*, two; *morph*, forms), that is, the two sexes can be distinguished. Males weigh several hundred pounds more than females. Fur seals are also polygynous (*poly*, many; *gyn*, female), meaning that one male mates with many females (usually between five and ten in a given breeding season). After their winter migration to low-latitude feeding grounds, the seals return to their breeding grounds. Older males are the first to arrive. On land they battle amongst themselves to stake out the boundaries of their territories. After a few weeks, the females arrive. They haul out on land and give birth almost immediately. Within several days of parturition (giving birth), a female mates with the male on her territory. The fertilized egg does not implant in the uterus or begin to grow for several months, however. This physiological adaptation, termed delayed implantation, allows breeding to take place when the population congregates on land and birth to take place at the appropriate season a year later, when the seals come ashore again. The younger, nonbreeding males arrive at the rookery after the adult females and gather at the outskirts in bachelor colonies.

Because of these characteristics, a population of breeding fur seals can easily be censused when it comes ashore, and the numbers of animals in each age and sex class can be determined (because of the size difference between the sexes and the spatial segregation of the bachelor males).

Furthermore, nonbreeding males can be removed from the population without decreasing the reproductive potential of the herd. These circumstances allow for a carefully monitored fur seal hunt. The effects of the harvest on the population can be assessed by comparing successive annual censuses (Baker *et al.* 1970).

Fur seal populations responded well to the protections afforded by the Fur Seal Treaty in 1911 (Chapter 1). After the population recovered, seal pelts were commercially harvested in the Pribilof Islands for several decades without apparent ill effects on the herd. During this period, the managed harvest provided a continuing supply of products without depleting the resource.

The Fur Seal Treaty expired in 1985 and was not renewed, however, primarily because of opposition to the killing of animals to obtain a luxury item (Weber 1985). Therefore, the U.S.A. no longer carries out a commercial hunt of northern fur seals (although Alaskan Natives still hunt seals for subsistence). We will return in Chapter 12 to the subject of the fate of northern fur seal populations in recent decades.

The risks of managing for sustained yield are relatively low in predictable environments. But if carrying capacity varies in unexpected ways, the assumptions underlying harvest management may not be met, and serious problems can ensue. Game animals and timber are often harvested from environments that have been simplified by managers, and thus they are relatively predictable. Not all environments are so stable, however. Rangelands, for example, are heterogeneous and experience variable and unpredictable weather. Furthermore, the relationships between soils, plant growth, and herbivory in rangelands are complex, and often they are poorly understood, especially in the western hemisphere, where domestic stock are not native and where exotic plants that are better adapted to livestock grazing have arrived as well. Plant species differ in their ability to tolerate grazing, and grazing animals preferentially graze some species over others. This makes it difficult to design a single grazing regime that will be appropriate for all the plants on a range (Dasmann 1945). If the level of grazing that is sustainable is misjudged, serious changes in soil fertility, erosion, and plant species composition may occur.

Management for sustained yield developed in response to overutilization of living natural resources. It has worked well to allow a continuous supply of wild products to be harvested under some circumstances. The greatest successes of this type of management occur in predictable environments where population dynamics and ecological interactions are fairly well understood, where populations respond to reduction with density-dependent compensation, where scientific advice is followed in setting regulations, and where there is compliance in following them. When these conditions are met, harvest management can provide a sustained flow of economically valuable products. But in cases where managers do not understand important natural variability, interactions, and processes; where excessive harvest levels are permitted; or where compliance is low, serious consequences can ensue.

In this chapter we have seen that managers concerned with sustained yield must estimate the productivity of a harvested resource and then adjust harvest levels accordingly. The other side of this coin involves manipulating habitats to maximize productivity of the harvested species. In the next chapter, we explore ways that utilitarian managers do this.

References

Baker, R. C., F. Wilke, and C. H. Baltzo (1970). *The Northern Fur Seal.* U.S. Department of the Interior, U.S. Fish and Wildlife Service, Bureau of Commercial Fisheries, Circular no. 336, Washington, DC.

Begon, M., M. Mortimer, and D. J. Thompson (1996). *Population Ecology: A Unified Study of Animals and Plants.* Oxford: Blackwell Science.

Bergerud, A. T. (1985). The additive effect of hunting mortality on the natural mortality rates of grouse. In *Game Harvest Management*, ed. S. L. Beasom and S. F. Roberson, pp. 345–366. Kingsville, TX: Caesar Kleberg Wildlife Research Institute.

Bergerud, A. T. (1988). Increasing the numbers of grouse. In *Adaptive Strategies and Population Ecology of Northern Grouse*, vol. 2, ed. A. T. Bergerud and M. W. Gratson, pp. 686–731. Minneapolis, MN: University of Minnesota.

Caughley, G. (1985). Harvesting of wildlife: past, present, and future. In *Game Harvest Management*, ed. S. L. Beasom and S. F. Roberson, pp. 3–14. Kingsville, TX: Caesar Kleberg Wildlife Research Institute.

Caughley, G. and A. R. E. Sinclair (1994). *Wildlife Ecology and Management.* Cambridge, MA: Blackwell Science.

Dasmann, W. (1945). A method of estimating carrying capacity of range lands. *Journal of Forestry* **43**:400–402.

Derocher, A. E. and I. Stirling (1995). Estimation of polar bear population size and survival in western Hudson Bay. *Journal of Wildlife Management* **59**:215–221.

Gottschalk, J. S. (1972). The German hunting system, West Germany, 1968. *Journal of Wildlife Management* **36**:110–118.

Gross, J. E. (1969). Optimum yield in deer and elk populations. *Transactions of the North American Wildlife Conference* **34**:372–386.

Highsmith, R. M., Jr., J. G. Jensen, and R. D. Rudd (1969). *Conservation in the United States*, 2nd edn. Chicago, IL: Rand McNally.

Jenkins, D., A. Watson, and G. R. Miller (1963). Population studies on red grouse, *Lagopus lagopus scoticus* (Lath.) in north-east Scotland. *Journal of Animal Ecology* **32**:317–376.

Kellert, S. R. (1993). Attitudes, knowledge, and behavior toward wildlife among the industrial superpowers: United States, Japan, and Germany. *Journal of Social Issues* **49**:53–69.

Leopold, A. (1966). *A Sand County Almanac*, 7th edn. New York, NY: Ballantine Books.

Little, R. M. and T. M. Crowe (1993). Hunting efficiency and the impact of hunting on greywing francolin populations. *Suid-Afrikaanse Tydskrif vir Natuurrnavorsing (South African Journal of Wildlife Research)* **23**:31–35.

Madsen, J. and A. D. Fox (1995). Impacts of hunting disturbance on waterbirds: a review. *Wildlife Biology* **1**:193–207.

Normile, D. (2000). Japan's whaling program carries heavy baggage. *Science* **289**:2264–2265.

Parker, I. and M. Amin (1983). *Ivory Crisis*. London: Chatto and Windus.

Rosenberg, A. A., M. J. Fogart, M. P. Sissenwine, J. R. Beddington, and J. G. Shepherd (1993). Achieving sustainable use of renewable resources. *Science* **262**:828–829.

Tormosov, D. D., Y. A. Mikhaliev, P. B. Best, V. A. Zemsky, K. Sekiguchi, and R. L. Brownell, Jr. (1998). Soviet catches of southern right whales *Eubalaena australis*, 1951–1971: biological data and conservation implications. *Biological Conservation* **86**:185–197.

Watson, A. and D. Jenkins (1968). Experiments on population control by territorial behavior in red grouse. *Journal of Animal Ecology* **37**:595–614.

Webb, W. J. (1960). Forest wildlife management in Germany. *Journal of Wildlife Management* **24**:147–161.

Weber, M. (1985). Marine mammal protection. In *Audubon Wildlife Report, 1985*, ed. R. L. Di Silvestro, pp. 180–211. New York, NY: National Audubon Society.

Wiley, J. J. Jr. (1959). Control techniques for managed even-aged stands. *Journal of Forestry* **57**:343–347.

5

Techniques – habitat management

Habitat management is a logical extension of the utilitarian focus on economically valued species. We have seen that one facet of utilitarian management involves managing harvests with the objective of guaranteeing a continuous supply of products. Another facet is the manipulation of habitats in ways that favor valuable species and discourage undesirable ones. This chapter explores some ways that utilitarian managers do that.

Managers can modify habitats directly or indirectly. Extending the analogy between agriculture and the production of wild "crops," managers can directly modify habitats by utilizing techniques from horticulture and farming such as planting, watering, fertilizing, pruning, thinning, and weeding, and by providing specific habitat requirements for selected species. Or, they can modify habitats indirectly by altering disturbance regimes, thereby speeding up or slowing down succession. We will look at examples of each of these approaches below.

5.1 Direct modification of plant communities

Managers alter habitat directly when they provide (or remove) specific habitat features, such as water, cover, or nest sites. The objective of this type of management is to increase the production of one or more products, such as wood or forage or selected wildlife species, usually by increasing the availability of resources that are limiting the growth of desired species. If scarcity of a resource is limiting the growth of a population, then increasing the supply of this limiting factor should allow the population to increase. If resources that are in short supply are correctly identified and effectively enhanced, this type of management can increase populations of desired species. (Note that, once again, this type of management is grounded in the assumption that populations are limited by resources; see Chapter 2).

5.1.1   Planting and fertilizing

Tree plantations

One of the most direct and obvious ways of modifying habitats is by planting desired species of plants. Foresters often attempt to speed up the re-establishment of trees on logged lands by means of reforestation, the planting of tree seeds or seedlings on a cutover area. European forestry has a long tradition of planting preferred tree species on logged forests. For example, early in the nineteenth century, most of Germany's mixed hardwood forests were converted to conifer plantations. Spruce was favored, because of its high yield, but pines were planted on sandy sites (Leopold 1936).

American foresters have often used a similar approach, taking advantage of the openings created by logging to plant fast-growing conifers or planting additional trees on lands that were not "fully stocked" (see Chapter 4). Utilitarian foresters see planting as a logical way to improve upon nature. This attitude is expressed in numerous textbooks and articles:

Harvesting methods usually provide for natural [tree] regeneration; but, unfortunately, nature does not always handle the job adequately. There is wide variation in quantity and quality of annual seed crop. . . . Moreover, when there is a good seed crop, rodents may consume the seeds or seasonal weather conditions may inhibit establishment of seedlings. Thus industry in its effort to shorten the growing cycle for economic reasons, is turning strongly to planting and to artificial reforestation through seeding. . . . In addition, it has been shown that the nation has a large area of commercial forest land that is poorly stocked for satisfactory crop production. . . . The obvious answer is that artificial reforestation is needed. (Highsmith *et al.* 1969:175–176)

This system [planting and fertilizing Douglas-fir] offers many opportunities to improve on natural regeneration processes and increase yields. Harvesting and site preparation techniques can be used to control competing vegetation, animal pressure, and the microclimate of the planted seedling. . . . Thus harvesting and regeneration practices increase Douglas-fir productivity about 30 percent. (Farnum *et al.* 1983:695)

In the 1930s, Civilian Conservation Corps crews planted red pines and jack pines on sites in the Midwest that had previously supported mixed deciduous forest (Alverson *et al.* 1994). In the Southeast, pines are favored for forest plantations; in the Pacific Northwest, Douglas-fir is the species of choice, and in the Rocky Mountains ponderosa pine, fir, and spruce plantations are common (Leopold 1978).

A cutover site may be burned, fertilized, drained, or prepared by chemical or mechanical means before seeds are sown or nursery stock is transplanted. Transplanting has a higher success rate, but it is labor-intensive and expensive. Seeding is much less expensive, but tree seeds are attractive, high-energy food sources for a variety of animals, including rodents and insects. Depredation by seed predators can seriously compromise the effectiveness of forest reveg-etation efforts. For this reason, seeds are often treated with chemicals that are toxic to rodents.

If the problem of seed depredation is solved, reforestation can be quite successful in terms of the objective of hastening the establishment of forest vegetation. If other objectives are considered, however, reforestation has some disadvantages. By planting only one or a few tree species, foresters create communities that are biologically and physically simple in comparison to forests that result from more natural regeneration. Because the stands that result from reforestation consist of trees that are all of the same age, this type of management creates homogeneous vegetation. Habitat with such a uniform physical structure supports relatively few species of animals or plants.

Food and cover crops for wildlife

One straightforward way to improve habitat for wildlife is by planting and fer-tilizing vegetation that will meet certain organisms' habitat requirements. This technique is widely used to increase the supply of wildlife food and cover on both public and private lands. For example, a variety of grasses and legumes, including clover, wheat, and ryegrass, are commonly planted to attract deer (Waer *et al.* 1997).

In the United States a number of federal programs have encouraged the revegetation of degraded lands or encouraged farmers to set aside cropland by withdrawing lands from production and planting cover crops instead. Between 1935 and 1943 the Forest Service planted more than 200 million trees in shelterbelts (barriers formed by woody vegetation to block wind and reduce erosion). Subsequently, the Food and Agriculture Act of 1965 estab-lished the Cropland Adjustment Program (CAP), which provided incentives for farmers to plant grasses and legumes as a means of reducing crop sur-pluses, and in 1985 passage of the Food Security Act allowed farmers to par-ticipate in a Conservation Reserve Program (CRP) by removing vulnerable land from production for 10 to 15 years. Under this program, in return for planting permanent cover on highly erodible cropland or other environmentally sensitive land, the government pays the landowner an annual fee and shares the cost of establishing the cover.

These programs were established to reduce crop surpluses and to lessen erosion and applications of agricultural chemicals on marginal land, but as a by-product they also benefit wildlife by diversifying cultivated landscapes with patches of uncultivated habitat. The benefits to wildlife from revegetation depend upon the type of crops that are planted in the set-aside lands and the diversity of planted species. In the Great Plains and the Midwest, substantial areas have been planted with native grasses, with concomitant benefits to wildlife. In the Southeast, however, pines are often planted on CRP lands. In a landscape that is already heavily forested, the conversion of open habitats to pine forests has probably been unfavorable for animals such as the northern bobwhite that require early successional habitats (Carmichael 1997).

Planting food crops for wildlife is an ancient practice. Marco Polo is said to have found in the Mongolian Empire

a valley frequented by great numbers of partridges and quails, for whose food the Great [Kublai] Khan causes millet, and other grains suitable to such birds, to be sown along the sides of it every season, and gives strict command that no person shall dare to reap the seed; in order that the birds may not be in want of nourishment. (Quoted in Leopold 1933:7.)

Planting food crops for wildlife is only one step removed from providing the food directly, that is from supplemental feeding. This too is a direct method of management that has been practiced for centuries. It is an effective way to build up local populations of game; but it is highly likely that the habitat will fail to meet all of the other requirements of such artificially high populations. This can lead to serious economic damage, a problem we will consider in the next chapter.

5.1.2   Removing unwanted vegetation

Controlling woody vegetation on rangelands
On rangeland in the American West, shrubs have increased in coverage at the expense of grasses and forbs. The reasons for this change are not entirely clear, but overgrazing and fire suppression have been suggested as causes. Because the removal of these plants is accompanied by an increase in the pro-

ductivity of grasses, utilitarian range managers have often sought to eradicate shrubs (or "brush"). In the western U.S.A., the principal target of these efforts has been sagebrush; in the Southwest, it has been mesquite.

A variety of methods are available for brush removal, including fire (see below), mechanical means, and chemical treatments. Mechanical means include cutting roots with bulldozer attachments, dragging cables between two tractors, disking, chopping, and mowing. Mechanical control disrupts the surface soil, however, increasing the chances of erosion and weed encroachment.

These disadvantages are avoided by chemical forms of shrub control. The earliest chemical treatments included kerosene and sodium arsenite, but these had to be applied to individual plants, and in the case of arsenite, chemical control was toxic to animals. Carbon-containing herbicides that affect plant physiological processes are more economical and less toxic (Stoddart *et al.* 1975).

The removal of shrubs to maximize livestock production often conflicts with management for wildlife. Shrubs offer food and cover to many species of wildlife in arid rangelands. Sagebrush, for example, provides forage for mule deer and bighorn sheep. Some species of specialized birds, particularly the sage-grouse and songbirds such as the sage sparrow, Brewer's sparrow, and sage thrasher, are able to meet their requirements only in habitats dominated by sagebrush. So, while shrub control may accomplish a range manager's objectives as far as forage production is concerned, from the point of view of the wildlife manager, it can have negative ecological effects.

In arid or semiarid climates, utilitarian range managers have also targeted riparian vegetation (vegetation growing along streams and rivers) for eradication (Bowser 1952). Many riparian trees and shrubs of dry regions have deep roots that contact underground water supplies; these species are termed phreatophytes. Range managers reasoned that the water consumed by these plants was wasted. In fact, one textbook written in the 1960s defined phreatophytes as "useless trees and shrubs whose roots reach the water table or the capillary fringe" (Highsmith *et al.* 1969:127). Removing riparian vegetation was seen as a way to make more water available for other uses, primarily forage production.

Riparian plant communities are very productive and diverse, however. The food and cover provided by riparian trees and shrubs are especially important in arid regions, where these zones provide critical resources that are in short supply for a variety of organisms, including many amphibians, bats, and birds. Again, we see that managing for some products – in this case, meat or wool – can negatively impact other resources.

Decreasing the density of forest stands

Taking out trees from forest stands is another form of habitat manipulation that involves vegetation removal. Foresters use techniques such as thinning, sanitation cutting, and salvage cutting to do this. Thinning decreases tree density throughout a stand by removing some of the smaller trees. The objective is to release the remaining trees from competition, allowing them to grow faster and to produce higher-quality timber (see Figure 5.1). Thinning does not necessarily increase productivity, but it increases the value of the wood produced, since it is concentrated in larger stems. Large trees are more valuable because they contain more wood, and the wood is of higher quality (Farnum *et al.* 1983). Sanitation cutting removes diseased or insect-infested trees in order to reduce the chance of these conditions spreading to other trees. Salvage cutting is the removal of damaged or dead trees that still have market value.

5.1.3   Managing water supplies

Where water is in short supply, its provision can be a straightforward method of improving a habitat's carrying capacity for some species. Many birds and mammals of arid environments, including javelina, bighorn sheep, deer, pronghorn, rabbits, sage-grouse, quail, and songbirds, often drink at artificial water sources, so providing these sources of water can potentially increase wildlife populations. Water developments in arid landscapes have potential drawbacks, however. A water source can provide a focal point for cattle or predators to congregate, resulting in overgrazing and trampling of the surrounding vegetation or increased mortality from predation.

Some areas are managed specifically to provide aquatic habitats and wetlands. (Wetlands are the interface between terrestrial and aquatic habitats; they may not always contain surface water, but their soils are wet for long enough during the growing season to support distinctive vegetation that can tolerate saturation.) Many U.S. national wildlife refuges were created specifically for the benefit of waterfowl and contain extensive areas of marsh habitat (see Chapter 1). Often a refuge's natural water regime has been eliminated, and water is controlled by means of dikes, pumps, and gates. By controlling the depth and timing of flooding, managers can indirectly manipulate the development of vegetation. Periodic drawdowns, in which water is removed from the management unit and mudflats are exposed, promote the development of desirable species of wetland plants that germinate in mudflats; these provide food and cover for waterfowl and other wetland birds (Kadlec and Smith 1992; Payne 1992; Fredrickson and Laubhan 1994).

Figure 5.1. Thinning of a stand of trees. (After Highsmith *et al.* 1969.)

In Europe, North America, and many other parts of the world, wetlands have decreased in area because of draining for agriculture and other land uses. In the southeastern U.S.A., loblolly pine wetlands have been drained to increase timber production (Farnum *et al.* 1983).

On the other hand, utilitarian management has been a force favoring wetlands protection. The U.S. national wildlife refuge system includes many significant wetlands that were set aside to protect waterfowl. Private conservation action by hunters has also led to the protection of large areas of habitat for ducks, geese, and swans. For example, Ducks Unlimited, a private organization founded in the 1930s, has acquired millions of acres of waterfowl breeding grounds in the U.S.A. and Canada. In the absence of such protection, many of these areas would have been drained for agriculture. Although these actions were motivated by a concern for game species, many nongame species have also benefited.

5.1.4 Providing special structures

Providing or safeguarding structures required by organisms of interest is another straightforward method of improving a habitat's carrying capacity for selected species. These may be natural features of the landscape or artificial structures.

Artificial structures

Artificial nest structures such as boxes or nest platforms are effective in enhancing the productivity of breeding birds. Where nest sites are in short supply, the provision of artificial structures can be critical. In managed forests of northern and eastern Europe, nest boxes have increased populations of cavity-nesting birds (Bruns 1960).

Another strategy for enhancing the productivity of nesting birds involves the creation of artificial islands where ground-nesting birds such as ducks, geese, and shorebirds can nest without being disturbed by mammalian predators. If predation by mammals limits the productivity of these species, then the creation of nesting islands can result in markedly increased productivity; however, the constructed islands will only deny access to mammalian predators if they are surrounded by water that is deep and wide enough to prevent them from swimming to the islands. Of course, avian predators will still be able to reach island nests.

Natural structures

Many species of wildlife require specific types of natural structures, such as cliffs, caves, talus, surface water, dead and down woody material, snags, or large organic debris (Maser et al. 1979a,b; Thomas et al. 1979a). Where this is the case, wildlife managers may try to insure that these structures are available. This can create a conflict with management for other resources, however. Many organisms – including woodpeckers, some songbirds, bats, and squirrels – feed, nest, or sleep in the decaying wood of standing dead trees (Figure 5.2), but because managing for cavity-dependent species involves leaving some old trees to die, and these trees can harbor insects or pathogens, this policy is inconsistent with management oriented toward maximum timber production (Leopold 1978).

5.2   Indirect modification of habitats: Modifying succession

Habitats can also be altered indirectly, through the modification of succession. As succession proceeds, early seral stages are replaced by later ones; this process is offset by periodic disturbances that set back succession and allow for the development of early seral stages (see Chapter 3). One of the main

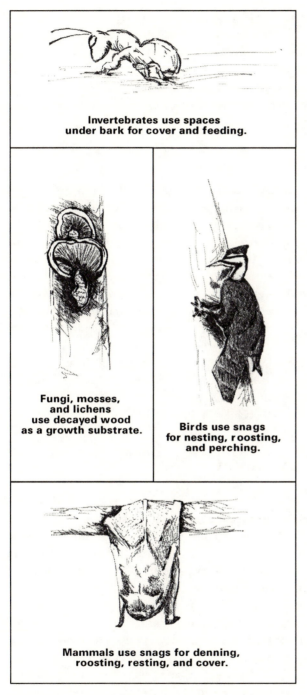

Figure 5.2. Some organisms that depend on snags. (After Thomas *et al.* 1979*a*.)

ways managers can modify habitat is by promoting or retarding disturbances. In general, the removal of vegetation promotes early seral stages. Burning, grazing, logging, flooding, applying herbicides, and mowing kill or remove vegetation and promote early seral communities. If the preservation of late successional stages is a goal, then managers will seek to decrease the frequency of disturbances that set back succession.

5.2.1 Flooding

In the absence of disturbance, sediments tend to build up in marshes. In temperate climates, where herbaceous vegetation dies each winter, the accumulation of dead plants and sediments can fill in a marsh. Natural disturbances – including abiotic disturbances such as floods and droughts and biotic disturbances such as herbivory from muskrats – tend to set back this process, destroying vegetation and replacing these patches with open water. In the absence of further disturbances, emergent vegetation (rooted vegetation that has stems and/or leaves extending above water or saturated soil into air) is likely to fill in the open areas. The plants then trap sediments, the marsh becomes shallower and more densely vegetated, and it begins to fill in again.

Wildlife managers sometimes choose to mimic natural disturbances in order to create openings in cattails or other emergents. They may use fire, flooding, or mechanical means to kill emergent vegetation and set back or retard succession (Weller 1978; Kadlec and Smith 1992).

5.2.2 Grazing management

Grazing by wild or domestic herbivores, especially large grazing mammals, has a major influence on the structure and composition of vegetation. Overgrazing can inhibit the reproduction of trees and shrubs and cause increases in unpalatable plants, loss of plant cover, soil compaction, and erosion. On the other hand, under some circumstances a moderate amount of grazing can prevent the accumulation of dead biomass, promote nutrient cycling, or enhance seedling recruitment. Thus, the manipulation of grazing pressure is an important tool at the disposal of the natural resource manager.

Plant species differ in the level of grazing they can tolerate. Heavy grazing results in the decline of species that are sensitive to grazing, termed decreasers, whereas increasers, species that are able to withstand heavy grazing, become more abundant. Unfortunately, in North America the more palatable and nutritious range plants are usually decreasers. As noted above, in steppes and deserts grazing often leads to a decline in the availability of grasses and an increase in shrub coverage.

The response of plant communities to grazing varies. Some types of vegetation seem to benefit from grazing, especially by native herbivores. On the other hand, some communities do not require grazing and are very vulnerable to negative impacts from grazers. (In Chapter 13, we will explore the reasons for these differences in sensitivity to grazing.)

5.2.3   Logging

The early successional stages that follow logging are beneficial to certain species of wildlife under some circumstances. The effects of logging on wildlife depend upon the size and method of the cut, the steepness of the site, the species of trees that are taken, and how a site is treated after the harvest. A few generalizations can be made, however.

The removal of trees makes light available and favors plant species that cannot reproduce in shade. Initially, logged sites – like burned areas – are invaded by grasses, forbs, and shrubs. These provide leaves, berries, and seeds that deer, elk, moose, bears, and many small mammals and songbirds feed upon. In addition, the deciduous trees and shrubs characteristic of recently logged sites provide nesting habitat for many species of birds.

On the other hand, the open environment of a clearcut has some disadvantages. It is hotter in the daytime and colder at night than the forest interior, and it provides poor hiding cover for large animals. Therefore, ungulates feeding in clearcuts must have hiding cover and thermal cover nearby (see Chapter 3). For this reason, deer and elk avoid the central portions of large clearcuts. If resource managers wish to manage for deer and elk as well as for timber, they must arrange cutover and forested areas to juxtapose these different requirements (see the discussion of arranging habitat components, below).

As succession proceeds after logging, additional species of plants and animals colonize a site. Some small mammals do not utilize clearcuts until several years after harvest. Similarly, a study in the oak woods of Burgundy reported that songbirds such as thrushes and warblers peaked in abundance in the first few decades after timber harvest (Figure 5.3).

Burning a site removes much of the debris left after clearcutting. This benefits large mammals by making travel easier, but it reduces hiding cover for small mammals. Here again, we see that there are tradeoffs; practices that favor some species inhibit others.

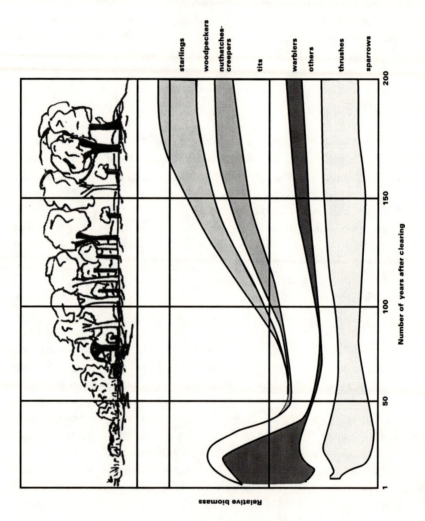

Figure 5.3. Biomass of bird populations in deciduous forests of Burgundy. (After Leopold 1978.)

5.2.4   Fire management

Fire suppression was part of the intensive management of European forests, but Native Americans used fire extensively as a tool for habitat modification – to create openings favorable to certain species and to clear fields for planting (Pyne 1982; Boyd 1999). Fires deliberately and accidentally set by Indians created a landscape mosaic of different successional stages. Plants and animals that utilized, and in some cases depended on, early successional stages were abundant in regions that were frequently burned. Many shade-intolerant plants that are favored by burning, such as blueberries, provide excellent food for wildlife and for people.

In contrast to this scenario, during the period of intensive logging of the frontier, conditions were created that favored large, hot fires. The woody debris left onsite after timber harvests furnished a source of dry fuel, and as a result, some extremely destructive fires occurred in the wake of the timber booms (see Chapter 1). Utilitarian forest managers responded to this situation by suppressing forest fires wherever possible. To them, fire was a destructive agent, and fire suppression a way to conserve timber. In 1910, the chief forester of the U.S. Forest Service declared that "the necessity of preventing losses from forest fires requires no discussion. It is the fundamental obligation of the Forest Service and takes precedence over all other duties and activities" (quoted in Pyne 1982:260).

The Forest Service's attitude toward fire was grounded in the Clementsian view of nature (see Chapter 3). In *Fire in America*, fire historian Stephen Pyne suggests that

the Clementsian concept of "nature's economy" . . . was a concept especially gratifying to foresters. . . . The Clementsian theory vindicated fire protection . . . as a means of assisting the succession of deforested land to forest climax and . . . as a means of promoting the innate and "natural" drive for successional climax. That lightning set many fires was really irrelevant: it was well known that nature suffered from waste and entropy, which human engineers could, and ought to, eliminate. (Pyne 1982:92)

Managers of wildlife and rangelands, however, saw fire as beneficial in some circumstances. Fire suppression changed the disturbance dynamics of North American forests, and favored late-successional habitats. This was detrimental to populations of some game species. In addition, in the Midwest fire had maintained steppe vegetation at the interface between steppe and forest, so fire suppression allowed trees to encroach upon grasslands. As noted above, fire may have also played a role in limiting shrubs in arid rangelands. Furthermore, in some types of unburned steppe communities, litter can build up and inhibit the development of grass and forb seedlings. For these reasons, wildlife and range managers were more inclined than foresters to use fire as a tool for modifying habitats (see Box 5.1).

Box 5.1   Managing red grouse habitat with fire at the turn of the century

From 1873 to 1910 a British royal committee studied red grouse (see Chapter 4) on the moors of England and Scotland in an effort to discover how mortality from parasitic diseases might be minimized. According to the committee's report, prior to 1850 the owners of large estates had their shepherds burn about one-tenth of their holdings annually, so that properties were burned on approximate 10-year rotations. In the mid-nineteenth century, however, landowners began appointing gamekeepers, in response to an increase in the value of grouse as game. The keepers wanted to increase cover to conceal hunters from their quarry, so they burned the moors less frequently. This, the commission concluded, led to a decline in the quality of the habitat for both grouse and sheep, as the amount of palatable heather and grass decreased. The solution, they argued, was to burn more frequently: "to avoid disease and heighten the average yield of the moor . . . the progressive landlord will . . . attempt to get the moor into good 'heart'" (Committee of Inquiry on Grouse Disease 1911:410).

The committee's observations led them to conclude that burning rejuvenated the food supply for grouse by removing old vegetation and stimulating the growth of nutritious young shoots. They recognized that the pattern of burning was as important as the amount of moor that was burned. By burning long, narrow swaths and alternating burned and unburned strips, gamekeepers could create a mosaic with small patches of young heather dispersed in a matrix of older patches. This arrangement, the committee suggested, would segregate the birds in separated strips of nutritious forage, thereby helping to lessen the transmission of parasites between birds. Because of their concern about disease transmission among individuals congregated in large habitat patches, the committee felt that there was no lower limit to patch size (Committee of Inquiry on Grouse Disease 1911:411). They recommended burning moors in a 15-year rotation. This regime, the committee felt, would be feasible for most landlords in most years and would allow three-fifths of a moor to be in good-quality heather at any one time.

The royal committee's insights were a significant contribution to game management in the British Isles at the turn of the century, because they were one of the first comprehensive attempts to develop a management plan that recognized and integrated the interrelationships between habitat quality, disturbance, patch configuration, and population dynamics.

5.3   Arranging habitat components

Wildlife managers are concerned not just with the amount and quality of habitat components, but with their juxtaposition as well (Giles 1978). For maximum effectiveness, all of a species' requisites should be accessible. The design of clearcuts used by deer and elk illustrates how the arrangement of habitat components can affect wildlife use of logged lands. Some of the negative impacts of clearcutting on ungulates can be minimized by cutting small areas so that animals feeding in the open can still be in close proximity to suitable thermal cover. If timber cuts are designed so that cutover areas are close to adequate thermal cover (Figure 5.4), then use by deer and elk can be maximized, but this comes at the expense of some harvestable timber (Thomas *et al.* 1979*b*).

From its beginnings, the discipline of wildlife management was concerned with maximizing edges between contrasting habitats: small patches of fields, forest, and brush on land; water and emergent vegetation in wetlands. For instance, habitat suitability models developed by the U.S. Fish and Wildlife Service incorporate interspersion of vegetation and open water as an important component of habitat quality for a variety of wetland birds, including western grebes, American coots, and red-winged and yellow-headed blackbirds (Schroeder 1982; Short 1984, 1985; Allen 1985).

The amount of edge present depends on the size and shape of habitat patches. If two patches are the same shape but differ in size, the smaller one will have a higher proportion of perimeter (that is, edge) in proportion to its area than the larger one (Figure 5.5A). If the area that can be managed as a given type of habitat is limited, a manager can maximize the edge of a patch by making it have a long, thin shape (Figure 5.5B) or an irregular border (Figure 5.5C). Thus a manager interested in increasing habitat edges can do so by interspersing small patches of contrasting habitats having irregular margins or thin shapes. This can be done by allowing for or mimicking patchy disturbances that create a mosaic of contrasting habitats.

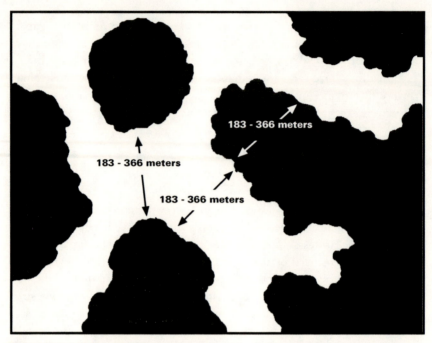

Figure 5.4. Arranging habitat components to maximize access to thermal cover. Shaded areas indicate cover; open areas indicate foraging habitat. (After Thomas *et al.* 1979*b*.)

5.4 Managing for multiple uses

The above discussion treats utilitarian forest, range, and wildlife managers as if they were in separate boxes, each concerned only with a single type of resource. In reality, utilitarian managers often try to manage habitats in ways that will favor the productivity of multiple resources. This typically involves tradeoffs between different objectives (Hall and Thomas 1979).

Since 1960 the U.S. Forest Service has been legally required to recognize four primary uses of national forests. In the years following World War II, social, economic, and political developments, including rising population and increased leisure time for recreation, resulted in pressure to adopt a broader approach to natural resource management. Americans became increasingly concerned that national forest management was too narrowly focused on timber production. In 1960, the Multiple Use Sustained Yield Act (MUSYA) was passed. Under this new policy, other uses, including recreation, wildlife

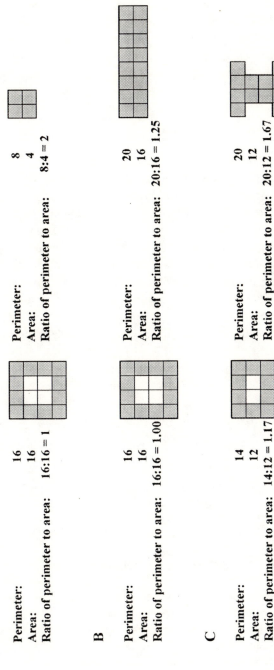

A

Perimeter:	16	8
Area:	16	4
Ratio of perimeter to area:	16:16 = 1	8:4 = 2

B

Perimeter:	16	20
Area:	16	16
Ratio of perimeter to area:	16:16 = 1.00	20:16 = 1.25

C

Perimeter:	14	20
Area:	12	12
Ratio of perimeter to area:	14:12 = 1.17	20:12 = 1.67

Figure 5.5. The influences of size and shape on the amount of edge (perimeter) of a habitat patch. In each case, the habitat patch on the right has a higher ratio of perimeter to area than the patch on the left. This ratio can be increased in three ways. (A) The two squares are the same shape, but the ratio of perimeter to area is greater for the smaller square than for the larger one. (B) The square and the rectangle have the same area, but the long and thin rectangle has more perimeter than the square. (C) The two figures have the same area, but the one with the irregular outline has more perimeter than the rectangle. Shading indicates the area of each patch that is influenced by the surrounding habitat. Note that the small, thin, and irregularly shaped patches have virtually no area that is not influenced by edge.

habitat, and watershed maintenance were to be considered by federal agencies. The focus of the MUSYA was still utilitarian, but by mandating the consideration of ecological values, the act initiated the beginning of a new approach to managing living natural resources. In 1976 Congress passed the National Forest Management Act, which elaborated on the principles set forth in the MUSYA.

5.5 Conclusions

Natural resource managers can manipulate habitat in a variety of ways to enhance the production of economically useful species of plants and animals. We have seen in this chapter that if the objective is to enhance the productivity of game animals, forage plants, or timber trees, a variety of methods are available to accomplish this. But, we have also seen that by concentrating on the utilitarian objective of maximizing production of a small number of species, resource managers may overlook some unintended and undesirable consequences of their management policies. Furthermore, managing to enhance production of one resource can conflict with managing for a different resource, so the choice of an appropriate management strategy depends upon our objectives. Both these points will come up repeatedly in the remainder of this book.

There is another side to utilitarian resource management. Reduction of species that are considered undesirable can complement management to enhance populations of and habitats for desirable species. In the next chapter we will consider management to reduce populations of species that harm people or prey on or compete with valuable species.

References

Allen, A. W. (1985). *Habitat Suitability Index Models: American Coot*. U.S. Department of the Interior Fish and Wildlife Service, Biological Report no. 82(110.115), Washington, DC.

Alverson, W. S., W. Kuhlman, and D. M. Waller (1994). *Wild Forests: Conservation and Biology*. Washington, DC: Island Press.

Bowser, C. W. (1952). Water conservation through elimination of undesirable phreatophyte growth. *Transactions of the American Geophysical Union* **33**:72–74.

Boyd, R. (ed.) (1999). *Indians, Fire, and the Land in the Pacific Northwest*. Corvallis, OR: Oregon State University Press.

Bruns, H. (1960). The economic importance of birds in forests. *Bird Study* **4**:193–208.

Carmichael, D. B., Jr. (1997). The conservation reserve program and wildlife habitat in the southeastern United States. *Wildlife Society Bulletin* **25**:773–775.

Committee of Inquiry on Grouse Disease (1911). *The Grouse in Health and in Disease,* vol. 1. London: Smith, Elder.

Farnum, P., R. Timmis, and J. L. Kulp (1983). Biotechnology of forest yield. *Science* **219**:694–702.

Fredrickson, L. H. and M. K. Laubhan (1994). Managing wetlands for wildlife. In *Research and Management Techniques for Wildlife and Habitats,* ed. T. A. Bookhout, pp. 623–647. Bethesda, MD: The Wildlife Society.

Giles, R. H., Jr. (1978). *Wildlife Management.* San Francisco, CA: W. H. Freeman.

Hall, F. C. and J. W. Thomas (1979). Silvicultural options. In *Wildlife Habitats in Managed Forests: The Blue Mountains of Oregon and Washington,* tech. ed. J. W. Thomas, pp. 128–147. U.S. Department of Agriculture Forest Service, Agriculture Handbook no. 553, Washington, DC.

Highsmith, R. M., Jr., J. G. Jensen, and R. D. Rudd (1969). *Conservation in the United States,* 2nd edn. Chicago, IL: Rand McNally.

Kadlec, J. A. and L. M. Smith (1992). Habitat management for breeding areas. In *Ecology and Management of Breeding Waterfowl,* ed. B. D. J. Batt, A. D. Afton, M. G. Anderson, C. D. Ankney, D. H. Johnson, J. A. Kadlec, and G. L. Krapu, pp. 590–610. Minneapolis, MN: University of Minnesota Press.

Leopold, A. (1933). *Game Management.* New York, NY: Charles Scribner's Sons.

Leopold, A. (1936). Deer and Dauerwald in Germany. I. History. *Journal of Forestry* **34**:366–375.

Leopold, A. S. (1978). Wildlife and forest practice. In *Wildlife and America,* ed. H. P. Brokaw, pp. 108–120. Washington, DC: U.S. Government Printing Office.

Maser, C., R. G. Anderson, K. Cromack, Jr., J. T. Williams, and R. E. Martin (1979*a*). Dead and down woody material. In *Wildlife Habitats in Managed Forests: The Blue Mountains of Oregon and Washington,* tech. ed. J. W. Thomas, pp. 78–95. U.S. Department of Agriculture Forest Service, Agriculture Handbook no. 553, Washington, DC.

Maser, C., J. E. Rodiek, and J. W. Thomas (1979*b*). Cliffs, talus, and caves. In *Wildlife Habitats in Managed Forests: The Blue Mountains of Oregon and Washington,* tech. ed. J. W. Thomas, pp. 96–103. U.S. Department of Agriculture Forest Service, Agriculture Handbook no. 553, Washington, DC.

Payne, N. F. (1992). *Techniques for Wildlife Habitat Management of Wetlands.* New York, NY: McGraw-Hill.

Pyne, S. (1982). *Fire in America.* Princeton, NJ: Princeton University Press.

Schroeder, R. L. (1982). *Habitat Suitability Index Models: Yellow-Headed Blackbird.* U.S. Department of the Interior Fish and Wildlife Service, Biological Report no. FWS-OBS-82/10.26, Washington, DC.

Short, H. L. (1984). *Habitat Suitability Index Models: Western Grebe.* U.S. Department of the Interior Fish and Wildlife Service, Biological Report no. FWS/OBS-82/10.69, Washington, DC.

Short, H. L. (1985). *Habitat Suitability Index Models: Red-Winged Blackbird*. U.S. Department of the Interior Fish and Wildlife Service, Biological Report no. 82(10.95), Washington, DC.

Stoddart, L. A., A. D. Smith, and T. W. Box (1975). *Range Management*, 3rd edn. New York, NY: McGraw-Hill.

Thomas, J. W., R. G. Anderson, C. Maser, and E. L. Bull (1979*a*). Snags. In *Wildlife Habitats in Managed Forests: The Blue Mountains of Oregon and Washington,* tech. ed. J. W. Thomas, pp. 104–127. U.S. Department of Agriculture Forest Service, Agriculture Handbook no. 553, Washington, DC.

Thomas, J. W., H. Black, Jr., R. J. Scherzinger, and R. J. Pedersen. (1979*b*). Deer and elk. In *Wildlife Habitats in Managed Forests: The Blue Mountains of Oregon and Washington,* tech. ed. J. W. Thomas, pp. 60–77. U.S. Department of Agriculture Forest Service, Agriculture Handbook no. 553, Washington, DC.

Waer, N. A., H. L. Stribling, and M. K. Causey (1997). Cost efficiency of forage plantings for white-tailed deer. *Wildlife Society Bulletin* **25**:803–808.

Weller, M. W. (1978). Management of freshwater marshes for wildlife. In *Freshwater Wetlands: Ecological Processes and Management Potential*, ed. R. E. Good, D. F. Whigham, and R. L. Simpson, pp. 267–284. New York, NY: Academic Press.

6

Techniques – management to minimize conflicts between pest species and people

In the utilitarian perspective on resource management, species are viewed as either good or bad. We have already seen how populations and habitats can be managed to favor species that are considered valuable. This chapter considers how utilitarian resource managers attempt to control populations of species that are considered overabundant. It begins with a consideration of the why and how of control efforts, then provides some historical background, and concludes by examining five case studies.

6.1 What is a pest?

Plants and animals that are perceived as detrimental to people or their interests are considered pests. Pests can be plants, fungi, invertebrates, fish, amphibians, reptiles, birds, or mammals. This chapter will emphasize attempts to control vertebrates that have been considered overabundant; insects and plant pests will be touched on only briefly. (See Chapter 5 for information about attempts to control woody vegetation on North American rangelands and Chapter 7 for a discussion of the ecological effects of chemicals used to control plants and insects.)

There are two main reasons why a species may be considered a pest: because it causes economic damage (directly, by killing a valued species or damaging property, or indirectly, by competing with valued species for limiting resources) or because it poses a threat to the health and safety of people or domestic animals. The first category includes (1) predators of domestic animals and game (wolves, cougars, bears, coyotes, seals, birds of prey, and some fish); (2) animals that damage crops or otherwise compete with people or livestock for resources (blackbirds, starlings, waterfowl, deer, elk, javelina, rodents, elephants); and (3) animals that damage property (pigeons, muskrats, nutria, beaver). Animals in the second category include those that potentially transmit diseases to people or domestic animals (skunks, raccoons, foxes, bats, deer, bison, deer mice, ground squirrels, bobcats, badgers), cause accidents (birds, deer, moose, reindeer, bears, elk); or attack people (wolves, bears, moose, cougars, elephants).

Pests are sometimes referred to as "weeds." Weedy species have high reproductive rates, are tolerant of people and therefore able to inhabit disturbed areas that have been impacted by people, and are good colonizers. Weeds do not have to be plants. To a livestock rancher, the coyote is a weedy species. Often, but not always, weeds or pests are not native to the region where they are considered pests (see Chapter 7).

Weeds have been described as organisms in the wrong place, that is species that that have become overabundant (perhaps because they have been transported to a region where their normal enemies are absent) or that conflict with the objectives of people in a given situation. This definition underscores the fact that weediness depends upon context and perspective. Pest status is in the eye of the beholder. Paradoxically, some animals, such as seals, deer, waterfowl, bears, and elephants are economically valuable and pests at the same time. They are harvested for sport or commerce but cause damage under certain circumstances. Polar bears and African elephants are valued by tourists and considered pests by local people (see below and Box 4.5). Grizzly bears and gray wolves were once relentlessly pursued by bounty hunters in North America and parts of Europe, but now they are endangered in parts of those regions, and large amounts of money are spent on recovery plans aimed at rebuilding their populations (Figure 6.1). Deer, elk, and waterfowl are valued as game, but when they damage grain fields or orchards they are pests.

Conflicting management objectives can exacerbate these problems. Managers seeking to maintain high populations for sport hunting or for recreational viewing may protect or even artificially feed populations that then build up to levels where they cause problems. For example, in intensively managed German forests of the nineteenth and early twentieth centuries, populations of artificially fed red deer increased to the point where they stripped bark from young trees and damaged nearby crops (Leopold 1936a,b; Webb 1960). In these situations, increasing the level of harvest can be a means of controlling damage, but this is likely to be controversial with birds and mammals that have a high degree of emotional appeal (Leopold 1955).

Pest control has social as well as biological dimensions. This is partly

Figure 6.1. Two views of wolves: villain and wilderness icon. (After Hunter 1996.)

because the definition of what constitutes a pest is a subjective one, and partly because human behavior affects the behavior of pests and often exacerbates problems. The examples below illustrate this.

6.2   How is damage from pests controlled?

To "control" a population of a pest means to reduce it to the point where it no longer poses a serious problem or to reduce the amount of damage it does. This can be done by lethal or nonlethal means. Lethal methods include shooting, trapping, gassing, and poisoning. The goal of lethal control is the reduction of population size. The principles underlying management to reduce populations are the same as the principles of harvesting for sustained yield (except that the harvest is generally not used). If the control operation causes deaths and emigration to exceed births and immigration, then the population will decline. If density-dependent compensation comes into play, however, then control efforts may not make any difference in population size (see Box 2.1).

Lethal animal damage control of protected species requires special permits, which are supposed to be issued only in situations where economic damage can be demonstrated. This type of activity is not to be confused with recreational hunting or trapping.

Recreational hunting is sometimes used to reduce population size. For example, it may be desirable to reduce the size of deer herds that exceed the carrying capacity of their habitat. Since deer are polygynous, however (like northern fur seals; Box 4.6), most females are bred by only a small percentage of males. This means that a large percentage of a herd's males can be killed without affecting its reproductive potential. If hunting is to be used as a tool to reduce herd size, therefore, some females must be removed. This is the justification for doe hunts. Although the biological arguments underlying doe hunts to reduce population size are sound, the idea that only bucks should be hunted is deeply ingrained with the public, and there is usually considerable resistance to doe hunts.

Some pest problems can be addressed by means of biological control, which uses natural ecological or biochemical processes to reduce pest populations. In one type of biological control, interactions between pest species and their predators or parasites are utilized to control unwanted organisms (Franz 1961). This technique makes use of disease-causing microorganisms (including viruses, bacteria, and fungi) and predatory, parasitic, or herbivorous insects. Typically biological control targets plant or insect pests. When exotic species are introduced into suitable habitat where they find few things that compete with, eat, or parasitize them, they are likely to prosper. Under these circumstances, the deliberate importation of predatory or parasitic insects from the homeland of the exotic pest may help to keep its population in check. This form of control avoids the ecological and health risks associated with chemical control (see Chapter 7), but it is not without risks. When biological control involves the deliberate introduction of exotic organisms, it can cause problems if it is not preceded by thorough studies showing that the aliens imported for control purposes will not adversely affect native flora and fauna. For example, a species of weevil that was imported into Canada in 1968, for the purpose of controlling alien musk thistles, is now attacking rare native thistles in the western U.S.A. (Louda *et al.* 1997).

Pest problems can sometimes be addressed without killing the pests, by removing problem individuals or altering their behavior, physiology, or environment. Nonlethal methods of control include live-trapping and relocating animals, modifying animal behavior through conditioning, using guard dogs to keep away predators, changing ranching or farming practices, sterilizing problem animals, vaccinating disease-carrying animals, reducing the food supply of problem populations, and using scare devices and lure crops. These techniques are designed to minimize wildlife mortality and other environmental impacts of control operations. Aversion conditioning is a nonlethal method of modifying the behavior of problem animals by exposing them to crops or carcasses treated with chemicals that are not poisonous but which sicken the animals that eat them. If the conditioning is effective, the affected individuals learn to avoid these foods.

Some forms of biological control use nonlethal chemicals that interfere with normal growth and reproduction, including (1) hormones that regulate growth and development and (2) pheromones, chemicals used in communication between members of a species. When pheromones are applied in unusual situations, the mating behavior of insect pests can be disrupted. For example, pheromones that function as species-specific sex attractants can be used to lure harmful insects into traps.

Nonlethal control sometimes involves modifying the behavior of people to reduce the frequency of encounters with animals or to change conditions that allow pest populations to build up. Problems with urban pigeons and with bears in national parks illustrate this in two very different settings (see below).

Pest control should identify objectives in terms of specific desired outcomes. These might include a reduction in economic losses or in disease transmission or accidents. The effectiveness of coyote control, for example, should be judged in terms of whether or not sheep losses are reduced, not in terms of how many coyotes are killed. We will see below that in populations with density-dependent compensation the number of animals removed may have little relationship to population size and therefore to the amount of damage. If objectives are not stated clearly at the outset of a control operation, managers may lose sight of the problem being addressed. When this happens, the control effort can become an end in itself regardless of whether it is effectively addressing the original problem.

6.3   Historical background

Historically, predators, rodents, and granivorous birds were among the principal targets of animal control efforts. Government-sponsored predator control in Europe dates back at least to the sixteenth century (Leopold 1936a). In England Henry VIII placed bounties on crows, choughs, and rooks, and Elizabeth I subsequently expanded the list of bounty species to include a number of other birds as well as stoats, weasels, wild cats, and polecats (Leopold 1933).

In Europe and the regions settled by Europeans, farmers and ranchers viewed predators as heinous villains. This attitude was not controversial; it was considered common sense and was shared even by biologists and conservationists until well into the twentieth century. For instance, William Hornaday, director of the New York Zoological Park, expressed concern about the decline of wildlife in his book *Our Vanishing Wildlife*, published in 1913. Although he was concerned about declines that resulted from the overexploitation of many animal populations, including bison (Chapter 1), Hornaday had no sympathy for large carnivores. Using the emotional language typical of his time, he wrote: "there are several species of birds that may at once be put under sentence of death for their destructiveness of useful birds. . . . Four of these are *Cooper's Hawk*, the *Sharp-Shinned Hawk*, *Pigeon Hawk* and *Duck Hawk*" (Hornaday 1913:80; emphasis in original).

In 1919, Lord Cranworth made a similar point about African game that might transmit diseases to domestic animals, using remarkably similar language:

there are at the present time certain animals, such as the eland and buffalo, which are under taint of suspicion of bringing in their train tsetse-fly or other obnoxious parasites, and therefore are inimical to stock raising. Should this suspicion develop into a certainty, these species must disappear from all settled lands. (Quoted in Collett 1987:140.)

In the early days of wildlife management, government predator control agents set out to exterminate predatory birds and mammals and justified doing so in the name of protecting game species as well as domestic stock. In 1917, the Chief of the U.S. Bureau of Biological Survey proclaimed to the International Association of Game, Fish, and Conservation Commissioners:

Everyone is aware that mountain lions, wolves, coyotes and other beasts of prey destroy vast numbers of game animals. For this reason, the destruction of the predatory animals, while primarily to protect live stock, at the same time is helping increase the amount of game. . . . There is little question that in five years we can destroy most of the gray wolves and greatly reduce the numbers of other predatory animals. In New Mexico we have destroyed more than fifty percent of the gray wolves and expect to get the other fifty percent in the next two or three years. (Quoted in Trefethen 1975:165.)

Until well into the twentieth century, state and local governments in the U.S.A. and some European nations paid bounties for wolves, foxes, cougars, weasels, hawks, owls, and any other animals suspected of killing game or domestic animals. The bounty system had numerous shortcomings, however. Its ecological consequences were not well thought out, and it was subject to fraud and abuse:

Some bounty claimants, with good insight into human psychology, presented for payment sacks of chicken heads well ripened in the sun, topped off with a few hawk heads. When one of these characters arrived at the county courthouse with his smelly trophies and offered to dump them for tally, the clerk almost invariably accepted the claim eagerly and paid promptly without examining the evidence. (Trefethen 1975:166–167)

In the U.S.A., western ranchers pay fees to graze their animals on federal land, which harbors predators that potentially threaten livestock. This combination of circumstances brings the U.S. government into the business of controlling predators on government-owned land, in response to the concerns of livestock growers. In 1931 the Animal Damage Control Act provided statutory authority for federal efforts to control animals thought to be harmful to crops and livestock. From the 1930s through the 1950s, there was little opposition to such federal programs to control animal damage (Di Silvestro 1985).

In 1949, however, Aldo Leopold published a now-famous essay entitled "Thinking like a mountain," in which he eloquently articulated a different viewpoint:

My own conviction on this score dates from the day I saw a wolf die. We were eating lunch on a high rimrock at the foot of which a turbulent river elbowed its way. We saw . . . a wolf [and a] half dozen others, evidently grown pups. . . .

In those days we had never heard of passing up a chance to kill a wolf. In a second we were pumping lead into the pack, but with more excitement than accuracy: how to aim a steep downhill shot is always confusing. . . .

We reached the old wolf in time to watch a fierce green fire dying in her eyes. I realized then, and have known ever since, that there was something new to me in those eyes – something known only to her and to the mountain. I was young then, and full of trigger itch; I thought that because fewer wolves meant more deer, that no wolves would mean hunters' paradise. But after seeing the green fire die, I sensed that neither the wolf nor the mountain agreed with such a view.

I now suspect that just as a deer herd lives in mortal fear of its wolves, so does a mountain live in mortal fear of its deer. And perhaps with better cause, for while a buck pulled down by wolves can be replaced in two or three years, a range pulled down by too many deer may fail of replacement in as many decades. (Leopold 1966:138–140)

At the time it was first published, Leopold's essay had little effect on predator control policies, but in the 1960s environmental awareness increased, and people began to question animal control programs on several grounds. Four types of concerns were raised: ecological concerns about the effects of poisons on the environment, humanitarian concerns about killing animals, political concerns about whether special interests exerted undue influence over government policy, and economic concerns about whether the results achieved by control programs justified the money the government spent on them.

In response to these concerns, a special committee was appointed by

Interior Secretary Stewart Udall to review federal animal control programs. The committee's report on "Predator and Rodent Control in the United States" became known as the Leopold Report after A. Starker Leopold, the committee's chair. A second committee review, known as the Cain Report, was completed in 1971. These reports advocated reforming control programs to make them more efficient and to lessen their ecological impacts by minimizing the broadcast application of poisons.

Some of the recommendations suggested by the Leopold and the Cain committees were implemented. Federal animal damage control programs now seek to target problem individuals rather than indiscriminately targeting entire populations of pest species. Nevertheless, animal damage control programs in the U.S.A., especially federally funded ones, remain highly controversial (Di Silvestro 1985). Some of the complexities of controlling pest species are illustrated in the examples that follow.

6.4　Case studies

6.4.1　Coyote control on rangelands in the western U.S.A.

With the disappearance of most large predators, especially the gray wolf and the grizzly bear, from most of the lower 48 states, coyote predation became a major concern in the U.S.A. Coyotes kill sheep on western ranges, but the magnitude of the damage is hotly debated. The North American sheep ranching industry has declined in recent decades because of economic and social factors, including the availability of synthetic fibers and falling markets for lamb and wool. In this context, ranchers argue that the additional burden of losses to coyotes is hard to bear, but critics suggest that the decline of the sheep industry is not due primarily to predation and should not be used to justify federal control programs.

Several nonlethal methods, including the use of guard dogs, llamas, and deterrents such as electric fences, sirens, or flashing lights, have been employed to control coyotes. In addition, a form of aversion conditioning in which dead sheep are laced with chemicals that cause vomiting has been used to condition coyotes not to eat sheep. One problem with this method is that it is likely to select for individuals that scavenge on carcasses rather than those who attack and kill sheep, but it is precisely the coyotes in the latter category that should be targeted for control. Lethal methods of coyote control are far more common. These include trapping, den hunting (the destruction of litters in the den, often by asphyxiation with carbon monoxide), shooting, and poisoning.

Studies of coyote populations in Texas showed that litter sizes were larger in areas where there was intensive control than in areas where control was light or moderate (Knowlton 1972). In other words, when coyote mortality increases, reproductive rate compensates by increasing, and the population quickly rebounds (Chapter 2). This density-dependent compensation makes coyote populations extremely resilient in the face of control. It is another reason why coyote control is controversial: critics point out that killing coyotes does little good because other individuals quickly replace the ones that are killed.

Coyote control is one of many cases in which substantial public opinion opposes lethal control. Disapproval is likely to be especially strong when the proportion of the population that must be killed to achieve management objectives is high. When that is the case, nonlethal control may be more acceptable as well as more effective and economical. This is illustrated by the following example.

6.4.2   Control of fox rabies in western Europe

Recently developed techniques for controlling the spread of rabies among wild and feral mammals provide an example of nonlethal control that effectively addresses a serious public health problem. Rabies is a well-known example of a disease that can be transmitted directly from wildlife to people or other warm-blooded animals. Because the rabies virus is shed in saliva, the disease is transmitted by bites from infected animals. Rabies kills tens of thousands of people annually. Most of its victims live in impoverished countries, where dogs are often unvaccinated and people who are bitten lack access to medical treatment or cannot afford it. Routine vaccination of dogs and cats can limit transmission from pets to humans, but the rabies virus is also present in some wild populations. In western Europe, foxes are the only significant wild reservoir of rabies, but in North America bats, skunks, foxes, and raccoons harbor the virus (Winkler and Bögel 1992).

In the 1950s, health workers tried to control rabies by reducing population density to the point where infected animals would have a low probability of transmitting the disease to another individual. This method was ineffective, expensive, and controversial. Biologists concluded that more than 60% of the target population would have to be removed to achieve the desired objective, and this level of mortality was not acceptable.

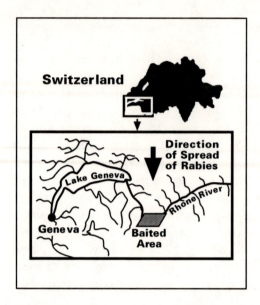

Figure 6.2. Map of area in Switzerland where the spread of rabies was controlled through bait-delivered vaccines. (After Winkler and Bögel 1992.)

Scientists next turned their attention to methods of vaccinating free-ranging animals. Since it would be impractical to trap animals, vaccinate them, and release them, interest in self-administered vaccines mounted. American researchers adapted the coyote getter, a device originally developed for poisoning coyotes, to this purpose. The apparatus consists of a small pipe stuck in the ground and baited with a tuft of scented wool. When an animal tugs on the bait, a cartridge fires a jet into the animal's mouth. Originally coyote getters fired the poison cyanide. It was a simple matter to substitute a dose of vaccine for the poison. Unfortunately, the device damaged the animal's mouth. While this was not a problem when the coyote getter was used for the purpose of killing the target animal, it was a serious drawback for the vaccine program, because it made the vaccinated animal unable to eat. Next, investigators developed a device consisting of a buried trigger pan that fired a vaccine-loaded syringe into the target animal's side when it stepped on the buried pan. This worked, but it was not economical and it was hazardous to nontarget animals.

These difficulties suggested that baits might be the best way to administer the rabies vaccine. This, too, presented technical challenges. Researchers had to develop a concentrated vaccine that would penetrate the mucous membranes of the mouth and throat, and they had to find a bait that would appeal to the target species but not to others. By the mid 1970s, these obstacles had been overcome, and field trials began on an island in the Aare River. When rabies spread among foxes on the eastern shore of Lake Geneva in 1978, scientists were ready to use baits to vaccinate the affected population. They succeeded in containing the outbreak by placing chicken heads laced with live rabies virus in a band across the Rhône River Valley (Figure 6.2). By 1989, bait-administered oral vaccines had successfully been used to control the spread of rabies in Switzerland as well as in 11 other western European countries (Winkler and Bögel 1992).

Delivering vaccines to wild animals through baits is a safe and relatively economical method of controlling rabies in feral or wild mammals. It offers a promising approach to reducing the number of human deaths from rabies in parts of the world where this disease continues to be a serious public health problem.

6.4.3   Human behavior and pigeon pests in urban environments

The control of wildlife pests in urban settings poses special challenges. Cemeteries, parks, college campuses, backyards, and even building ledges provide habitat for wildlife in and around residential and business areas; but wild animals in cities are prone to becoming pests for two reasons. First, wild organisms that are found in urban areas tend to be species that are capable of reaching high population densities around people. Pigeons, starlings, rats, and house mice are familiar examples. Second, in cities and suburbs, wild animals find themselves with little habitat. Under these circumstances, they may damage buildings, make noise, and spread diseases to people or their pets. This is true even of native species that are not usually regarded as pests. For instance, Canada geese become a nuisance when large flocks congregate in parks and on golf courses. In some instances, deer or other wild animals in cities or suburbs face starvation because of inadequate resources.

On the other hand, wildlife in metropolitan areas has positive recreational and educational values. People who dwell in cities typically value contact with wildlife and have few opportunities to see wild animals. Thus, although it is often desirable for economic, esthetic, or humanitarian reasons to reduce wildlife populations in or near urban areas, city residents are likely to react negatively toward control operations that kill wildlife. As a consequence, urban wildlife control becomes a matter of public relations as well as biology, and managers must strive to find methods of controlling unwanted organisms that are acceptable to the public.

This is certainly the case in cities where people feed pigeons. High densities of street pigeons (rock doves) occur in many North American and European cities. These birds have both negative and positive values in urban settings. They provide city dwellers with an opportunity to see and interact with wild animals, but they transmit diseases and parasites to people and domestic animals, and their droppings damage buildings and statues. The result is a paradoxical situation in which city governments kill pigeons on the one hand, while city residents feed them on the other.

Efforts to control street pigeons by lethal means are expensive and often ineffective. Between 1961 and 1985, game inspectors in the city of Basel, Switzerland, trapped and shot about 100000 street pigeons, but these measures had no lasting effect on the population. Like coyotes, street pigeons are resilient in the face of population reductions. Where they are not controlled, pigeon populations have high reproductive rates and high rates of juvenile mortality. When adult mortality increases, there is a compensatory decline in juvenile mortality, and juveniles quickly replace individuals that disappear from the population (Haag-Wackernagel 1995).

Faced with the ineffectiveness of lethal control by itself, city officials in Basel decided to add another component to their strategy. In addition to removing individuals from the population, they reduced the birds' food supply. Since pigeon feeding was a popular activity that provided substantial food to the birds, it was necessary to educate the people of Basel about the effects of high pigeon populations. Pamphlets and posters that showed young pigeons suffering from density-dependent diseases and parasitic infections were distributed, with text explaining that excessive pigeon densities were bad for pigeons as well as people. But pigeon feeding can fulfill an important social function. It often provides people who have no one to care for with emotional ties and a sense of being useful. Recognizing this, the Pigeon Action project of the University of Basel decided to provide opportunities for pigeon feeding by maintaining a small number of flocks in supervised lofts where population density was carefully controlled.

This project met with mixed success. Unregulated pigeon feeding declined markedly, as did the number of pigeons in the city and the amount of damage they caused, but the designated pigeon encounter areas were less successful. Few people used them, in part because the project had been so successful in getting its antifeeding message across that it created social pressure against any pigeon feeding (Haag-Wackernagel 1995). This example illustrates the complexities of controlling animal damage caused by a popular species while providing opportunities for positive interactions between people and wildlife.

6.4.4   Attacks by black bears and grizzly bears on visitors to U.S. national parks

Encounters between bears and people in American national parks are another situation where the behavior of people is an important component of the problem. High densities of visitors in parks with bears lead to frequent interactions between people and bears. Some of these result in attacks on people, and occasionally the attacks are fatal. Black bears are far more common and are considered less dangerous than grizzlies; however, because interactions with black bears are so common, and because people fear them less than grizzlies, the rate of injuries from black bear encounters exceeds the rate of injuries from grizzlies.

In 1967 two young women were unexpectedly killed by grizzly bears in separate, unprovoked attacks during the same night in Glacier National Park. In response to this situation, the National Park Service sought to identify factors that increase the likelihood of bear attacks. A dramatic increase in the number of backcountry visitors in recent decades had set the stage for the attacks by increasing the likelihood of encounters between people and bears. In addition, grizzlies were attracted to food in the campgrounds and at garbage dumps, which increased the likelihood of encounters even further and created a dependence upon artificial sources of food. In 1969 the dumps were closed. Strict regulations for disposal of all food and dishwater have been implemented. As a result of these measures, bears no longer frequent the park campgrounds. Films, signs, and pamphlets educate the public about the importance of garbage control and about what to do if they see a bear. Park bears are monitored, and areas where the probability of an attack is considered high may be closed to camping or hiking. If problem bears can be identified, these individuals may be moved to other locations. If this is not successful, problem bears are sometimes killed (Wright 1992).

These changes in the management of people and bears have raised the level of awareness about how to prevent bear attacks, but problems have not been completely eliminated. This is partly because there are always some visitors who do not comply with regulations, and partly because grizzly bears are notoriously unpredictable. Even when they are not attracted by food and they are not provoked, on rare occasions attacks with no obvious cause occur.

Although the rate of grizzly attacks on people is extremely low (less than one injury per million visitors to parks with grizzlies), some risk remains as long as people and grizzly bears use the same habitat. For some visitors this risk actually enhances the experience of visiting parks with grizzlies, not just because of the excitement associated with danger but also because they are put in a situation in which they are no longer the dominant species.

6.4.5   Elephants and crop damage in Africa

In the last two examples, animal control was complicated by the fact that wild animals simultaneously have both negative and positive values. This creates a situation where managers need to provide opportunities for people to inter-act with wildlife while minimizing risks and damage. The situation of the African elephant is even more complicated, because there are three conflict-ing values. In addition to destroying crops and being a valuable tourist attrac-tion, elephants are a source of ivory, a commercially valuable product. Conventional solutions to the problem of agricultural damage are compli-cated by the enormous emotional appeal of elephants on the one hand and by the enormous profits that can be obtained from marketing ivory on the other.

In parts of Africa where elephant populations are "overabundant," that is, where there are more elephants than the habitat can support, the animals come into conflict with farmers because they damage crops (Parker and Graham 1989*a,b*). In this type of situation, it would usually be considered appropriate to reduce populations of the pest species. This can be done by shooting problem elephants or by "culling," in which an entire group is systematically removed. The deliberate killing of an entire group of such appealing and intelligent animals is a disturbing prospect, however, that is vehemently opposed by conservation and humanitarian groups.

The lucrative trade in ivory complicates matters still further. Ivory has fas-cinated people throughout the world for millennia (Parker and Amin 1983). In the 1970s, trade in ivory escalated, and elephant-killing became more effi-cient because of the availability of automatic weapons. In 1989 the Ivory Trade Review Group (established by the International Union for the Conservation of Nature) issued a report on the elephant trade. It shed light on several important components of the problem in addition to the number of tons of ivory that were sold. Although the yield of ivory had been fairly stable for about a decade, more and more elephants were being killed to provide the same amount of ivory. This was due to the fact that most of the larger bulls, with the biggest tusks, had been killed, so females and smaller and younger elephants were being killed to supply an equivalent amount of ivory. This reduction in adult males resulted in lost reproductive opportunities because on some areas females were unable to find mates. The killing of females also caused mortality among orphaned infants and disrupted the ele-phants' matriarchal social organization (Lewin 1989).

These developments fueled concern both within and outside of Africa about the effects of the ivory trade on the continent's elephants. When the

Convention on International Trade in Wild Fauna and Flora (CITES) met later in 1989, it banned trade in African ivory. (For more information about CITES, see Chapter 9.) This decision was controversial, however. Opponents of legalized trade in ivory say that the ban is necessary because as long as there is an ivory trade, poachers will be able to sell ivory from illegally killed elephants. But the nations of southern Africa argue that well-managed harvests are the best way to conserve elephants, protect habitat, and minimize damage. They point out that damage could be controlled in this way and that the profits generated by the sale of ivory are used to fund conservation programs and provide local people with incentives for conservation.

The question of how best to control damage from African elephants remains unresolved. In 1997, CITES temporarily lifted the ban on trading ivory. Not surprisingly, this was a highly controversial decision.

6.5   Conclusions

It should be clear from these examples that there are a variety of situations in which people seek to minimize conflicts with organisms that cause economic harm or threaten health or safety. It should also be clear that whether and how to do this is subject to debate and depends upon our objectives. In the past, utilitarian managers actively sought to exterminate unwanted species such as predators. Pest control efforts were often wasteful and ecologically harmful. More recently, the emphasis has shifted to developing control efforts that are efficient and specifically target problems while minimizing ecological impacts, but pest control remains controversial.

Programs to control populations that are regarded as overabundant can have unforeseen impacts. I have already alluded to some of these: if the control program is not carefully designed and monitored it may waste money and labor and accomplish little; chemicals administered to control pests may poison nontarget species and accumulate in the food chain; and successful control may reduce or eliminate species that perform significant ecological functions. Also, pest control may have unintended evolutionary consequences, because it creates selective pressures for certain characteristics. The evolution of resistance to pesticides is particularly worrisome.

This chapter and the two preceding chapters describe techniques of utilitarian management that address important societal needs and that strive to maximize the productivity of valued wild plants and animals while minimizing problems from pests. But when pursued single-mindedly, harvest management, habitat management, and pest control can overlook subtle ecological interactions. This theme – that utilitarian management dramatically simplifies landscapes and has other unforeseen ecological impacts – has been repeated often in the last few chapters. In Parts II and III we will see how two alternative approaches to resource management developed in response to the limitations of management to maximize the production of selected species.

The next four chapters consider management to preserve species and habitats from human impacts. This approach was developed in part as a reaction to problems with utilitarian management, but its roots actually go back as far as those of utilitarian management, and the two approaches developed simultaneously. The preservationist approach reached its heyday, however, as problems stemming from resource utilization became apparent after World War II. These problems will be considered in Chapter 7.

References

Collett, D. (1987). Pastoralists and wildlife: image and reality in Kenya Maasailand. In *Conservation in Africa: People, Policies, and Practice*, ed. D. Anderson and R. Grove, pp. 129–148. Cambridge: Cambridge University Press.

Di Silvestro, R. L. (1985). The federal animal damage control program. In *Audubon Wildlife Report, 1985*, ed. R. L. Di Silvestro, pp. 130–148. New York, NY: National Audubon Society.

Franz, J. M. (1961). Biological control of pest insects in Europe. *Annual Review of Entomology* **6**:183–200.

Haag-Wackernagel, D. (1995). Regulation of the street pigeon in Basel. *Wildlife Society Bulletin* **23**:256–260.

Hornaday, W. T. (1913). *Our Vanishing Wildlife: Its Extermination and Preservation*. New York, NY: Charles Scribner's Sons.

Hunter, M. L., Jr. (1996). *Fundamentals of Conservation Biology*. Cambridge, MA: Blackwell Science.

Knowlton, F. F. (1972). Preliminary interpretations of coyote population mechanics with some management implications. *Journal of Wildlife Management* **36**:369–382.

Leopold, A. (1933). *Game Management*. New York, NY: Charles Scribner's Sons.

Leopold, A. (1936a). Deer and Dauerwald in Germany. I. History. *Journal of Forestry* **34**:366–375.

Leopold, A. (1936b). Deer and Dauerwald in Germany. II. Ecology and policy. *Journal of Forestry* **34**:460–466.

Leopold, A. (1966). *A Sand County Almanac*, 7th edn. New York, NY: Ballantine Books.

Leopold, A. S. (1955). Too many deer. *Scientific American* **193**(110):101–108.

Lewin, R. (1989). Global ban sought on ivory trade. *Science* **244**:1135.

Louda, S. M., D. Kendall, J. Connor, and D. Simberloff (1997). Ecological effects of an insect introduced for the biological control of weeds. *Science* **277**:1088–1090.

Parker, I. and M. Amin (1983). *Ivory Crisis*. London: Chatto and Windus.

Parker, I. S. C. and A. D. Graham (1989*a*). Elephant decline: downward trends in African elephant distribution and numbers (Part I). *International Journal of Environmental Studies* **34**:287–305.

Parker, I. S. C. and A. D. Graham (1989*b*). Elephant decline: downward trends in African elephant distribution and numbers (Part II). *International Journal of Environmental Studies* **35**:13–26.

Trefethen, J. B. (1975). *An American Crusade for Wildlife*. New York, NY: Winchester Press and the Boone and Crockett Club.

Webb, W. J. (1960). Forest wildlife management in Germany. *Journal of Wildlife Management* **24**:147–161.

Winkler, W. G. and K. Bögel (1992). Control of rabies in wildlife. *Scientific American* **266**(6):86–92.

Wright, R. G. (1992). *Wildlife Research and Management in the National Parks*. Urbana, IL: University of Illinois Press.

The figure opposite illustrates some characteristics of a landscape managed primarily with a preservationist approach. High-elevation and low-elevation forest reserves are connected by a wooded corridor. Clearcuts have been consolidated, leaving large blocks of mature or old-growth forest. This forest has few openings (because of fire suppression). Disturbances have been prevented in the woodlot, so there are no significant openings and all trees are the same age and size. Meanders have been restored to the stream channel, as well as shrubs along its banks.

PART TWO

Protection and restoration of populations and habitats – a preservationist approach to conservation

7

Historical context – the rise of environmental concerns after World War II

In Part I we saw how management of living natural resources developed in response to unregulated exploitation and how the disciplines of utilitarian forestry, wildlife management, and range management emphasize regulated exploitation of selected species. This kind of management was a great improvement over unregulated exploitation and sometimes succeeded in sustaining the take of harvested species, in controlling unwanted organisms, and in manipulating habitats for the benefit of certain organisms, but it has not been entirely satisfactory. Utilitarian management simplifies managed habitats and sometimes fails to foresee some of the consequences of this simplicity. Furthermore, changing conditions in the middle of the twentieth century created new pressures on species and habitats, and changing values dictated a broader focus, one that considered the needs of all species. This chapter describes the evolution of an approach to resource management that seeks to preserve living things regardless of their utilitarian values.

The roots of this approach go back to a nineteenth-century movement to preserve wild places for their intrinsic beauty and spiritual value (see the Introduction). A century later, awareness of environmental problems and the accelerating loss of species provided additional motivation for preserving wild things from human impacts.

This chapter examines the rise of environmental concerns about several interrelated problems: introduced organisms, toxins in the environment, impacts of human resource use, and the loss of species. It does not contain an exhaustive review of environmental problems. Instead, it focuses on the emerging consciousness, after 1950, of several key problems. We shall see in the next three chapters that this emerging awareness had important consequences for how resource managers think about and implement management strategies. Chapter 8 explores some of the scientific concepts underlying this new approach, and Chapters 9 and 10 examine techniques for preserving habitats and species.

7.1   Economic and demographic changes

In the wake of World War II, fundamental demographic and economic changes took place in both the developed and the developing worlds. These changes resulted in new pressures on habitats and species. Industrialized nations experienced simultaneous population growth and rising standards of living, both of which raised levels of resource consumption. In the United States, agriculture became increasingly mechanized. Because mechanized farming required less labor, much of the rural population was displaced and migrated to urban centers. At the same time, people began moving out of cities to escape social unrest, crime, and congestion. Instead of returning to rural areas, they settled in suburban "bedroom communities" and commuted to urban jobs. These demographic changes profoundly altered the landscape. Extensive monocultures replaced small, diverse farms, and discrete cities surrounded by farms were replaced by suburban sprawl.

In developing countries, on the other hand, a different set of forces came into play in the late twentieth century. In different ways, foreign investors and international aid programs both put pressure on living natural resources. Because the commercial exploitation of resources usually requires capital and sophisticated technology, developing nations depended upon foreign investments and expertise. Timber harvest, in particular, requires large amounts of capital; this is usually provided by "mega-corporations" backed by "financial muscle" in developed countries (Myers 1979:193). This dependence tended to make host countries timid about imposing environmental regulations that could scare off foreign investors. Because of this unbalanced relationship, foreign investors faced few restraints on how they managed resources. At the same time, they were "driven to apparently reckless forms of exploitation" by interest rates on their substantial investments (Myers 1979:193). Needless to say, these conditions favored short-term profits rather than sound ecological stewardship.

At the same time, rising standards of living in the developed world created a demand for resources from the developing world, such as timber, paper pulp, and beef. For instance, imports of tropical hardwoods by the developed world increased by a factor of 13 between 1950 and 1973. Furthermore, logging in the tropics opened up inaccessible forests to subsistence cultivators, who exerted a new set of pressures upon local resources (Myers 1979).

In addition, through development projects, international aid agencies created novel pressures on resources in Asia, Africa, and Latin America. These projects typically encouraged people to abandon traditional modes of resource use, and in some cases replaced them with inappropriate management strategies based on western models and conditions (Homewood and Rodgers 1987; Oba *et al.* 2000). These agencies also built roads, dams, and other forms of infrastructure that had unforeseen environmental impacts.

By the 1950s, many people began to be concerned about widespread environmental changes. Interest in nonconsumptive uses of wildlife grew, as did public awareness of the connection between environmental quality, the well-being of wild organisms, and human health and welfare. This increase in environmental awareness was accompanied by heightened concern for species and habitats that do not provide commodities. As the pace of resource extraction quickened, people began to notice more and more undesirable effects of resource consumption, even when it was regulated by utilitarian management.

7.2   Awareness of ecological problems

7.2.1   Invading species

In 1958 the British ecologist Charles Elton published a book on *The Ecology of Invasions by Animals and Plants*. (Organisms that are accidentally or deliberately introduced and successfully colonize places they did not previously inhabit are considered non-native, alien, or exotic.) Invasions of exotics often pose a dire threat to native organisms and communities, but until Elton's book the seriousness of this problem was seldom recognized. "Nowadays we live in a very explosive world," he wrote:

It is not just nuclear bombs and wars that threaten us, . . . there are other sorts of explosions . . . ecological explosions. I use the word 'explosion' deliberately, because it means the bursting out from control of forces that were previously held in restraint by other forces. . . .

Ecological explosions . . . can be very impressive in their effects, and many people have been ruined by them. . . .

We are living in a period of the world's history when the mingling of thousands of kinds of organisms from different parts of the world is setting up terrific dislocations in nature. We are seeing huge changes in the natural population balance of the world. (Elton 1977:15,18)

Where organisms invade or are released into suitable environments, introductions can be dramatically successful. This is especially likely if the newcomers encounter few antagonists – such as predators, grazers, parasites, or competitors – in their new home. Organisms in the latter group often become exotic pests, or invaders. (Fictional versions of this phenomenon have been humorously treated in numerous science fiction movies, such as *Invasion of the Body Snatchers* and *Little Shop of Horrors*.) Successful invaders often undergo an explosive, exponential increase in population size and geographic range (Mack 1981). As a result, they often have negative impacts on native species and habitats. Exotic species may compete with native organisms, eat them, or parasitize them. Exotics can also indirectly alter ecological processes such as nutrient cycling, thereby causing changes in community structure and function (Vitousek 1986).

For an alien organism to become established, it must be able to get to its new homeland, and it must survive and reproduce when it arrives. Dispersal is limited by geographic and climatic barriers, but many organisms are aided in their dispersal by the activities of people. We introduce plants, animals, and microorganisms both deliberately and inadvertently. Contaminated batches of grain and ballast water in ships have unintentionally transported dozens of species to new environments. Species that can take advantage of these means of transportation are more likely to invade new regions than species that cannot. This is why beetles, which hitchhike in stored grain and ship ballast, and Old World rats and mice, which stow away on ships, are well represented among invading fauna.

Wildlife managers and sport hunters have often tried to introduce game animals to locations outside their native range or to restock areas from which game has disappeared. In a surprising number of cases, these attempts to enhance opportunities for recreational hunting have failed. For instance, between 1883 and 1950, 23 introductions of four species of foreign grouse into the United States were unsuccessful, and repeated attempts to introduce ring-necked pheasants from China were initially unsuccessful as well. Often the introductions failed because the new environment was unsuitable.

But when they succeed in becoming established, introduced species can have dramatic effects. A few of the examples Elton recounted are described below:

- The muskrat. The muskrat is an aquatic furbearer native to the New World. In 1905, a Czechoslovakian landowner introduced five muskrats.

Supplemented by subsequent introductions, this rodent spread throughout Eurasia. Within a few decades, it numbered in the millions.

- Chestnut blight. The American chestnut tree once dominated deciduous forests throughout much of eastern North America. As a result of the introduction of chestnut blight, a fungus that was apparently brought to New York City on nursery stock from Asia around the turn of the century, the American chestnut was virtually eliminated within a few decades.
- The sea lamprey. The sea lamprey spends most of its life in the North Atlantic but spawns in fresh water. For millennia, Niagara Falls formed a natural barrier that prevented lampreys from entering the Great Lakes except for Lake Ontario. In 1829 the completion of the Welland Ship Canal gave lampreys access to Lake Erie. In the 1930s they entered Lake Huron and Lake Michigan. Larval lampreys are free-living filter feeders, but adults parasitize other fish. They attach themselves by means of their jawless mouths to the bodies of host fish, secrete an anticoagulant, and feed on blood and other body fluids sucked from the host. The introduction of sea lampreys into the Great Lakes devastated populations of native fishes, especially lake trout.

The fossil record provides numerous examples of natural arrivals of alien species that occurred when geographic changes allowed two previously isolated biotas to come into contact. Typically such events are followed by the disappearance of many species from one or both of the newly joined areas. For instance, about 3 million years ago North and South America were joined when a land connection, the Isthmus of Panama, developed between the two continents. After millions of years of separation, organisms from North America expanded their range into South America and vice versa. The fossil record shows that in both continents the number of families of mammals increased initially, as new species arrived, but subsequently the number of mammalian families declined in both North and South America. Although the fossil record does not allow us to test hypotheses about these events directly, it has been suggested that the arrival of new competitors and predators was a factor in the extinctions that followed. "Could it not be," asked Elton "that this intermingling of species that had not evolved into ecological balance led to dislocations as catastrophic as the entry of the sea lamprey into the inner Great Lakes, or the spread of the Asiatic chestnut fungus in America?" He noted, however, that natural invasions occur much more slowly than anthropogenic ones: "the scale of time is totally different – one in millions [of years] and the other in decades" (Elton 1977:40).

Many deliberate introductions, Elton suggested, are motivated by utilitarian concerns, the desire to simplify ecological relationships in ways that maximize production and minimize losses:

with land in cultivation, whether pastoral, ploughed, or gardened, the earnest desire of man has been to shorten food-chains, reduce their number, and substitute new ones for old. We want plants without other herbivorous animals than ourselves eating them. Or herbivorous animals without other carnivorous animals sharing them.

Some of the profoundest changes in food-chains have come about through the introduction and spread of domestic grazing animals. A hundred years ago, the grass plains of North America were still occupied by huge roaming herds of bison.... The bison was the chief grazing animal in the centre of the continent, but in a comparatively few years was completely replaced by cattle and sheep, as well as by other kinds of farming. The structure and composition of the prairie vegetation also changed. (Elton 1977:127–128)

Although Elton expressed concern about the ecological consequences of introduced species, for the most part the general public remained unaware of this problem. In fact, even today, many people know little about this serious environmental issue. That is not the case with environmental problems that directly affect the well being of people or appealing species of wildlife. At about the time when *The Ecology of Invasions* appeared, concern was mounting about the widespread use of pesticides and other forms of pollution.

7.2.2   Harmful substances in the environment

Early warnings

Elton viewed the widespread application of synthetic pesticides as an outgrowth of the same attitude that led to the replacement of bison by cows and sheep: the desire to simplify ecological relationships. "The applied biologist," wrote Elton (1977:138), "seeks to bypass all the irritating and complex interactions of natural populations, in fact simply sweep away natural food-chains altogether, leaving only the crop plants to give an ordered and useful appearance to the landscape." This goal produced a strategy of "incredibly massive use of insecticides" and an "astonishing rain of death upon . . . much of the world's surface" (Elton 1977:137–138,142).

By the early 1960s, a number of prominent scientists had come to the conclusion that synthetic chemicals in air, water, and soil posed a serious threat to the health of people and wildlife. In 1962 and 1963, three books appeared which sounded the alarm about the proliferation of these substances in the environment: Barry Commoner's *Science and Survival*, Murray Bookchin's *Our*

Synthetic Environment, and Rachel Carson's *Silent Spring* (Carson 1962; Commoner 1966; Bookchin 1974; see also Commoner 1975). The most influential of these was the award-winning *Silent Spring*, but all three authors voiced similar concerns.

Radioactive substances emit radiation that is capable of damaging cells in ways which cause genetic damage and disrupt cellular growth and reproduction. Studies in the aftermath of aboveground tests of atomic weapons in the 1950s revealed that wind and water currents distributed radioactive fallout around the globe. The radioactive isotopes strontium-90 and cesium-137 quickly accumulated in tundra lichens, which absorb nutrients and moisture from the air because they lack roots. The atmospheric pollutants that became concentrated in tundra lichens were subsequently ingested by caribou and then passed on to Inuit people.

After World War II, the production of pesticides expanded. Large amounts of insecticides, rodenticides, herbicides, and fungicides were used to control agricultural pests, but pesticides were also developed for use in homes, offices, and factories.

Chlorinated hydrocarbons are a widely used group of insecticides. DDT is the best known of these. During World War II, powdered DDT was used to control lice, and thus typhus, among soldiers, prisoners, and refugees. Because powdered DDT is not readily absorbed through the skin, it was thought to be safe; however, DDT actually attacks the central nervous system of insects and affects liver function in birds. This can lead to reproductive failure in birds, because liver function controls calcium transport in the oviduct and therefore eggshell formation. As a result of this disruption of normal reproductive physiology, the eggs laid by affected birds contain insufficient calcium in their shells and are too thin and soft to support the weight of the incubating parent. Because DDT is not very specific, it affects a variety of wildlife species. Pelicans, grebes, terns, and several species of raptors (birds of prey), including ospreys and bald eagles, experienced dramatic declines in reproduction following exposure to DDT.

In addition to their toxic effects, chlorinated hydrocarbons have two other properties that cause problems. First, they are long-lasting; therefore, they accumulate in the tissues of carnivorous animals, reaching concentrations that are several orders of magnitude greater than their original concentrations in the environment (see Box 7.1). Second, they are soluble in fat and stored in fatty tissues. This means that they are further concentrated by being dissolved in a small amount of tissue. To make matters worse, if an animal uses up most of its fat reserves at one time, there will be less fat for the toxin to dissolve in, and it will become even more concentrated. This is exactly what happens to many birds during their annual cycle. Fat is stored in preparation for reproduction and, in some cases, also for migration. When most of this stored fat is mobilized to meet the energy demands of these activities, fat-soluble pesticides concentrate in the remaining fatty tissue.

Box 7.1   Passing toxins up the food chain: Secondary poisoning and biomagnification

Because toxic substances are often introduced in small amounts and appear to be rapidly diluted in air, water, and soil, people initially thought that pollutants such as radioactivity and pesticides would not reach harmful concentrations in the environment. In the 1950s and 1960s, however, it became apparent that complacency about the vastness of the earth was not justified, because under certain circumstances physical and biological processes can concentrate toxins by several orders of magnitude.

First of all, toxins can be passed from the tissues of one organism to those that feed on it. This means that a given amount of toxin can kill more than once. This phenomenon, termed secondary poisoning, occurs when an animal is poisoned by feeding on another poisoned animal. Secondary poisoning has resulted in mortality among predaceous insects, trout, amphibians, reptiles, nestling and adult birds, rodents, coyotes, foxes, and domestic animals (Rudd 1964). If toxicity is high, it can occur even when the pesticide is short-lived. For example, organophosphates, which break down rapidly, have been implicated in secondary poisoning of gulls, owls, and other carnivorous birds.

Stable toxins present an even more serious problem, however, because in addition to being passed from one organism to another, they can become increasingly concentrated with each step in the food chain, a process termed biomagnification (or bioconcentration). Because energy transfers are rather inefficient, herbivorous animals must eat many grams of plant matter to gain one gram of biomass. The same is true for carnivores feeding on herbivores. If a toxin breaks down rapidly, then it will not concentrate as it is passed up the food chain. Stable pesticides, however, are not broken down and excreted. As biomass is transferred from one trophic level to another, toxins become increasingly concentrated, even if they are present at low concentrations initially (Figure 7.1). As a result of this bioconcentration, carnivores and scavengers accumulate high concen-

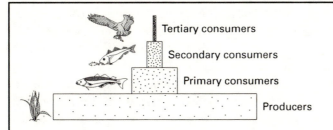

Figure 7.1. Bioconcentration. Dots represent molecules of a persistent toxin such as DDT. The amount of toxin stays the same at each trophic level, but biomass decreases with each successive level, so that the toxin becomes more concentrated at the top of the biomass pyramid. (See Figure 3.1.) (After Woodwell 1967.)

trations of chlorinated hydrocarbons and other persistent toxins in their tissues.

Biomagnification affects both aquatic and terrestrial communities. Clear Lake in northern California is a shallow, warm, highly productive body of water with a soft, mucky bottom. Its high productivity produces a good crop of sport fish, making it a favorite spot for anglers. Unfortunately, however, it is also favored by the Clear Lake gnat, whose larvae develop in the mud of the lake bottom. Although this insect is related to mosquitoes, it does not bite people, but because of its sheer numbers it was considered a serious threat to the tourist business at Clear Lake (Hunt and Bischoff 1960; Carson 1962; Rudd 1964).

Efforts to control gnats at Clear Lake began in 1916, but until chlorinated hydrocarbons were developed, the gnats continued to be a problem. Studies of DDT and DDD, a closely related compound, demonstrated that these insecticides were effective in killing gnat larvae. Of the two, DDD appeared to be less harmful to fish. Preliminary studies indicated that DDD caused relatively low fish mortality when applied at a concentration of one part of insecticide per 70 million parts of water (0.014 parts per million, ppm). The lake bottom was surveyed, lake volume was calculated, and in September 1949, DDD was applied to the lake at an estimated dilution of one part insecticide to 70 million parts water.

Initially, the control operation seemed to be very successful. Follow-up studies indicated that 99% of the gnat larvae were killed, and very few gnats were observed for two years following the treatment. But in 1951 larvae were found again, and by 1952 they had reached problem levels.

DDD was reapplied in 1954, this time at a higher rate: one part insecticide to 50 million parts water (0.02 ppm). Again the treatment appeared to be effective, but the gnat population rebounded, this time even more quickly, so that the lake was treated for a third time in 1957.

The first sign of possible problems from gnat control came in December 1954, when 100 western grebes were reported dead on the lake. Investigators who examined the carcasses found no signs of infectious disease. More dead grebes were found in March of 1955 and in December of 1957, and again none of the birds examined showed any evidence of dying from disease. But chemical analysis of the birds' fat tissue revealed extraordinarily high concentrations of DDD (up to 1600 ppm, an 80000-fold increase over the application rate of 0.02 ppm).

What had happened to bring about this extraordinary increase? Additional tissue studies of fish, frogs, and birds revealed that DDD was present in all animals that were sampled, and concentrations increased with trophic level, according to the following progression:

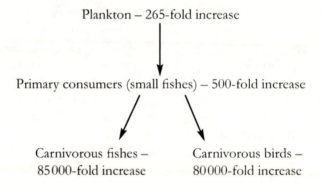

Plankton – 265-fold increase

Primary consumers (small fishes) – 500-fold increase

Carnivorous fishes –
85000-fold increase

Carnivorous birds –
80000-fold increase

A comparison of DDD levels in fishes with different diets provides further evidence that feeding patterns were responsible for the observed tissue concentrations. Fish that fed on plankton, algae, or plants had lower levels of DDD in their tissues than those that fed on other fishes. Furthermore, younger fishes had lower levels of pesticide than older fishes, indicating that the effects of ingesting DDD-laden food were cumulative.

Although about 1000 pairs of western grebes nested on the lake prior to the first treatment, less than 25 pairs were found during nesting surveys in 1958 and 1959. Grebes were not found in areas where they had previously nested, and pairs that did attempt to nest were apparently unsuccessful.

Thus, even after all traces of DDD had vanished from the water itself, DDD remained part of the "fabric of life" of Clear Lake (Carson 1962:52). Although the poison was gone from the lake's water, it persisted and continued to exert its effects on the lake's organisms.

Other types of insecticides break down rapidly but are highly toxic. For example, organophosphates, such as Parathion, are highly toxic to vertebrates. Like chlorinated hydrocarbons, compounds in this group work by interfering with central nervous system functioning. Their toxicity and lack of specificity stem from the fact that they inhibit an enzyme that is necessary for the proper functioning of vertebrate nerves and muscles. The short life of organophosphates is clearly an advantage over DDT, but this must be weighed against the fact that compounds in this group are more toxic, and hence more likely to directly poison people or wildlife, than chlorinated hydrocarbons.

Herbicides, or "weed killers," vary greatly in their effects on animal tissues. Some have relatively little effect on animals, while others are toxic or cause tumors, birth defects, or mutations. They can also degrade wildlife habitat by removing vegetation.

Polychlorinated biphenyls (PCBs), dioxin, and heavy metals such as mercury and lead are examples of industrial pollutants (toxic by-products of industrial processes). PCBs, which like DDT are chlorinated hydrocarbons, are used in the construction of electrical transformers, where they function as coolants. They have been implicated in developmental abnormalities, reproductive disorders, and immune suppression in wildlife, and because they are very persistent, PCBs accumulate in the tissues of carnivores.

Radiation, insecticides, herbicides, and industrial pollution are worrisome because of their toxicity. There are other forms of pollution that are not toxic but nevertheless have harmful ecological consequences. For example, when fertilizers, sewage, or high-phosphate detergents enter surface waters, they add massive infusions of nutrients that cause algae to proliferate. In a process termed eutrophication, the resulting algal blooms deplete waters of dissolved oxygen, which creates conditions unfavorable for many other organisms.

Some synthetic substances cause problems because they resist decay. The earliest synthetic plastics were nonbiodegradable. To create long-lasting materials, chemists joined molecular units together with linkages that naturally occurring enzymes are unable to attack. Garbage has been dumped on land or into the ocean for centuries, but in the past it dissolved, decayed, or sank. Long-lived plastics constitute a new type of solid waste, however. Although this was considered primarily an esthetic problem, it was also hazardous to wildlife. Barry Commoner noted that when a plastic ring connecting a six-pack of beer is "tossed aside, it nevertheless persists until it comes to float on some woodland lake where a wild duck, too trustingly innocent of modern technology, plunges its head into the plastic noose, [a] fatal conjunction [of] some plastic object and some unwitting creature of the earth" (Commoner 1975:164).

Commoner (1966, 1975) also warned of another consequence of pollution: global atmospheric changes with the potential to alter profoundly the distribution and composition of the earth's biota. The effects of atmospheric changes are of special concern because no place on earth, no matter how remote or how well protected, is insulated from these effects. Commoner identified two examples: global climate change (the "greenhouse effect") and ozone depletion. He pointed out that between 1860 and 1960 increased rates of fossil fuel combustion were associated with a rise in atmospheric carbon dioxide, which, like glass, transmits visible light but absorbs infrared rays. "Carbon dioxide makes a huge greenhouse of the earth," therefore, "allowing sunlight to reach the earth's surface but limiting radiation of the resulting heat into space" (Commoner 1966:11). This, Commoner suggested, could lead to a rise in the earth's temperature, melting of the antarctic ice cap, and catastrophic flooding.

Ozone, a form of oxygen in which each molecule consists of three atoms (O_3) instead of the usual two (O_2), in the stratosphere (the upper level of the atmosphere) forms a protective shield that screens out most harmful solar ultraviolet radiation. Ozone depletion results from the release of chlorine into the stratosphere. At cold temperatures, chlorine is a catalyst in chemical reactions that break down ozone. A catalyst enables a chemical reaction to take place but is not used up in that reaction; hence, a single molecule of a chlorinated catalyst can catalyze the destruction of many molecules of ozone. Commoner pointed out that "the continued existence of terrestrial life is dependent on the layer of ozone in the stratosphere – a protective device that itself is the product of life. Should the ozone in the stratosphere be reduced, terrestrial life would be seriously threatened by ultraviolet radiation. It is unfortunate that some human activities raise this threat" (Commoner 1966:31).

The specific concerns of Carson, Commoner, and Bookchin about harmful substances released into the environment are tied together by a more general worry – a sense that the anthropogenic changes wrought by synthetic chemicals upset natural balances and present conditions that life cannot adapt to. Like Elton, these writers expressed concern about the accelerated pace of change. "It took hundreds of millions of years to produce the life that now inhabits the earth," wrote Carson (1962:17):

– eons of time in which that developing and evolving and diversifying life reached a state of adjustment and balance with its surroundings. . . . For time is the essential ingredient; but in the modern world there is no time. . . .

The rapidity of change and the speed with which new situations are created follow the impetuous and heedless pace of man rather than the deliberate pace of nature. Radiation is no longer merely the background radiation of rocks, the bombardment of cosmic rays, the ultraviolet of the sun that have existed before there was any life on earth; radiation is now the unnatural creation of man's tampering with the atom. The chemicals to which life is asked to make its adjustment are no longer merely the calcium and silica and all the rest of the minerals washed out of the rocks and carried in rivers to the sea; they are the synthetic creations of man's inventive mind, brewed in his laboratories, and having no counterparts in nature.

To adjust to these chemicals would require time on the scale that is nature's; it would require not merely the years of a man's life but the life of generations.

Commoner expressed concern that both the magnitude and duration of the changes brought about by modern technology are unprecedented:

The new hazards are neither local nor brief. Air pollution covers vast areas. Fallout is worldwide. Synthetic chemicals may remain in the soil for years. Radioactive pollutants now on the earth's surface will be found there for generations, and, in the case of carbon 14, for thousands of years. Excess carbon dioxide from fuel combustion eventually might cause floods that could cover much of the present land's surface for centuries. (Commoner 1966:28)

One consequence of this growing concern was that ecologists turned their attention to the process of evolution, in an effort to understand how changes in the modern world affect the adaptive potential of species. This topic is explored in Chapter 8.

Acid precipitation

Acid precipitation, another form of atmospheric pollution, became evident in the 1960s. Although this phenomenon is usually referred to as acid rain, both rain and snow are affected, so it is more correctly termed acid precipitation. Swedish scientist Svante Odén pointed out in 1968 that precipitation in Europe had become more acidic since the European Atmospheric Chemistry Network began keeping records in the 1950s. The affected area spread from the Netherlands, Belgium, and Luxembourg to include northern France, Germany, southern Scandinavia, and the eastern British Isles by the late 1960s. Much of the northeastern United States was also receiving acid precipitation by 1955 (Odén 1976; Likens et al. 1979).

Acid precipitation results when oxides of nitrogen and sulfur are released into the atmosphere by the burning of fossil fuels. The pH scale measures hydrogen ion concentration. It reflects whether a solution is acidic, neutral, or basic. The lower the pH, the greater the hydrogen ion concentration, and the more acidic the solution. A pH of 7 indicates a neutral solution. Acids have values below 7, whereas pH values greater than 7 indicate basic solutions. The scale is exponential (see Chapter 2), so that a liquid with a pH of 5 is 10 times more acidic than a liquid with a pH equal to 6 and 100 times more acidic than a solution with a pH of 7.

In regions where the underlying bedrock contains limestone, a form of calcium carbonate, acid precipitation may be neutralized when it reacts with carbonate in the substrate. But in areas of granitic bedrock, this buffering does not take place, and soils and surface waters become acidified. Forest-dwelling species are affected indirectly by loss of habitat resulting from extensive tree mortality. In aquatic communities, on the other hand, organisms are affected both directly and indirectly by acid precipitation. Aquatic invertebrates, fishes, and amphibians are affected directly when the levels of acidity in their environment exceed their physiological tolerances (see Chapter 3). In addition, acidity interferes with calcium metabolism in fishes, thereby producing skeletal deformities. Consumers are also affected indirectly by declining food supplies. The rate at which organic matter decomposes also slows with acidification, presumably because of changes in the microbial communities of acidified waters.

7.2.3   Extinctions

A third environmental problem that sparked alarm during this period was the disappearance of species. The plights of a number of dwindling species, such as pandas, gorillas, whales, whooping cranes, birds of prey, elephants, and tigers, became well known. These "charismatic megafauna" are large, well-known species that people tend to regard with affection. But the idea that all forms of life on earth deserve protection, regardless of their emotional appeal, was a new development. This was epitomized by the controversy over the fate of the snail darter, a small, previously unknown species of perch thought to occur only in water near the site of a proposed dam on the Little Tennessee River (see Chapter 9).

As concern about the fate of endangered species mounted, scientists echoed a familiar theme – people are bringing about changes at an unprecedented rate, with potentially dangerous consequences. The rate at which species are going extinct is thought to have increased dramatically in the last few hundred years. Ehrlich et al. (1977) estimated that the extinction rate of birds and mammals between 1600 and 1975 was five to 50 times higher than it was throughout most of the rest of evolutionary history and predicted that it would escalate still further. In some cases, we have evidence that extinctions after 1600 were caused by exploitation, habitat loss, or introduced species. In others, too little information is available to determine the cause of extinction, although it seems likely that people contributed to the demise of many of those species as well (Caughley and Gunn 1996).

Because of this accelerated loss of species, scientists such as the entomologist E. O. Wilson and others argued that we should be concerned about the erosion of biological diversity or "biodiversity." Initially, the term biological diversity meant the total number of species. (A more precise term for this is species richness.) It has been expanded, however, to encompass different levels of organization. Most biologists now use the term biodiversity to denote the diversity of life at all levels of organization, including genetic material, species, and communities (Reaka-Kudla *et al.* 1997). (We will see in Chapter 9, for example, that one facet of biodiversity is the genetic diversity within species.)

No one can predict with certainty the effects of accumulated extinctions, but many scientists argued that it would be prudent to try to minimize extinctions. The Ehrlichs used the metaphor of an airline mechanic removing rivets from a plane wing to illustrate the potential dangers of species loss:

As you walk from the terminal toward your airliner, you notice a man on a ladder busily prying rivets out of its wing. Somewhat concerned, you saunter over to the rivet popper and ask him just what the hell he's doing.

"I work for the airline – Growthmania Intercontinental," the man informs you, "and the airline has discovered that it can sell these rivets for two dollars apiece."

"But how do you know you won't fatally weaken the wing doing that?" you inquire.

"Don't worry," he assures you. "I'm certain the manufacturer made this plane much stronger than it needs to be, so no harm's done. Besides, I've taken lots of rivets from this wing and it hasn't fallen off yet. . . ."

You never *have* to fly on an airliner. But unfortunately, all of us are passengers on a very large spacecraft – one on which we have no option but to fly. And, frighteningly, it is swarming with rivet poppers behaving in ways analogous to that just described. (Ehrlich and Ehrlich 1985:xi–xii; emphasis in original)

Extinctions are triggered primarily by four things: overexploitation, habitat modification, harmful substances in the environment, and exotic species. The latter two have been touched on already; the former are discussed below.

Exploitation

We have already seen that unregulated market hunting was responsible for the depletion and extinction of numerous species. In response to the excesses of wildlife exploitation in North America, regulations were adopted that gave full protection to some species (such as the herons, egrets, and terns that had been exploited by the plume trade) and regulated the harvest levels of game species such as ungulates, waterfowl, and upland game birds. These measures were fairly successful in maintaining viable populations of the managed species. Today there are still numerous situations in which exploitation poses a threat to species, however. The international trade in wild plants and animals and commercial fishing are two examples.

International commerce in wild organisms involves dozens of countries and thousands of species and their products, some of which are threatened or endangered. Some familiar species such as crocodilians, rhinoceroses, and elephants have declined markedly as a result of this contemporary form of market hunting. The regulation of this trade is complex and difficult. The international trade in wild organisms is diverse. It includes legal and illegal activity, trade in products for everyday use as well as luxury items, and products obtained by killing wild animals and plants as well as by taking live specimens.

Wildlife products have been traded for centuries. But the international wildlife business has reached an unprecedented level for several reasons. First, improvements in transportation have promoted commerce in wildlife. Rapid air transport, in particular, facilitates the shipping of live organisms. Second, techniques of hunting and capturing have improved. Third, the prosperity that followed World War II provided many people with disposable income to spend on wild plants and animals.

The regulation of international wildlife commerce is discussed in Chapter 9. Much of this trade is carried out legally, but poaching wildlife for profit is an international problem. Some animals are killed to obtain body parts that are believed to have medicinal value or to be powerful aphrodisiacs. For instance, elk are killed to obtain the velvet from their antlers, and several species of bears are killed for their gall bladders, which sell for hundreds of dollars per gram. These products can be sold at enormous profit in parts of Asia. Other animals are killed for trophies or to obtain valuable skins or ivory. Unfortunately, as populations of wild plants and animals decline, scarcity drives prices up and increases the economic incentives for poaching. Also, officials involved in regulating trade in harvested products are more likely to be corrupted in situations where the potential financial rewards are enormous.

Fisheries managers have attempted to apply the principles of maximum sustained yield to the harvest of fish populations, but this has been less successful than managing harvests of terrestrial animals. The reasons for this are both biological and political (see Chapter 4). Fish stocks mix at sea, making it difficult to get the information needed to set harvest limits, and management of international waters is politically complex. As a result, many oceanic fish stocks have been depleted.

Habitat modification

Many forms of resource exploitation exert their effects indirectly by modifying habitats. The most obvious forms of habitat modification involve landscapes that are paved over for development. This clearly has serious impacts on biodiversity: few wild organisms survive in habitats dominated by concrete, steel, and glass. But most habitat modification involves more subtle transformations that result from resource consumption. Farms, pastures, and tree plantations may look "natural," in the sense that they support wild organisms and are green and open, but they are not necessarily biologically diverse. Some effects of resource use on habitats are very briefly reviewed below.

Consumption of nonrenewable resources.   Nonrenewable resources are concentrated in ore bodies or pools beneath the earth's surface. Since deposits of oil, coal, natural gas, and minerals occur far underground, they do not participate in soil formation and are not used by living organisms. Thus, their loss is of no direct consequence to the habitats under which they are found. But mining nonrenewable resources exerts strong indirect negative effects on aboveground communities, because it disturbs and pollutes them. Bare areas are created that may take centuries to recover, especially if soil is removed and pollutants are left on the surface. Heavy metals and acids may also enter aquatic systems as a result of mining. Arctic environments are particularly vulnerable to any activities that involve excavation (see Box 13.2). Nonrenewable resource use also affects biodiversity indirectly when oil spills pollute marine environments or when fossil fuel combustion causes acid rain.

Consumption of renewable resources.   One of the principal ways that we alter habitats is by using renewable resources. Some of these effects are briefly summarized below. Unlike nonrenewable resources, under the appropriate conditions renewable resources can grow back after they are removed. Nevertheless, the harvest of renewable resources, even when it is conducted on a sustained yield basis, impacts habitats both directly (by removing resources) and indirectly (by creating disturbances in the process of extracting resources).

Timber harvest.   The harvest of wood for fuel, pulp, or lumber can irreversibly alter forest communities and be devastating to biodiversity, it can improve habitat for some species (see Chapter 5), or it can have effects that lie somewhere in between these two extremes. The effect of timber harvest may be irreversible, or recovery can proceed relatively rapidly after wood is removed. The types of impacts depend upon the amount of timber removed, the methods used, and the type of forest community. Temperate and tropical forests are very different in their responses to disturbance (see Boxes 13.3 and 13.6).

Careless or excessive exploitation of temperate forests (see Chapter 1) has serious ecological consequences, including irreversible soil loss and changes in community composition. Furthermore, when a single tree species is planted to hasten the regeneration of woody vegetation on a logged site, monolithic forests with low biodiversity develop. On the other hand, if appropriate pre-cautions are taken – such as harvesting at a moderate rate, avoiding steep or highly erodable sites and the habitats of rare plants or animals, leaving special habitat features and biological legacies (Chapter 13), and designing cuts that will not overly fragment the landscape – the negative impacts of logging on biodiversity of temperate forests can be minimized.

Removing timber from tropical forests can have severe negative impacts. Moist tropical forests are slow to recover from disturbances that remove veg-etation, because a high proportion of their biomass is located above ground and because the removal of vegetation can cause irreversible changes in their soils (see Box 13.6).

This sensitivity to resource use is cause for concern because of the high biological value of tropical forests. First of all, tropical forests contain a great many different kinds of organisms. Although these forests have high biolog-ical diversity, the geographic ranges of many tropical forest species are quite small. Thus, when tropical forests are cleared, many species are eliminated. Second, many animals that summer in North America or Eurasia migrate to the tropics in winter. Loss of wintering habitat could mean the extinction of these species. Third, because of their great biomass, tropical forests provide important ecological services including flood control, maintenance of atmos-pheric quality, climate moderation, nutrient recycling, and disposal of wastes. Fourth, the pace of tropical forest destruction escalated in the late twentieth century, for the reasons mentioned at the beginning of this chapter.

Grazing. Livestock grazing is one of the principal land uses in dry climates where the dominant vegetation is chaparral, steppe, desert, savanna, or dry forest. (See Boxes 13.3–13.6 for more information on these ecosystems.) Grazing is particularly attractive where the land is unsuitable for other uses such as agriculture or timber harvest. Many plant species are able to cope with a certain amount of grazing. Natural selection favors those individual plants that have modifications allowing them to withstand grazing. If grazing is not excessive, leaves and shoots that are removed by an herbivore are replaced; in fact, under some conditions grazing may actually stimulate regrowth. In some parts of the world, such as Africa and India, nomadic pastoral tribes have subsisted on grazing for centuries. These grazing practices appear to be sustainable, in the sense that they have supported people and their livestock over long periods of time (Collett 1987).

On the other hand, grazing can cause serious habitat degradation, especially where the effects of grazing are not well understood, where grazers have been introduced, and where grazing systems that are not well suited to local environmental conditions are imported. Excessive grazing exceeds plants' ability to recover and promotes the growth of increaser species at the expense of decreasers (see Chapter 5). Where the native vegetation is especially sensitive to overgrazing, dramatic changes in plant cover and increased erosion may result.

Agriculture. At least three types of activity associated with agriculture potentially affect biodiversity: (1) habitat alteration, (2) water diversion, and (3) application of agricultural chemicals. During the last six decades, North American agriculture has been characterized by increased mechanization and the consolidation of family farms into large businesses. In contrast to small farms, which frequently increase habitat diversity, in large farming operations hundreds or thousands of hectares of cropland are typically planted to a single crop. The resulting monocultures reduce habitat diversity markedly, and allow populations of insect pests to build up. In addition, large-scale cultivation may influence habitat potential by causing soils to erode or soil fertility to decline.

In the tropics, both large- and small-scale farming operations pose problems for biodiversity. In moist tropical forests, shifting cultivation (also known as slash-and-burn agriculture, milpa, or swidden agriculture) exhausts the soil's fertility within a few years. A family will cut down the trees on a few hectares of forest and burn them before planting root crops, such as manioc (also known as cassava) and fruits. The fields are weeded several times a year. After two or three years, the site is abandoned because of declining soil fertility.

Water use. Another way in which human activities inadvertently impact biodiversity is by changing patterns of water use. This phenomenon is an ancient one. Water can be manipulated by dams, impoundments (structures that collect, confine, or store water), diversion of surface waters from one place to another for irrigation or other purposes, drainage of subsurface waters, and straightening of stream channels. Dams were once considered an ecologically benign method of generating energy. Since water does work as it moves downhill, this energy can be harnessed to provide power, without generating nuclear waste, acid rain, or greenhouse gases. Thus hydroelectric power is a renewable resource that lacks many of the drawbacks associated with other common methods of power generation.

Dams are not without their negative environmental impacts, however. They submerge scenic canyons and riverine habitats. (For an early example of a controversy regarding the scenic costs of building a dam, consult the Introduction.) Furthermore, the construction of dams brings about drastic changes in the ecology of flowing waters. When water is prevented from moving downstream by a dam, its velocity and temperature are altered. The cool, fast-moving waters of a river or stream are replaced by the warmer, slower, deeper waters behind a reservoir. In addition to interrupting water flow, dams interfere with the movements of organisms such as anadromous fishes (Nilsson and Berggren 2000).

Dams interrupt the natural disturbance regimes of rivers. Some riparian trees such as cottonwoods require floods in order to reproduce. On undammed rivers, seasonal floodwaters remove vegetation from the shoreline and deposit silt as they recede, leaving behind moist, scoured bars that provide ideal sites for the germination of pioneer species adapted to colonize such sites, such as willows and cottonwoods. Thus, the altered flow regimes created by dams interfere with the regeneration of riparian tree species that provide valuable habitat for wildlife (Rood and Mahoney 1993).

It is not just the amount of water in a channel that is important. Some organisms do well in dry years while others prosper in wet years, so in order to sustain biodiversity it is important to retain at least some of the natural variation in flow regimes, with high flows in some years and low flows in others (Poff *et al.* 1997). Dams tend to do just the opposite. By decreasing the variability of flow dynamics, they decrease environmental heterogeneity and diversity.

Dams are often associated with dikes and levees built to straighten river channels and control floods. The diverse and productive ecosystems of natural floodplains adjacent to lowland rivers depend on periodic flooding, however. Flood control structures disrupt the important connection between rivers and floodplains, cutting off inputs of sediment, water, and nutrients (Allan and Flecker 1993). Because of their gentle topography, scenic views, and rich, moist soils, floodplain lowlands attract agricultural and residential developments, so most temperate zone floodplains have been extensively modified by water control structures. Ironically, by confining flowing waters to a straight channel, these measures have often actually exacerbated downstream flooding.

Water diversion poses another serious threat to the biodiversity of riverine habitats. In deserts and the drier portions of steppes, agriculture depends upon irrigation, but when irrigation waters are withdrawn from rivers and streams, the resulting drop in water levels can negatively impact organisms that depend upon habitats associated with rivers. Islands and sandbars in rivers furnish crucial nesting habitat for many species of ground-nesting birds, including waterfowl and shorebirds. Because the young of ground-nesting species are extremely vulnerable to predators, productivity in these species tends to be highest on islands that are surrounded by water deep enough to make them inaccessible to mammalian predators (Chapter 5), but when river water is diverted, nesting islands may become accessible to terrestrial predators.

Islands are also threatened by erosion in dammed rivers. In a naturally flowing river, islands are continually being remodeled by deposition and erosion. When dams are constructed, the processes that create and maintain islands and bars are altered. Silt is trapped behind a dam instead of being deposited downstream, but erosion continues. The net result is the loss of island nesting habitat (Palmer 1991).

In many parts of the world substantial areas of wetland have been drained to create farmland and sites for urban, residential, and industrial development. By the early 1980s, over half the wetlands in the contiguous United States had been drained for agriculture and development. This has had negative effects on biodiversity. As a result of the extensive loss of prairie potholes, waterfowl production in the Midwest has declined to about one-third of its former level.

Habitat fragmentation.   Many of the activities discussed in this chapter cause habitats to become fragmented into small, isolated patches. Even if these fragments of habitat are undisturbed, their small size and isolation compromise their suitability for many kinds of wild things. Animals that require large tracts of undisturbed habitat may avoid smaller patches, or they may inhabit them but fail to reproduce in them. A familiar example is the northern spotted owl, which thrives in large expanses of old-growth forest in the Pacific Northwest.

Habitat fragmentation has serious impacts on Neotropical migrant birds, songbirds that breed in North America and winter in Central and South America. These birds have declined markedly in recent decades. The loss of habitat on their wintering grounds has contributed to their decline, but reduction and fragmentation of nesting habitat in North America has also played a role. Although small lots containing deciduous trees still remain in many eastern states, it appears that many of these patches are too small to support breeding populations of many Neotropical migrants (Galli *et al.* 1976; Whitcomb *et al.* 1981; Lynch and Whigham 1984; Askins *et al.* 1990).

Why are these birds so vulnerable to fragmentation of their breeding habitat? If a large forest is carved up into smaller bits as a result of suburban development, why don't Neotropical migrants continue to breed in the remaining patches, or just move to one of the remaining large patches?

Neotropical migrants have several characteristics that make them especially vulnerable when their breeding habitat is fragmented. First, they are unlikely to move to a new breeding site. Each spring these birds return to the site where they bred the previous year. If that site no longer contains suitable habitat, Neotropical migrants rarely settle and breed elsewhere. Paradoxically, they are both migratory and sedentary. Although they move thousands of kilometers between their wintering grounds and their breeding grounds, they are fairly inflexible about where they breed.

Secondly, Neotropical migrants usually require habitat found in the interior of forests. Forest edges are unsuitable. Many permanent residents of temperate forests – like chickadees, nuthatches, and woodpeckers – lay their eggs in cavities, where they are fairly secure from predators. Neotropical migrants often construct open, cuplike nests on the ground, however. Since small patches of forest have a large proportion of edge (see Chapter 5), they receive many visits from humans and their pets. Ground nests near edges are often destroyed, either by trampling or by predation from cats. Brood parasitism (a special type of parasitism in which the parasitizing species lays its eggs in the nest of a host species, which then feeds and rears the young of the parasite) by cowbirds also increases in fragmented forests (Brittingham and Temple 1983). Brown-headed cowbirds feed in open, grassy areas such as pastures and lawns. From these sites they move as much as several hundred meters into forests, where they lay their eggs in the nests of other species (see Box 9.1).

Finally, Neotropical migrants are typically characterized by low biotic potential (see Chapter 2). Because of the energy demands of migration, they often have only enough fat stores left over to lay a single clutch of eggs. If the first nesting attempt is unsuccessful, they are therefore unable to renest.

Some of the effects of habitat fragmentation are not easily observed. Habitat modifications that split populations into small units can set in motion a host of genetic changes. For instance, they can cause closely related species to hybridize. This happens when habitat changes bring together two related species that were formerly segregated in different habitats. For example, the clearing of deciduous forests in the eastern U.S.A. brought black ducks into contact with mallards, allowing the two species to hybridize (Brodsky and Weatherhead 1984). The likely result of such interspecific hybrids is that the rarer species will be genetically swamped by genes from its more abundant relative. Other genetic consequences of small population size are discussed in Chapter 8.

Nonconsumptive resource use.   Camping, hiking, swimming, horseback riding, boating, snowmobiling, birding, whale watching, and wildlife photography involve the enjoyment of wild plants and animals through activities not aimed at harvest. Although these forms of recreation do not have the taking of plants or animals as their primary objective, losses may nevertheless result indirectly from such activities.

Areas that have remained free of development are attractive for recreation. Unfortunately, however, these sites also provide important habitat for plants and animals; in fact, that is part of their recreational value. Vehicular traffic poses the greatest threat, but even foot traffic can have serious impacts. The negative impacts potentially associated with recreation include outright mortality and lowered reproduction (Boyle and Samson 1985).

7.3   Diagnosing the problem

Is there an underlying cause of this host of daunting environmental problems that threaten biodiversity? According to historian Lynn White, the underlying cause is an anthropocentric (human-centered) philosophical tradition that encourages the exploitation of nature. In an influential article published in the journal *Science* in 1967, White suggested that the ethic of western Christianity encourages environmental domination:

Especially in its Western form, Christianity is the most anthropocentric religion the world has ever seen. . . . Christianity . . . not only established a dualism of man and nature but also insisted that it is God's will that man exploit nature for his proper ends. . . .

The present increasing disruption of the global environment . . . cannot be understood historically apart from distinctive attitudes toward nature which are deeply grounded in Christian dogma. The fact that most people do not think of these attitudes as Christian is irrelevant. . . . Hence, we shall continue to have a worsening ecologic crisis until we reject the Christian axiom that nature has no reason for existence save to serve man. (White 1967:1205,1207)

White's interpretation has been disputed (see, for example, Passmore 1974; Kellert 1995), but for our purposes the important point is that even if White is correct about the content and consequences of the Christian ethic, his argument tells us little about why environmental problems came to a head in the middle of the twentieth century.

Others saw the problem in terms of the amount of resources used by countries in the developed world, especially the United States, in the period following World War II. As early as 1953, Samuel Ordway suggested that Americans' high levels of consumption were straining resources (Ordway 1953, 1956). The economist John Kenneth Galbraith (1958*a,b*) picked up this theme, pointing out that since World War II American "consumption of most materials has exceeded that of all mankind through all history before that conflict" (Galbraith 1958*b*:90) and arguing that such levels of resource use were not sustainable.

Probably the best-known explanation for late-twentieth-century environmental problems is population growth, however. Extending the views of the Reverend Thomas Malthus (1986) (see Chapter 2), Paul Ehrlich (1968) and Garrett Hardin (1968) argued that the human population had reached a level which put excessive pressure on the earth's resources. They argued that unregulated population growth was the inevitable result of people acting in their own short-term self-interest. Hardin took as his starting-point an argument articulated by the amateur nineteenth-century mathematician William Lloyd. In 1833 the question of whether there should be public relief for the poor was being hotly debated in Britain. In this context, Lloyd published a pamphlet suggesting that people will add additional calves to a common pasture or bear additional children, since it is in their short-term interest to do so and there are no immediate negative incentives for restraint. (This is the problem of externalities discussed below.)

In a famous essay on "The tragedy of the commons," Garrett Hardin elaborated on this idea:

The tragedy of the commons develops in this way. Picture a pasture open to all. It is to be expected that each herdsman will try to keep as many cattle as possible on the commons. Such an arrangement may work reasonably satisfactorily for centuries because tribal wars, poaching, and disease keep the numbers of both man and beast well below the carrying capacity of the land. Finally, however, comes the day of reckoning, that is, the day when the long-desired goal of social stability becomes a reality. At this point the inherent logic of the commons remorselessly generates tragedy.

As a rational being each herdsman seeks to maximize his gain. Explicitly or implicitly, more or less consciously, he asks, "What is the utility *to me* of adding one more animal to my herd? This utility has one negative and one positive component.

(1) The positive component is a function of the increment of one animal. Since the herdsman receives all the proceeds from the sale of the additional animal, the positive utility is nearly +1.

(2) The negative component is a function of the additional overgrazing created by one more animal. Since, however, the effects of overgrazing are shared by all the herdsmen, the negative utility for any particular decision-making herdsman is only a fraction of -1.

Adding together the component partial utilities, the rational herdsman concludes that the only sensible course for him to pursue is to add another animal to his herd. And another; and another. . . . But this is the conclusion reached by each and every rational herdsman sharing a commons. Therein is the tragedy. (Hardin 1968:1244; emphasis and ellipsis in original)

The only alternatives to the tragedy of the commons, Hardin concluded, are privatization or state regulation. Privatization should increase individual responsibility because the individual bears the costs and reaps the benefits of his or her actions. Government control, on the other hand, is a necessary evil: "The alternative of the commons is too horrifying to contemplate. Injustice is preferable to total ruin" (Hardin 1968:1247).

When Hardin described the consequences of grazing a "pasture open to all," he was apparently thinking of what economists term an open-access resource, that is, a resource that no one has defined rights to use. Since no one has rights to use open-access resources, restrictions cannot be placed on their utilization. A "commons" on the other hand, is land used in common by the people of a community. For example, the "commons" of pre-sixteenth-century England was a communally owned village pasture (see Chapter 1). Economists term this common property. Since community members have rights to use a commons, the community can also limit the extent of that use. By applying the term "commons" to open-access resources, Hardin seemed to imply that small communities are incapable of regulating the use of resources that are owned in common (Jensen 2000). (We shall return to this point in Chapter 12.)

The issue of the commons has surfaced repeatedly in different forms. According to Hardin, the answer to overuse of communal food sources is enclosure of farmland, pastures, and hunting and fishing grounds. Pollution is a different kind of abuse of the commons, and "matters of pleasure," including reproduction, are a third. These tragedies, Hardin argued, must be averted though government regulation.

In *The Population Bomb*, Ehrlich compared population growth to a malignancy:

A cancer is an uncontrolled multiplication of cells; the population explosion is an uncontrolled multiplication of people. Treating only the symptoms of cancer may make the victim more comfortable at first, but eventually he dies – often horribly. A similar fate awaits a world with a population explosion if only the symptoms are treated. We must shift our efforts from treatment of the symptoms to the cutting out of the cancer. The operation will demand many apparently brutal and heartless decisions. The pain may be intense. But the disease is so far advanced that only with radical surgery does the patient have a chance of survival. (Ehrlich 1968:166–167)

To Ehrlich, it was not just population size that created environmental problems. Like Ordway and Galbraith, he recognized patterns of resource distribution as part of the problem, and he drew attention to the enormous disparity in consumption patterns in rich and poor nations. "At the moment the United States uses well over half of all the raw materials consumed each year," he wrote in 1968. "Think of it. Less than 1/15th of the population of the world requires more than all the rest to maintain its inflated position" (Ehrlich 1968:133). A few years later Paul and Anne Ehrlich published *The End of Affluence*, in which they articulated this critique more forcefully:

Overdeveloped countries [ODCs] are those in which population levels and per capita resource demands are so high that it will be impossible to maintain their present living standards without making exorbitant demands on global resources and ecosystems. . . .

People in ODCs, especially the United States, aren't eating more *food*; they are eating more meat, poultry, and dairy products. Americans (6 percent of the world's population) not only consume about 30 percent of the world's natural resources, they also consume 30 percent of the world's meat.

The protein-rich, highly varied diet of the average American requires nearly *five times* the agricultural resources (such as land, water, fertilizers, and pesticides) that are needed to feed a citizen in a UDC [underdeveloped country]. (Ehrlich and Ehrlich 1975:21 footnote,23; emphasis in original)

Others saw the misuse of science and technology as the root cause of environmental problems. Barry Commoner argued that neither population growth nor increased affluence (as measured by the gross national product of the U.S.A.) was sufficient to account for the level of environmental degradation (Commoner *et al.* 1971). Rather, the problem was the misuse of science and technology, particularly the "erosion of the principles which have long given science its remarkable capability to understand nature." This "erosion" occurred in the face of pressure from military and industrial interests during and after World War II, so that "even *basic* scientific work" often came to be "controlled by military and profit incentives" (Commoner 1966:48,61; emphasis in original). Commoner's conclusion that increasing affluence was a minor contributing factor was based on domestic production; imports were explicitly excluded (Commoner *et al.* 1971). Thus, he did not take into consideration the flow of resources from less developed to more developed countries.

In spite of Commoner's insistence that affluence was not the culprit, by the late 1970s, ecologists began to scrutinize western import patterns and their influence on resource use in the developing world, particularly the tropics. "Species disappear because of the way we prefer to live, all of us," wrote Norman Myers in his preface to *The Sinking Ark*. "For example, the expanding appetites of affluent nations for beef at 'reasonable,' i.e. non-inflationary prices," encourage "the conversion of tropical moist forests into cattle ranches" (Myers 1979:xi; 1981). Likewise the "booming demand on the part of the developed world for tropical hardwoods" results in increased pressure on tropical forests (Myers 1979:158).

To understand why the workings of the marketplace had failed to prevent such serious environmental problems, scientists turned to the discipline of economics. Ansley Coale concluded that pollution and other forms of environmental degradation proliferate because the free market does not require people to be responsible for the environmental consequences of production. It pays people for the "goods" they produce but does not "make them pay for the bads":

The way our economy is organized is an essential cause, if not *the* essential cause, of air and water pollution, and of the ugly and sometimes destructive accumulation of trash. I believe it is also an important element in such dangerous human ecological interventions as changes in the biosphere resulting from the wholesale use of organic fertilizers . . . [and] the accumulation in various dangerous places such as the fatty tissue of fish and birds and mammals of incredibly stable insecticides. . . .

The economist would say that harmful practices have occurred because of a disregard of what he would call *externalities*. An externality is defined as a consequence (good or bad) that does not enter the calculations of gain or loss by the person who undertakes the economic activity. It is typically a cost (or a benefit) of an activity that accrues to someone else. . . . Air pollution created by an industrial plant is a classic case of an externality; the operator of a factory producing noxious smoke imposes costs on everyone downwind, and pays none of these costs himself – they do not affect his balance sheet at all. This, I believe, is the basic economic factor that has a degrading effect on the environment: we have in general permitted economic activities without assessing the operator for their adverse effects. There has been no attempt to evaluate – and to charge for – externalities. (Coale 1970:132; emphasis in original)

The writings of scientists like Ehrlich, Hardin, and Commoner outline specific policy recommendations involving regulations and incentives aimed at changing the behavior of people and businesses. To some people, this did not go far enough. In an article published in 1973, the Norwegian philosopher Arne Naess suggested that conventional approaches to environmental problems were inadequate. He coined the term "deep ecology" as an alternative to a "shallow ecology" concerned solely with "improving the health and afflu-ence of people in the developed countries" by minimizing resource depletion and pollution (Naess 1973:95). The key features of the deep ecology "eco-philosophy" developed by Naess are "a deep-seated respect, or even venera-tion, for ways and forms of life," which have an *"equal right to live and blossom"* (Naess 1973:95,96; emphasis in original). In this view, which Naess called "ecological egalitarianism" and others have referred to as biocentrism (in con-trast to anthropocentrism), the root of our ecological crisis is an arrogance toward the natural world. This attitude should be replaced by "a keen, steady perception of the profound *human ignorance* of biospherical relationships" (Naess 1973:97; emphasis in original).

Although Naess agreed with Lynn White that the fundamental problem was philosophical arrogance, he went further than White in outlining social and economic remedies. "Ecologically inspired attitudes," wrote Naess, "favour diversity of human ways of life, of cultures, of occupations, of econ-omies. They support the fight against economic and cultural, as much as mil-itary, invasion and domination, and they are opposed to the annihilation of seals and whales as much as to that of human tribes or cultures" (Naess 1973:96). He concluded that the earth's environmental problems can only be solved by a radical shift to an economic system characterized by decentraliza-tion and local autonomy that fosters ecological and cultural diversity and mini-mizes class distinctions.

7.4   The response to the problem: The rise of preservationist management

These concerns and insights led to a critical re-examination of utilitarian resource management. In response to these concerns, a preservationist approach to managing living natural resources emerged. This approach seeks to maintain or prevent the loss of biodiversity by preserving and restoring species and habitats threatened by the activities of people.

In the early 1980s, an academic discipline termed "conservation biology" was initiated in response to concerns about declining biodiversity. Conservation biology is the application of scientific theory to the task of maintaining biodiversity. Conservation biologists are professionals trained in various branches of the biological sciences, including zoology, botany, ento-mology, herpetology, and so on, but not necessarily in the management of wildlife or other natural resources. Thus, conservation biology encompasses individuals with shared concerns from diverse disciplines, and provides an alternative to the traditional utilitarian disciplines of forestry, wildlife management, and range management (Soulé and Wilcox 1980).

Conservation biology is not synonymous with preservationist management, but preservationist management is a major concern of conservation biologists. (The concerns of ecosystem managers, discussed in Part III, also fall within the discipline of conservation biology.)

In this and the previous chapter we have seen that utilitarian management of renewable natural resources failed to protect many species of wild organisms from threats posed by overexploitation, habitat alteration, pollution, and alien species. The seriousness of the environmental problems described in this chapter led scientists to search for new approaches to managing renewable natural resources. In particular, they turned their attention to the causes of extinction, in an effort to respond to the earth's escalating environmental problems. The next chapter considers this area of study.

References

Allan, J. D. and A. S. Flecker (1993). Biodiversity conservation in running waters. *BioScience* **43**:32–43.

Askins, R. A., J. F. Lynch, and R. Greenberg (1990). Population declines in migratory birds in eastern North America. *Current Ornithology* **7**:1–57.

Bookchin, M. (1974). *Our Synthetic Environment*, 1st Harper Colophon edn. New York, NY: Harper and Row.

Boyle, S. A. and F. B. Samson (1985). Effects of nonconsumptive recreation on wildlife: a review. *Wildlife Society Bulletin* **13**:110–116.

Brittingham, M. C. and S. A. Temple (1983). Have cowbirds caused forest songbirds to decline? *BioScience* **33**:31–35.

Brodsky, L. M. and P. J. Weatherhead (1984). Behavioral and ecological factors contributing to American black duck–mallard hybridization. *Journal of Wildlife Management* **48**:848–852.

Carson, R. (1962). *Silent Spring*. Greenwich, CT: Fawcett Publications.

Caughley, G. and A. Gunn (1996). *Conservation Biology in Theory and Practice*. Cambridge, MA: Blackwell Science.

Coale, A. (1970). Man and his environment. *Science* **170**:132–136.

Collett, D. (1987). Pastoralists and wildlife: image and reality in Kenya Maasailand. In *Conservation in Africa: People, Policies, and Practice*, ed. D. Anderson and R. Grove, pp. 129–148. Cambridge: Cambridge University Press.

Commoner, B. (1966). *Science and Survival*, 3rd edn. New York NY: Viking Press.

Commoner, B. (1975). *The Closing Circle*, 4th edn. New York, NY: Alfred A. Knopf.

Commoner, B., M. Corr, and P. J. Stamler (1971). The causes of pollution. *Environment* **13**(3):2–19.

Ehrlich, P. (1968). *The Population Bomb.* New York, NY: Ballantine Books.

Ehrlich, P. and A. Ehrlich (1975). *The End of Affluence.* Rivercity, MA: Rivercity Press.

Ehrlich, P. and A. Ehrlich (1985). *Extinction: The Causes and Consequences of the Disappearance of Species,* 2nd Ballantine printing. New York, NY: Ballantine Books.

Ehrlich, P. R., A. H. Ehrlich, and J. P. Holdren (1977). *Ecoscience: Population, Resources, Environment.* San Francisco, CA: W. H. Freeman.

Elton, C. S. (1977). *The Ecology of Invasions by Animals and Plants,* Science Paperback edn. London: Chapman and Hall.

Galbraith, J. K. (1958*a*). *The Affluent Society.* Boston, MA: Houghton Mifflin.

Galbraith, J. K. (1958*b*). How much should a country consume? In *Perspectives on Conservation: Essays on America's Natural Resources,* ed. H. Jarrett, pp. 89–99. Baltimore, MD: Johns Hopkins University Press.

Galli, A., C. F. Leck, and R. T. T. Forman (1976). Avian distribution patterns in forest islands of different sizes in central New Jersey. *Auk* **93**:356–364.

Hardin, G. (1968). The tragedy of the commons. *Science* **162A**:1243–1248.

Homewood, K. and W. A Rodgers (1987). Pastoralism, conservation, and the overgrazing controversy. In *Conservation in Africa: People, Policies, and Practice,* ed. D. Anderson and R. Grove, pp. 111–128. Cambridge: Cambridge University Press.

Hunt, E. G. and A. I. Bischoff (1960). Initial effects on wildlife of periodic DDD applications to Clear Lake. *California Fish and Game* **46**:91–106.

Jensen, M. N. (2000). Common sense and common pool resources. *BioScience* **50**:638–644.

Kellert, S. R. (1995). Concepts of nature east and west. In *Reinventing Nature?: Responses to Postmodern Deconstruction,* ed. M. E. Soulé and G. Lease, pp. 103–121. Washington, DC: Island Press.

Likens, G. E., R. F. Wright, J. N. Galloway, and T. J. Butler (1979). Acid rain. *Scientific American* **241**(4):43–51.

Lynch, J. F. and D. F. Whigham (1984). Effects of fragmentation on breeding bird communities in Maryland, USA. *Biological Conservation* **28**:287–324.

Mack, R. N. (1981). Invasion of *Bromus tectorum* L. into western North America: an ecological chronicle. *AgroEcosystems* **7**:145–165.

Malthus, T. R. (1986). *An Essay on the Principle of Population,* 7th edn. Fairfield, NJ: Augustus M. Kelley.

Myers, N. (1979). *The Sinking Ark.* Oxford: Pergamon Press.

Myers, N. (1981). The hamburger connection: how Central America's forests become North America's hamburgers. *Ambio* **10**:2–8.

Naess, A. (1973). The shallow and the deep, long-range ecology movement: a summary. *Inquiry* **16**:95–100.

Nilsson, C. and K. Berggren (2000). Alteration of riparian ecosystems caused by river regulation. *BioScience* **50**:783–792.

Oba, G., N. C. Stenseth, and W. J. Lusigi (2000). New perspectives on sustainable grazing management in arid zones of sub-Saharan Africa. *BioScience* **50**:35–51.

Odén, S. (1976). The acidity problem: an outline of concepts. *Water, Air, and Soil Pollution* **6**:137–166.

Ordway, S. H., Jr. (1953). *Resources and the American Dream: Including a Theory of the Limit of Growth*. New York, NY: Ronald Press.

Ordway, S. H., Jr. (1956). Possible limits of raw material consumption. In *Man's Role in Changing the Face of the Earth*, vol. 2, ed. W. L. Thomas, Jr., pp. 987–1009. Chicago, IL: University of Chicago Press.

Palmer, T. (1991). *The Snake River: Window to the West*. Washington, DC: Island Press.

Passmore, J. (1974). *Man's Responsibility for Nature: Ecological Problems and Western Traditions*. London: Duckworth.

Poff, N. L., J. D. Allan, M. B. Bain, J. R. Karr, B. D. Richter, R. E. Sparks, and J. C. Stromberg (1997). The natural flow regime. *BioScience* **47**:769–784.

Reaka-Kudla, M. L., D. E. Wilson, and E. O. Wilson (1997). *Biodiversity II*. Washington, DC: Joseph Henry Press.

Rood, S. B. and J. M. Mahoney (1993). River damming and riparian cottonwoods. In *Riparian Management: Common Threads and Shared Interests*, tech. coord. B. Tellman, H. J. Cortner, and M. G. Wallace, pp. 134–143. U.S. Department of Agriculture Forest Service, General Technical Report RM-226, Ford Collins, CO.

Rudd, R. L. (1964). *Pesticides and the Living Landscape*. Madison, WI: University of Wisconsin Press.

Soulé, M. E. and B. A. Wilcox (ed.) (1980). *Conservation Biology: An Evolutionary Ecological Perspective*. Sunderland, MA: Sinauer Associates.

Vitousek, P. M. (1986). Biological invasion and ecosystem properties: can species make a difference? In *Ecology of Biological Invasions in North America and Hawaii*, ed. H. A. Mooney and J. A. Drake, pp. 163–176. New York, NY: Springer-Verlag.

Whitcomb, R. F., J. F. Lynch, M. K. Klimkiewicz, C. S. Robbins, B. L. Whitcomb, and D. Bystrak (1981). Effects of forest fragmentation on avifauna of the eastern deciduous forest. In *Forest Island Dynamics in Man-Dominated Landscapes*, ed. R. L. Burgess and D. M. Sharpe, pp. 125–205. New York, NY: Springer-Verlag.

White, L., Jr. (1967). The historical roots of our ecologic crisis. *Science* **155**:1203–1207.

Woodwell, G. M. (1967). Toxic substances and ecological cycles. *Scientific American* **213**(3):24–31.

8

Central concepts – the causes of extinction

We saw in the previous chapter how events that followed World War II directed attention toward threats to biodiversity. Because of this situation, it became critical for scientists and resource managers to understand the processes that give rise to species and that allow them to persist on the one hand or cause them to become extinct on the other. This knowledge is a prerequisite to doing something about the situation.

In this chapter we consider how genetic changes in populations occur and how they result in local adaptation and evolution. We will also briefly consider how species arise, some of the consequences of small population size, and circumstances that make certain species and populations more vulnerable to extinction than others. More information on genetic variation, evolution, and the risks of small population size can be found in Frankel and Soulé (1981), Caughley and Gunn (1996), Futuyma (1997), Meffe and Carroll (1997), and Primack (1998).

8.1 Speciation: The formation of species

8.1.1 What is a species?

A species is generally regarded as a kind of organism that is reproductively isolated from other groups in nature. Thus, the species cannot normally interbreed with other species, and the members of a species share a common evolutionary history. Gene flow, the exchange of genetic information, can occur between members of a species but not between members of different species.

This definition works well in many cases, but there are situations where it is problematic. Where hybridization is common (as in many groups of plants), where asexual reproduction (reproduction without the recombination of genetic material from two parents) occurs (as in microorganisms and many plants), and where we are dealing with fossils of extinct forms, it is difficult or impossible to apply the criterion of reproductive isolation. The important point, however, is that since gene flow does not normally occur between species, each species is a repository of unique genetic information. Once that species becomes extinct, its genetic information is lost forever. This is one reason why the loss of species is of such concern. (For a more detailed discussion of different concepts of species and their implications for conservation consult Rojas (1992).)

8.1.2   The theory of natural selection

In Chapter 2 we saw that all organisms are theoretically capable of producing more offspring than can survive. Not all individuals are equally likely to survive, however. In the mid-nineteenth century, Charles Darwin and Alfred Russel Wallace pondered these points and independently proposed the theory of natural selection, which states that individuals that are better adapted to their environment will have a greater probability of surviving and reproducing than those that are not so well adapted and that these reproducers will pass on their hereditary characteristics to their offspring. Darwin summarized this "doctrine of Malthus, applied to the whole animal and vegetable kingdoms" succinctly in the introduction to *The Origin of Species*:

As many more individuals of each species are born than can possibly survive; and as, consequently, there is a frequently recurring struggle for existence, it follows that any being, if it vary however slightly in any manner profitable to itself, under the complex and sometimes varying conditions of life, will have a better chance of surviving, and thus be *naturally* selected. From the strong principle of inheritance, any selected variety will tend to propagate its new and modified form. (Darwin 1958:29; emphasis in original)

Darwin and Wallace held that this mechanism produces (1) gradual changes within populations and (2) new species. Individuals that leave more surviving offspring pass on more of their genes to the next generation; they are said to have greater genetic fitness. This is an important concept. In evolutionary biology the word "fitness" has a different meaning from everyday usage. An individual's fitness is defined in terms of its success in passing its genes on to succeeding generations; this need not have anything to do with physical fitness in the sense the term is used in athletics. The mistaken idea that fitness is synonymous with strength was promoted by "social Darwinists," who misinterpreted Darwin's concept of "survival of the fittest" to mean that the more successful or powerful members of society owed their positions to biological superiority.

In addition to providing an explanation for the origin of species, the theory of natural selection has a number of practical applications. It predicts that if a large population with a high reproductive rate is exposed to a toxin and not all individuals are killed outright, then resistance to the toxin will evolve. This is exactly what happens when insecticides are broadcast to control insect pests. (See Chapter 7: Recall that at first DDD appeared to be spectacularly successful when it was used to kill gnats at Clear Lake, but problems surfaced three years after the first treatment and even sooner after the second application.) The theory of natural selection also allows us to predict which functional insect groups, herbivores or predators, will be most likely to evolve resistance (see Box 8.1).

Box 8.1   Evolution of resistance to pesticides

When a pesticide is applied to a population, it affects those individuals that are sensitive to its mode of action. But unless the target population is genetically uniform, individuals are likely to differ in their susceptibility to toxins. If a few individuals have the ability to break down the pesticide, they will be at an enormous selective advantage. These individuals will have superior ability to survive and reproduce after being exposed to the pesticide, and they will therefore leave more offspring than susceptible individuals will leave. In the next generation, a higher proportion of individuals will be resistant to the pesticide. Thus, the application of a pesticide creates a strong selective pressure for resistance to it. When the process is repeated over several generations, a population in which all or most individuals are resistant to the pesticide can be created. This typically leads to a search for newer, even more toxic products.

Insects occupying a variety of ecological niches can be found in an untreated field. Some are herbivores that feed on the crop; others are carnivorous insects that feed on the herbivores. From a utilitarian standpoint, the herbivores are harmful, and the carnivores are beneficial because they attack the herbivores. To achieve optimum pest control, it would be best if the herbivorous insects could be reduced while leaving the beneficial carnivores to assist in pest control by eating any remaining herbivores.

Evolutionary theory predicts that pesticide resistance is more likely to evolve among herbivorous insects than in carnivores, however. Plant-eating insects have had to contend throughout their evolutionary history with plant defenses produced by the species they feed on. Natural selection favored those individual insects that possessed mutations allowing them to break down or detoxify naturally occurring plant chemical defenses. For this reason, herbivorous insects typically possess the metabolic machinery to deal with some of the naturally occurring insecticides that plants produce. Although when they are first exposed to a new pesticide they may not be able to break down the particular toxins it contains, they are likely to evolve the metabolic machinery they need relatively rapidly. To evolve the ability to break down a novel toxin may require only one or two changes in metabolic pathways these insects already possess. In other words, because of their long history of evolution in response to plant defenses, herbivorous insects may be only a few steps away from being able to deal with a new poison. Predatory insects, on the other hand, have little evolutionary experience with breaking down toxins. For them, evolution of the required biochemical pathways typically requires the development of complex new adaptations starting from scratch. This will require many mutations. Consequently, the application of pesticides is likely to eliminate beneficial predaceous insects while producing resistant strains of the most harmful crop-destroying insects. Clearly, this is just the opposite of what resource managers would like pesticides to accomplish.

In nature, as herbivorous insects evolve the ability to break down plant toxins, the plants that are fed upon will evolve counter-defenses. These adaptations will in turn be favored by natural selection, and so on. Over time the plant species and the herbivore species that feed on them undergo coevolution, that is, reciprocal evolutionary changes between interacting populations (Thompson 1994). This process has been likened to an arms race, with each side escalating its arsenal against its evolutionary antagonists in a stepwise fashion. The evolution of defensive adaptations in plants leads to counter-adaptations in herbivores; this leads to further counter-adaptation, and so on. Similar coevolution takes place between predators and prey.

Wallace and Darwin knew nothing about how genetic changes arise nor how inherited traits are passed on (see Box 8.2). But they did identify natural selection, acting on the raw material provided by genetic variation, as the driving force of evolution. We shall see later in this chapter that this has important implications for the conservation of rare species.

Box 8.2   Inheritance and genetic change

We now know that genetic material is located on chromosomes and that chromosomes are composed of deoxyribonucleic acid (DNA) and protein. (DNA occurs in cell nuclei and in organelles such as mitochondria and chloroplasts, which can reproduce themselves within cells.) The functional unit of genetic information is the gene, a segment of DNA. Genes occur in definite positions (loci [singular: locus]) on chromosomes. Genes, specifically the DNA they contain, control the synthesis of proteins.

Most animals are diploid (*di*, two; *ploid*, sets of chromosomes) throughout most of their life cycle; that is, their cells normally possess two sets of chromosomes, a maternal set inherited from the female parent and a paternal set inherited from the male parent. Gametes (eggs and sperm) normally contain one set of chromosomes; they are haploid. (The social insects – ants, bees, and wasps – are a notable exception. In this group and some other kinds of invertebrates, females, except the queen, are haploid; that is, they possess only one set of chromosomes.) When a haploid egg and a haploid sperm are united in the process of fertilization, they form a diploid zygote, with two sets of chromosomes. In plants, a condition termed polyploidy (*poly*, many; *ploid*, sets of chromosomes), in which multiple sets of chromosomes are present, is common.

Different forms of a gene are termed alleles. If both alleles at a given locus are the same, an individual is said to be homozygous (*homo*, same; *zygous*, zygote) for the trait controlled by that allele. If two different alleles are present at a locus, the individual is heterozygous (*hetero*, different; *zygous*, zygote) for the trait in question. Normally, a diploid individual has two copies of each allele. (An exception occurs with the X chromosome in male mammals. Since the X chromosome is paired with a shorter Y chromosome in males, some alleles on the X chromosome do not have a corresponding partner on the Y chromosome.) Within a population, many alleles for a given locus may be present, but an individual normally does not have more than two alleles at a locus. An allele is said to be recessive if it is not expressed when it occurs with a dominant allele. Thus, for a recessive allele to be expressed in a diploid situation, an individual must have two copies of the recessive allele, that is, it must be homozygous at that locus.

Mutations are changes in genetic material. A mutation can involve a change in the structure of a DNA molecule, a rearrangement of genes on a chromosome, or a change in the number of chromosomes. These changes in genetic material occur at random. The genetic variants introduced by mutations provide the raw material on which natural selection acts; thus, they permit populations to evolve in response to environmental changes.

Mallards and black ducks, two inhabitants of shallow waters in eastern North America, probably diverged from a single ancestral species. At the end of the last ice age, about 10 000 years ago, glaciers that persisted along the Appalachian Mountains divided the mallard's geographic range into two segments. East of the mountains they adapted to forested conditions; to the west, the ducks encountered a steppe, or prairie, environment. Several other species of shallow-water ducks also nested in the steppe. In this open habitat, male mallards with bright plumage are more likely than dull-colored males to be recognized by females of their own species, to mate with them, and to pass their genes on to the next generation. As a result, natural selection in the steppes, where there were many species of ducks, favored mallard drakes with bright plumage. In the forests to the east, however, where closely related species were absent, there was no selective premium on bright males. Under these conditions, natural selection (exerted by predators) favored drab, protective coloration. This difference, combined with other adaptations to the forest environment, led to the evolution of a new species, the black duck, in which males resemble females. These two species are still so closely related that they can hybridize under certain conditions (Chapter 7).

There are other mechanisms that can produce species. For example, in plants, new species can arise when an entire set of chromosomes is duplicated. Because many plant species can reproduce through self-fertilization, an individual with an extra set of chromosomes may be able to fertilize itself. If that is the case, finding a mate with matching chromosomes will not be a problem.

8.1.3   The tempo of speciation

Many new forms appear abruptly in the fossil record. Most major groups of animals actually appear rather suddenly (in geologic terms) without transitional forms. Those who believe that evolution must occur gradually have argued that the absence of transitional forms occurs because the fossil record is imperfect, that is, the transitional forms evolved, but we have not found them. In this view, new species are produced by gradual changes accumulating slowly over long periods of time, but because of the low probability of fossilization, many of the transitional forms are missing from the fossil record.

An alternative explanation for our failure to find transitional forms in the fossil record is that they are generally not there. Paleontologists Niles Eldredge and Stephen Jay Gould argue that species appear suddenly in the fossil record because in many cases they appeared relatively suddenly in real life; this suddenness is real, not just an artifact of an imperfect fossil record (Eldredge and Gould 1972; Gould and Eldredge 1977). (A relatively "sudden" transition in the fossil record is not sudden in the everyday sense of the term. In geological terms, 50000 years is sudden. If a transition took place over a period of 50000 years, we would not be able to distinguish it in the fossil record; the new form would appear to have arisen abruptly.)

In this view, termed punctuated equilibrium, there are long periods of time during which speciation does not occur; these periods of "equilibrium" are "punctuated" by the relatively sudden appearance of new species. According to this theory, species are well adapted and do not change much throughout most of their existence. These are the forms we find in the fossil record. New species appear fairly quickly, often during periods of rapid environmental change, in small populations at edges of the ancestral species' range, when a splinter population is subjected to rather strong selective pressure. Once the new species has adapted to its environment, it stays that way for a long time. Species can also arise fairly abruptly through changes in chromosome number.

A key piece of evidence that led to the theory of punctuated equilibrium is the fact that few transitional forms are found in the fossil record. Eldredge and Gould remind scientists that they failed for so long to understand that speciation can occur relatively suddenly because they refused to believe what the fossil record was telling them. Instead, scientists argued that species arose gradually but the fossilized transitions were missing. Gould and Eldredge reasoned that the absence of transitional forms in the fossil record is an important piece of data, one that should be understood instead of being explained away. Their insistence on believing what the fossil record had to say instead of trying to make it fit preconceived concepts led to an important new development in evolutionary theory.

8.1.4   Adaptive radiation, isolation, and endemism

Darwin and Wallace were impressed with the numbers of endemic species (species found nowhere else) that they encountered on oceanic islands. This phenomenon results from the fact that isolated islands provide habitat that is free from competitors. An original immigrant or small number of immigrants to an island can evolve to fill vacant ecological niches in a burst of speciation

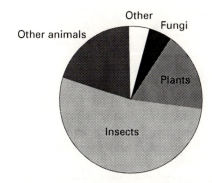

Figure 8.1. Approximately half of all species that have been described are insects. The "other" category includes bacteria, blue-green algae, algae, and protists (single-celled organisms with a membrane-bound nucleus, such as the familiar *Amoeba* and *Paramecium*).

known as adaptive radiation. Darwin observed the results of this phenomenon on the Galapagos Islands, where he found many specialized forms of finches and tortoises that had resulted from such evolutionary radiations. Isolated patches of habitat such as the tops of mountains contain many endemic forms for similar reasons.

Because they contain so many endemics, the native flora and fauna of isolated habitat patches and oceanic islands are unique. These special biotas have high extinction rates as a result of their restricted geographic range and vulnerability to introduced competitors, predators, diseases, and herbivores (see below). This is a serious problem, because when an endemic species becomes locally extinct, it also becomes globally extinct as well, since it doesn't occur anywhere else.

8.2 How many species are there?

About 1.4 to 1.7 million species have been described by scientists to date. Familiar organisms make up only a small portion of this (Figure 8.1). Only about 43 000 species are vertebrates, and nearly half of the vertebrates are fishes. Only about 4000 species of mammals are known to science. By contrast, about 250 000 species of plants and 751 000 species of insects, including over 400 000 species of beetles have been described. There are many more species that have not yet been described, especially in the tropics. Some scientists estimate that the total number of species on earth probably exceeds

Table 8.1. *Classification of the coyote,* Canis latrans

Category	Example	Description
Kingdom	Animalia	Multicellular heterotrophs
Phylum	Chordata	Vertebrates and close relatives
Class	Mammalia	Vertebrates with mammary glands
Order	Carnivora	Flesh-eating mammals
Family	Canidae	Carnivores adapted for running
Genus	*Canis*	Dogs and their close relatives
Species	*latrans*	Coyote

5 million and some believe the number may be much higher, although this is controversial (Erwin 1991; Gaston 1991).

8.3   Classification

8.3.1   Nomenclature

The science of classifying organisms is termed taxonomy or systematics. A taxon (plural: taxa) is any taxonomic category, for instance, a genus, species, or subspecies. Each species is designated by a two-part Latin name, or binomial, and a taxon recognized by science is classified in a series of ranked categories. These are listed in Table 8.1. At any level in the hierarchy, an organism can belong to only one rank.

This classification scheme organizes species into a framework that expresses the relationships between them. Species that are closely related to each other are grouped in the same genus (plural: genera). The gray wolf (*Canis lupus*), the coyote (*Canis latrans*), and the domestic dog (*Canis familiaris*) are all placed in one genus. Genera that are closely related are placed in the same family. Similarly, families are grouped together in the same order, related orders are grouped in a single class, and so on. This can be thought of as a branching diagram or family tree. Both existing (or extant) species and extinct species are included on the tree. Thus, the classification of species is not only a necessary tool for naming and keeping track of the different kinds of living organisms; it is also a powerful conceptual framework that summarizes scientists' understanding of the relatedness of different forms of life.

Scientists continue to study the interrelationships between species that have already been described. As new species are discovered and new information about relationships is uncovered, new hypotheses are formed, and our understanding of our family tree is revised and refined. Taxonomy is not just an academic exercise. In order to manage biodiversity, scientists need to know what they are managing (see Box 8.3).

Box 8.3   What is a red wolf?

The red wolf (usually classified as *Canis rufus*), is a small-bodied, reddish canid that was once abundant in densely vegetated bottomland forests, swamps, and coastal marshes from central Texas to southern Pennsylvania. As its habitat dwindled and its range shrank, that of the closely related coyote, a canid that is far less specialized in its habitat preferences, expanded. By the 1980s, the red wolf was extinct in the wild, and its fate depended upon 19 captive individuals. After several years, the captive population had increased to several dozen wolves, and biologists decided to try reintroducing the animals to portions of their natural habitat, a costly and time-consuming endeavor (Cohn 1987).

On the basis of samples of mitochondrial DNA derived from red wolf blood and pelts obtained prior to 1975, geneticists R. Wayne and S. Jenks proposed that the red wolf originated through hybridization between the gray wolf, *Canis lupus*, and the coyote, *Canis latrans* (Wayne and Jenks 1991). Other authorities suggest it is a subspecies of the gray wolf; still others continue to regard it as a species that is closely related to the gray wolf but has hybridized with the coyote at some time in the past (Dowling *et al.* 1992; Nowak 1992; Phillips and Henry 1992).

Currently the red wolf's taxonomic status is the subject of heated debate. The resolution of this question has important practical and political ramifications. The Endangered Species Preservation Act has been interpreted to mean that hybrids between endangered species should not be protected (O'Brien and Mayr 1991). Thus, if it is determined that the red wolf is a hybrid between coyotes and gray wolves, it could lose its endangered listing. Some conservation biologists have suggested that, regardless of the red wolf's "undeniable cuddliness," the resources expended on reintroducing red wolves should be redirected if it proves to be a hybrid (Gittleman and Pimm 1991:525), and members of the sheep industry have argued that the red wolf should no longer be listed under the Endangered Species Preservation Act because it is not a separate species.

8.3.2   Genetic differentiation within species: Subspecies and local adaptation

A species consists of many populations. These populations differ from each other in some characteristics. When the differences are noticeable and consistent, scientists sometimes designate distinct subspecies, or races. Although they are different enough to be distinguished, subspecies are not reproductively isolated from one another; members of a subspecies are capable of interbreeding with members of other subspecies in the same species.

This taxonomic category is different from the other taxonomic categories we discussed above for two reasons. First, unlike species, the boundaries between subspecies are not fixed because, by definition, a member of a subspecies can interbreed with members of other subspecies that belong to the same species. Thus, gene flow is possible between subspecies. Second, the category need not be used at all. Scientists designate subspecies as a matter of convenience, but that designation is always somewhat subjective. Every organism must belong to a species; every species is part of a genus; every genus belongs to a family; and so on, but an organism does not have to belong to a subspecies, and the boundaries between subspecies are somewhat arbitrary. Many taxonomists have abandoned altogether the practice of dividing species into subspecies. Instead they use quantitative techniques of describing geographic variation. This practice is more rigorous and objective. Nevertheless, the practice of designating subspecies persists.

Regardless of whether scientists recognize separate subspecies, species consist of one or more populations. Each population possesses adaptations to its local environment. (We shall see in the next chapter that this becomes an important consideration when managing rare populations.) As a consequence, populations of the same species differ genetically, even though they are capable of interbreeding.

8.4   Extinction

When all representatives of a group have died, leaving no living descendants, the group is said to be extinct. Extinction can affect a population, a subspecies, or a species. If all populations of a species in a local area die out, the species is said to be locally extinct, although other populations may exist elsewhere. If all populations of a species go extinct, it is globally extinct. Sometimes some subspecies but not others become extinct. When that happens, the genetic makeup of a particular subspecies is gone, but other subspecies of the same species persist.

Extinction occurs when losses from a population consistently exceed gains, that is when death and emigration are greater than birth plus immigration, so that new members are not recruited into the population (see Chapter 2). This may happen because a species is not adapted to its environment (perhaps because the environment has changed in some way) or because of excessive exploitation. Another kind of extinction occurs when a species evolves into a new species, and the original form no longer exists.

8.4.1   Extinctions in the fossil record

There have always been extinctions. Darwin recognized that the number of species on earth has not increased markedly in spite of the ongoing process of speciation and pointed out that for this to be the case some species must become extinct as new ones arise. In addition to isolated extinctions, the geological record is marked by numerous mass extinctions, in which large numbers of organisms disappeared at about the same time. Perhaps the most familiar example is the dramatic disappearance of the dinosaurs 65 million years ago. Actually, the extinction of these ruling reptiles was accompanied by the disappearance of many other groups of terrestrial and marine organisms as well. An earlier and even more extensive mass extinction occurred 225 million years ago. During that episode 95% of the species of marine invertebrates disappeared. There have been numerous other mass extinctions, and there is considerable debate about their causes. The relatively sudden extinction of so many forms suggests a drastic, widespread environmental change. Continental drift, climatic change, and collision with meteors have been suggested as possible causes.

If mass extinctions have occurred repeatedly in the past, why are conservation biologists so concerned about contemporary extinctions? One reason is because under current conditions of development and habitat fragmentation, it is highly improbable that the earth's biodiversity would be able to recover from massive losses. While the earth's extinction rate has escalated, the rate of speciation has not shown a corresponding increase. In fact speciation may be occurring less frequently today than in the past. Paul and Anne Ehrlich compared this situation to a sink: the earth's species are disappearing down the "drain," but the flow of new species through the "faucet" is not increasing (Ehrlich and Ehrlich 1985: 32).

At the end of the last ice age, about 12000 years ago, many large mammals, including mammoths as well as giant beaver and ground sloths, became extinct within a relatively short period of time. While other simultaneous mass extinctions were characterized by the disappearance of many unrelated groups, this instance was unique in that only large mammals were affected. It has been suggested that prehistoric hunters may have caused, or at least contributed to, the demise of this group (Martin and Klein 1984). Scientists do not by any means agree on this issue, however, and the causes of these extinctions remain controversial (Caughley and Gunn 1996). On the one hand, the selective disappearance of large mammals is striking. On the other hand, prehistoric hunters lacked the kinds of weapons that allowed modern, market hunters to drive innumerable species to extinction (see Chapter 1 for examples), leading some scientists to question whether or not prehistoric hunting could have caused the extinction of large numbers of mammalian species.

8.4.2   Why are some species more vulnerable than others?

Why do some species go extinct while others in the same environment do not? Why do cockroaches, dandelions, thistles, starlings, magpies, pigeons, and house mice proliferate as development progresses, while many other organisms disappear? The factors that predispose a species to decline and ultimate extinction can be divided into two categories: ultimate factors and proximate factors. Ultimately, the vulnerability of a species depends on a number of characteristics of its reproduction, ecology, anatomy, and behavior (Wolfheim 1976). These factors are genetically determined and result from a species' evolutionary history. We cannot do anything about them. Superimposed on these conditions are short-term factors – such as the kinds of threats we reviewed in Chapter 7: exploitation, habitat change, pollution, and the introduction of exotic species – that affect the balance between mortality and reproduction. These factors may be considered the proximate (immediate) causes of extinction.

Although managers cannot change ultimate factors that affect vulnerability to extinction, an understanding of these factors can be used to identify species that are at risk. Species that are especially likely to become extinct typically have one or more of the following characteristics: low biotic potential, specialized habitat or food requirements, limited geographic range, limited ability to disperse, and limited experience with evolutionary antagonists. These traits increase the likelihood that habitat alteration will make it impossible for a species to meet its requirements, and they also decrease the likelihood that individuals will colonize new habitats (see below).

All other things being equal, species with a high biotic potential are able to replace individuals removed by exploitation more rapidly than are species with low reproductive rates (although, as indicated in the next paragraph, species with high reproductive rates are not necessarily secure). Specialization, limited geographic range, and limited dispersal ability are interrelated (see Box 9.1 for a discussion of efforts to aid an endangered species that has extremely specific habitat requirements and a very small geographic range.)

Many insects feed on only one or a few species of plants. The fate of these insect specialists depends on the fate of their food plants, regardless of whether the insects have high reproductive rates. Thus, specialists may be vulnerable even if their biotic potential is high. In other words, a single vulnerability factor is enough to boost a species' extinction risk.

In some geographic settings, organisms evolve with little exposure to certain kinds of evolutionary antagonists. This is often the case on oceanic islands. Because large land mammals are incapable of traversing large bodies of salt water on their own, island plants typically evolved without selective pressure from large herbivores. Consequently, they lack defenses against deer, goats, or cows. This makes them extremely vulnerable to herbivory by livestock or other introduced mammalian herbivores. Such evolutionary naiveté is not restricted to islands. Even biotas on continents may, under some circumstances, lack experience with certain types of evolutionary antagonists (see Chapter 13).

For animals, high trophic level and a large home range are associated with risk. Animals that are high on the food chain may decline because of a reduction in their prey, because toxic substances accumulate in their tissues, or because they are targeted by animal control programs. For these reasons, carnivorous birds and mammals are vulnerable. In addition, each individual in a population of large animals often requires a large area, or home range. Animals that move through a large area are more likely to have part of their habitat altered and to come into conflict with people than are those with small home ranges. Many large birds and mammals, such as grizzly bears, bald eagles, giant pandas, California condors, and tigers, have large home ranges and have declined because of habitat loss. (Note that either a *small geographic* range or a *large home* range can contribute to vulnerability.)

Finally, traits that increase the likelihood of intensive exploitation pose additional risk. These include conspicuousness and high population density, because organisms with these characteristics are typically easy to find and to harvest. In addition to being easy to find, some brightly colored organisms such as flowers and butterflies are also attractive to people, and this increases their likelihood of being exploited. Likewise, individuals that concentrate in dense groups are vulnerable to exploitation. Some species congregate only at certain times of year, for example during the breeding season, but at these times they are especially susceptible to exploitation. This is the case with the breeding aggregations of some invertebrates, amphibians, waterfowl, shorebirds (such as those that were exploited by the plume trade), seals, and bats. Because they occurred in large groups, both the passenger pigeon and the bison were overexploited, even though they were initially present in enormous numbers. Defensive behaviors also can increase vulnerability. Remaining with a wounded individual or attempting to defend young may succeed against animal predators, but these behaviors increase susceptibility to human predators armed with weapons.

Some organisms are at risk for several different reasons. For example, carnivores are high on the food chain and therefore face threats from bioconcentration of toxic substances, reductions in prey, and control programs. Furthermore, carnivores often have low biotic potential and large home ranges. The grizzly bear, gray wolf, red kite, peregrine falcon, and bald eagle are examples of organisms that have been reduced to the brink of extinction throughout much of their former geographic range by a combination of these factors. Similarly, plants and animals on oceanic islands also face a combination of risk factors. They have limited geographic ranges, they are often highly specialized, and they lack adaptations against predators or herbivores.

8.4.3 Why are some populations more vulnerable than others?

Even within a given species, some populations may be more vulnerable to extinction than others. The outlook for small populations is especially precarious. They are more likely to go extinct than large populations, because they face several kinds of risks that are unlikely to be problems for larger populations.

Genetic risks

Causes of reduced genetic variation. The genetic diversity of a population is the amount of variation in the genetic material of that population. Genetic diversity declines in small populations. There are two principal mechanisms responsible for this erosion of genetic diversity in small populations: genetic drift (especially when it results from founder effects or bottlenecks) and inbreeding.

Genetic drift. Small populations (or even populations that have been small at some time in the past) are likely to have fewer alleles than populations that have remained large throughout their evolutionary history. This is because some alleles lose out in the "intergenerational genetic lottery" (Caughley and Gunn 1996:177). This chance loss of alleles that occurs from one generation to the next is termed random genetic drift. The alleles in the gametes of a given generation can be thought of as a "gene pool." A finite number of these form the zygotes that give rise to the next generation. Some of the alleles in the gene pool are in this sample that is passed on to the progeny, but some may not be. (To see how this works, suppose that you have a jar consisting of 500 beans of a dozen colors. If you remove a handful of beans, the sample in your hand will probably contain only a few of the 12 colors present in the original population of beans.) Thus, with each successive generation, some alleles are likely to be lost as a result of this random sampling, which is totally independent of any selective pressures. Once an allele is lost (that is, it is not passed on to the next generation) it is gone forever, unless it arises again by a mutation (an unlikely event), or it is reintroduced by an immigrant.

The smaller the sample, the greater the likelihood that some forms will fail to occur in the sample. Therefore, this kind of sampling "error" is likely to lead to a substantial loss of genetic diversity any time that a population is reduced suddenly. This occurs when a few colonists form a new population, a phenomenon termed the founder effect. It can also occur when a population experiences a sudden and dramatic reduction in numbers, for example if there is high mortality because of a natural disaster. When a population declines to a low level it is said to have gone through a bottleneck. The individuals that survive a pronounced decline are only a small sample of the original population, and they are likely to have only some of the alleles that were represented in the population before it declined. Even if the population rebounds to higher levels after going through a bottleneck, its allelic diversity will remain low. Recall from Chapter 1 that Scandinavian brown bears went through a bottleneck around 1930, when the population was reduced to about 130 individuals. Box 8.4 presents another example of a population that apparently went through a bottleneck in the past.

Box 8.4   Low genetic diversity in the cheetah

Research on captive cheetahs from southern Africa suggests that genetically these individuals are nearly identical, like a highly inbred strain of laboratory mice. Four kinds of evidence point to this conclusion in captive cheetahs (O'Brien *et al.* 1985):

- **Sperm quality.** Examinations of ejaculate from zoo cheetahs reveal a high level of abnormally shaped sperm as well as low sperm concentrations and poor motility. These findings would be expected for a genetically depauperate population.
- **Biochemical studies.** Electrophoresis, a method of determining how many forms of a protein are present in a sample, reveals exceedingly low frequencies of variant proteins in captive cheetahs. Since genes direct the synthesis of proteins, the similarity in the proteins produced by different individuals suggests genetic similarity among captive cheetahs.
- **Skull morphology.** Measurements of skull characteristics in cheetahs reveal a high level of asymmetry when left and right sides are compared. This type of asymmetry is associated with a reduction in the ability of an animal to buffer itself against environmental changes during development, and thus it is believed to reflect reduced fitness. It is characteristic of inbred populations.
- **Tissue compatibility.** Cheetahs do not respond to tissue grafts from other individuals as foreign tissue. When 14 reciprocal skin grafts were performed, 11 were accepted for as long as grafts from another part of the cheetah's own body. In contrast, skin grafts from domestic cats were rejected. This suggests that different cheetahs have the same proteins in their skins and is further evidence for genetic uniformity.

In small populations the proportion of individuals that are heterozygous for a given trait declines over time; thus, homozygosity tends to be high in small populations. If two individuals that are heterozygous for a particular trait mate, on average one-quarter of their offspring will be homozygous for the recessive allele, one-quarter will be homozygous for the dominant allele, and one-half of the progeny will be heterozygous. In the next generation, the proportion of heterozygotes will continue to decline. Thus, going through a bottleneck is likely to result in reduced heterozygosity.

Inbreeding.   Inbreeding, the mating of close relatives, can occur in a large population (if individuals do not disperse far from the site where they were born and they mate with other individuals born nearby), but it is far more likely in small populations. Like genetic drift, inbreeding causes a loss of heterozygosity.

To summarize, genetic drift and inbreeding are exacerbated in small populations. Genetic drift causes genetic diversity to decline within populations (loss of alleles), and both genetic drift and inbreeding cause genetic diversity to decline within individuals (loss of heterozygosity). Both of these types of lowered genetic variation may increase a population's risk of extinction. The reasons for this are explained below.

Effects of reduced genetic variation.   Reduced genetic diversity is worrisome for two reasons: It reduces a species' evolutionary potential to adapt to environmental changes, and it is associated with reduced fitness. Natural selection acts on the raw material provided by genetic differences. Since natural selection acts on genetic variation, favoring those differences that confer an advantage in a changing environment, the loss of alleles from a population means the loss of raw material for natural selection to act upon. Under normal circumstances, when a species' environment changes, those individuals with advantageous alleles will be favored by natural selection. But if few genetic variants are present, there will not be many possibilities to select from. For instance, if a population is exposed to a novel disease, only those individuals with the alleles for resistance to the disease will survive and reproduce. If the population is genetically uniform and the allele conferring resistance has been lost, extinction may result unless the allele for resistance arises through a mutation. Since mutations arise at rates that are much slower than rates of environmental change, and since most mutations are harmful, new mutations are unlikely to produce favorable variations rapidly enough to allow for adaptation to a changing environment.

A second problem stemming from reduced genetic diversity is reduced fitness resulting from low heterozygosity. There are a number of examples in animal populations where heterozygous individuals have higher fitness than homozygotes. For instance, heterozygosity has been shown to have positive effects on survival and on resistance to disease in some species (Allendorf and Leary 1986). Perhaps the best-known example is sickle-cell anemia. People who are heterozygous for the allele that is responsible for the production of sickle-cell hemoglobin have higher resistance to malignant falciparum malaria than people who are homozygous for the normal allele at that locus (Allison 1961), while people homozygous for the sickle-cell allele have severe, often fatal, anemia.

Inbreeding depression is a related, but distinct, phenomenon. Like low heterozygosity, it is associated with reduced fitness (Frankel and Soulé 1981). One manifestation of inbreeding depression occurs when the progeny of closely related parents display inherited deleterious traits. Most harmful mutations are recessive. They are also rare, so it is unusual for an individual to have two alleles with the same harmful mutation. Close relatives are likely to share some alleles, however, because they have one or more common ancestors. If the parents share alleles for harmful recessive mutations, they are likely to produce some offspring that are homozygous for these traits. Because the alleles for these harmful traits are recessive, they will be expressed in the homozygous progeny even though they were not evident in the parents.

Inbreeding does not always result in inbreeding depression, however. Some organisms can contend with extreme inbreeding without any evident problems. (The extreme case is self-fertilizing plants.) Furthermore, mating between individuals that are too dissimilar genetically can also have disadvantages for wild organisms. Local populations are usually adapted to local conditions; they possess alleles for traits that are advantageous in their particular environment. If they mate with individuals from another population of the same species, the favorable alleles may be diluted. This is something that needs to be considered in the conservation of rare species. If managers transplant individuals from one region to another, for example to supplement a declining population, they may do more harm than good unless the genetic consequences of such translocations are understood. We will return to this point in Chapter 9.

It is hard to document cases in which wild populations have gone extinct as a result of low genetic diversity, because usually genetic data are not available for populations before they disappear. In fact, the connection between genetic diversity and population persistence is not fully understood. There is evidence that genetic diversity is extremely low in some organisms, such as elephant seals (Bonnell and Selander 1974) and cheetahs (Box 8.4). These results suggest that the ancestors of contemporary cheetahs and elephant seals went through one or more genetic bottlenecks in the past. (In the case of the elephant seal, this was due to heavy exploitation in the nineteenth century; see Chapter 1.) The significance of this for the long-term future of these species is unclear, however. It appears that wild populations of cheetahs have persisted for several thousand years after losing most of their genetic diversity. Although cheetahs in zoos have a host of problems that have been attributed to low genetic diversity (including poor quality sperm, low resistance to disease, and high infant mortality), problems stemming from low genetic diversity have not been documented in wild cheetah populations, so the effects of reduced genetic diversity in wild cheetahs are unresolved (Laurenson et al. 1995).

Other kinds of risks facing small populations

Small populations are also vulnerable to chance fluctuations in birth and death rates and in sex ratio. In addition, they are vulnerable to environmental fluctuations and to catastrophes (Figure 8.2). Recall from Chapter 1 that a combination of these risk factors came into play when only a single, small population of heath hens remained.

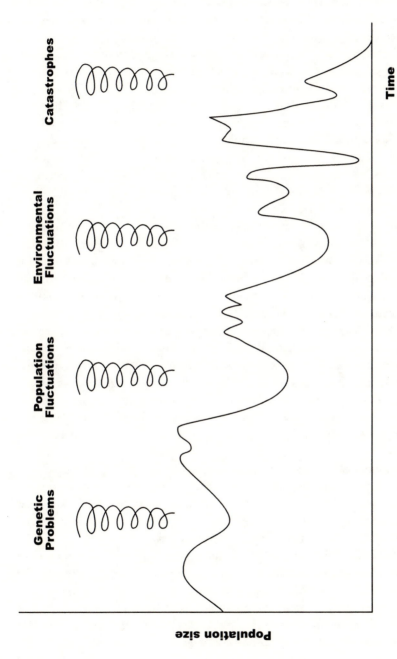

Figure 8.2. Four causes of downward spirals to extinction. Small populations are vulnerable to genetic problems, demographic fluctuations, environmental fluctuations, and catastrophes. Any one of these factors or a combination of them can drive a population to extinction.

Risks to populations on islands and in fragmented habitats

Many of the risk factors that face small populations are exacerbated on islands and habitat patches where populations are small. To make matters worse, if a population does go extinct on an island or in an isolated habitat patch, it may be difficult for colonists to get there to establish a new population. In the 1960s the ecologists E. O. Wilson and Robert MacArthur attempted to discover why an area of a given size on an island generally contains fewer species of plants and animals than an area of the same size and type of habitat on a nearby mainland (MacArthur and Wilson 1967; MacArthur 1972). They proposed that colonization and local extinction are the processes responsible for this phenomenon. If we consider only short-term events (so that we exclude the development of new species through speciation), the number of species on an island must equal the number of species that colonize the island by arriving and establishing successful populations minus the number of species that disappear from it by going extinct.

Wilson and MacArthur suggested that the number of species on an island is influenced by two geographic features: the island's size (Figure 8.3) and its distance from a source of colonists. Their field observations, and those of other ecologists such as F. W. Preston, indicated that for a particular group of organisms within a given climatic region, large islands typically contain more species than smaller islands (Figure 8.4). There are two reasons for this pattern. First, larger islands support larger populations, which have relatively low extinction rates. MacArthur and Wilson hypothesized that large populations have lower extinction rates because they are less vulnerable to fluctuations in their birth and death rates; however, they also recognized that habitat diversity plays a role. Large islands are likely to encompass more different kinds of habitats than small islands; therefore, more kinds of organisms will be able to fulfill their habitat requirements on large islands than on small islands.

The number of species on an island also reflects the rate of immigration to it (Figure 8.4). It is obvious that most terrestrial organisms cross land more easily than water; thus they colonize islands infrequently. Islands located near a source of colonists are more likely to be reached and successfully colonized than islands that are far from a source of colonists.

MacArthur and Wilson synthesized these observations into what came to be known as the equilibrium theory of island biogeography, or the MacArthur–Wilson (MW) model. This theory proposes that the number of species on an oceanic island will eventually reach an equilibrium, a point at which colonization and extinction rates balance each other out. At equilibrium, extinctions and colonizations continue to occur, but the number of

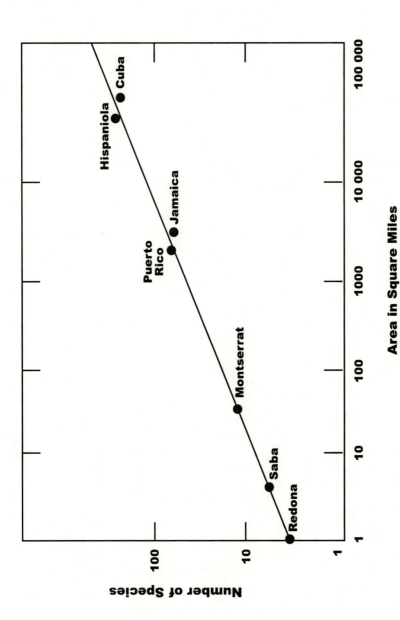

Figure 8.3. The relationship between area and species richness for amphibians and reptiles on islands of the West Indies. (After MacArthur and Wilson 1967. Copyright 1967 by Princeton University Press. Reprinted by permission of Princeton University Press.)

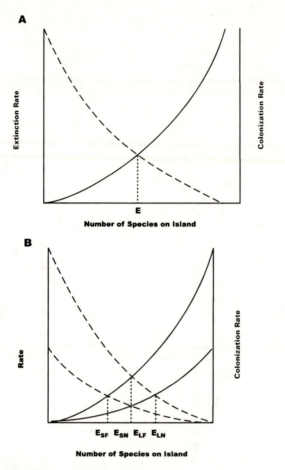

Figure 8.4. The dynamics of extinction and colonization on islands according to MacArthur and Wilson's equilibrium model of island biogeography. Solid curves indicate extinction rates; dashed curves indicate colonization rates. (A) The equilibrium model for a single island. The model predicts that colonization and extinction balance each other out in a dynamic equilibrium. E is the number of species on the island when equilibrium is reached. (B) The equilibrium model for several islands. E = equilibrium number of species; S = small; L = large; N = near; F = far. The model predicts that E is higher for large islands and islands that are near a source of colonists because they have larger populations (and thus lower extinction rates) and higher rates of immigration. According to the model, E_{SN}, the number of species at equilibrium on a small island located near a source of colonists, is lower than E_{LN}, the equilibrium number of species for a large island located the same distance from a source of colonists. Similarly, E_{SF}, the equilibrium number of species on a small island that is far from a source of colonists is less than E_{LF}, the number of species at equilibrium on an island that is also far from a source of colonists but large in area. Note the similarity between these curves and Figure 2.5, which also depicts a model of two processes (reproduction and mortality) at equilibrium. (After MacArthur and Wilson 1967. Copyright 1967 by Princeton University Press. Reprinted by permission of Princeton University Press.)

species on the island remains fairly constant because the extinction rate and the colonization rate are equal. The change in species composition that occurs on an island as some species disappear and others arrive is referred to as turnover.

Thus, islands contain fewer species than mainlands because they have few episodes of colonization and relatively frequent extinctions. The theory predicts that on small, isolated islands colonization will be rare and extinction will be common, so at equilibrium these islands will contain few species. Large islands located near sources of colonists will be colonized relatively frequently and will have relatively low extinction rates, so when they reach equilibrium they will have more species. Islands that are large and isolated or small and close to a source of colonists will be intermediate in species richness (Figure 8.4).

Of course, rates of extinction and colonization are different for different groups of organisms. Bats, birds, some insects, and some plants (for instance those with seeds or fruits that can tolerate long periods of immersion in salt water, such as coconuts), are fairly good at crossing oceans to colonize islands; large mammals are not. This has significant consequences in terms of the vulnerability of island biotas to introduced species.

Evidence for the MW model is summarized in Box 8.5.

Box 8.5 Testing the predictions of the MW model of island biogeography

A report published in 1917 lists species of birds known to breed at that time on the Channel Islands off the coast of southern California. Half a century later, the ecologist Jared Diamond surveyed the same islands and compared his results to the earlier census (Diamond 1969). The MW model predicts that we would find evidence of repeated colonizations and extinctions on the islands during the period between the two censuses, but that extinctions and colonizations would balance each other out, so that the total number of species on each island would be fairly stable. As predicted, Diamond did not find some of the species that had been observed earlier and found others that had not been reported; furthermore, for most islands the number of reported species did not change very much. He concluded that extinctions and colonizations were in equilibrium for the bird fauna of the Channel Islands and that substantial turnover occurred, as predicted by the MW model.

Diamond inferred extinction and colonization from census data; he did not observe these phenomena directly. There are other possible explanations for the discrepancies between successive censuses. (See Chapter 12 for additional discussion of Diamond's interpretation of the Channel Islands data.) To address this problem, E. O. Wilson and his colleague Daniel Simberloff designed an experiment that allowed them to do just that (Simberloff and Wilson 1969, 1970; Wilson and Simberloff 1969). At the start of their experiment, they censused the terrestrial arthropods, such as insects, spiders, and mites, on four small islands of red mangroves (see Box 13.9) in Florida Bay. Then they "defaunated" the islands by encasing them in plastic sheeting and spraying them with insecticide. If the defaunation was thorough, any arthropods that were found on the islands in subsequent censuses had to have arrived there as colonists, and any colonists that disappeared after they reached the island must have gone extinct. Of course, it is possible that some of the animals found on the islands had not really established breeding populations before they disappeared, so they could not really be said to have become extinct. Nevertheless, even when they interpreted their data conservatively to account for this possibility, Simberloff and Wilson estimated that considerable turnover was occurring on the islands (about 1.5 extinctions per island per year) (Simberloff 1976).

Not long after the MW model was proposed, ecologists realized that MacArthur and Wilson's insights about the processes of extinction and colonization might apply to patches of habitat on a mainland as well as to islands. Ecologists studied the dynamics of naturally occurring discontinuous habitats, such as caves, lakes, mountaintops, or clumps of plants, and applied the insights gained from these studies to habitat fragments created by agriculture, logging, development, or other forms of habitat alteration (see Chapter 10).

We noted in Chapter 7 that contemporary landscapes are highly fragmented by habitat destruction and development. Many types of habitat are confined to discrete patches surrounded by "oceans" of inhospitable habitat. A habitat patch surrounded by a matrix of dissimilar habitats can be considered analogous to an island, if the surrounding matrix is hard for organisms to cross. (The matrix is comparable to water surrounding the island.) A habitat patch can be a natural feature that is discontinuous across a landscape, or it can be artificially created. If organisms cannot cross the habitat between patches, colonization will cease. Extinctions continue to take place, however. This means that there is no longer an equilibrium between extinction and colonization for the biotas of these patches. Small, isolated habitat patches would be expected to have fewer episodes of colonization and more extinctions than large patches that are not isolated. Of course, the number of species that live in a patch will also depend upon the quality and the variety of habitats that it contains.

Whether or not the land separating habitat fragments acts as a barrier and prevents colonization – that is, whether or not a habitat patch functions as an island – depends upon a number of factors, including dispersal abilities, physiological tolerances, behaviors, and sensitivity to the presence of humans and vehicles. If habitat fragments are separated by large expanses of unsuitable habitat, movement between patches is unlikely. Different groups of organisms vary tremendously in their willingness and ability to move through habitats. Even a two-lane highway presents an impenetrable barrier for some organisms, whereas others can easily fly over hundreds of kilometers of unsuitable habitat. If movement between patches is not possible, then extinctions cannot be offset by recolonization.

If the habitat surrounding a patch does not function as a barrier, "colonization" occurs frequently and the patch cannot really be said to function like an island for the organisms in question. If the patch is somewhat isolated from other areas of suitable habitat, so that population size is constrained by the size of the patch, and colonization occurs sometimes but not very often, then extinctions and colonizations will occur, and they may balance each other, like islands at equilibrium. If, on the other hand, a patch really is totally isolated, that is, if the organisms in question cannot or will not move through the intervening habitat, then colonization cannot occur but extinction will continue (Weddell 1991). If an area that was formerly surrounded by suitable habitat becomes totally isolated, then we would expect the number of species in the patch to decline, as extinctions continue and colonizations cease.

The fates of small mammals in alpine habitats at high elevations in the mountains of the southwestern U.S.A. and the Great Basin illustrate this (Figure 8.5). Studies of the distribution of small mammals in the cool environments on mountaintops have shown that large habitat patches do indeed support more species than small ones. Because small mammals adapted to a cool climate cannot move very far through the warm valleys separating mountaintops, small alpine mammals cannot colonize mountaintops under current conditions, but extinctions continue. For small mammals in this environment, patches of suitable habitat function like islands, that is, extinction rates are higher in small patches than in large ones. Unlike the islands described by MacArthur and Wilson, colonization cannot be said to be related to distance from a source of colonists in this case, because under existing climatic conditions colonization is impossible for small mammals on mountaintops (Brown 1971; Patterson 1984).

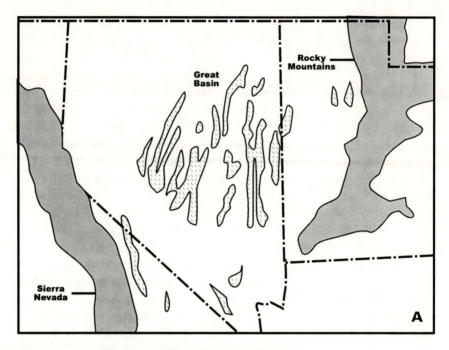

Figure 8.5. (A) Isolated mountaintops in the Great Basin between the Rocky Mountains and the Sierra Nevada Mountains. The stippled areas have elevations greater than 3000 m. The hot, dry desert valleys between the cool habitats on mountain peaks form barriers that small mammals are unable to cross. (Reprinted from Brown 1971. © 1971 University of Chicago Press.) (B) Differences in the relative isolation of mountaintops for two different species of alpine environments. The gray areas represent patches of alpine habitat. Solid line represents characteristic dispersal distance of a hypothetical

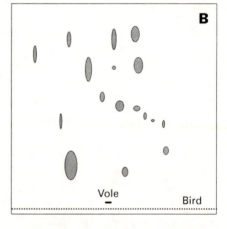

small mammal; dashed line represents characteristic dispersal distance of a hypothetical bird. The bird is capable of long-distance flight and can cross lowland barriers between mountaintops with ease, so that for it, mountaintops are not like islands. The distances between patches of alpine habitat are inconsequential; distance has no effect on colonization, and the probability of occupancy is high. The small mammal, on the other hand, can only disperse relatively short distances and cannot tolerate the climate at lower elevations. For it, patches of alpine habitat on mountaintops are so widely separated that they cannot be colonized. The probability of patch occupancy is low because extinctions are not offset by colonizations. (After U.S. Army Corps of Engineers 2001.)

On the other hand, studies of bird populations on mountaintops in the same region do not show a strong relationship between area and species diversity. This is not really surprising because many species of birds can readily fly from one mountain peak to another; the low elevation habitat is not a barrier to them. So mountaintops do not function like islands for birds. In other words, for species that can move readily between mountaintops, extinction and colonization rate are not related to patch size or distance between patches, as we would expect for island-like habitats (Johnson 1975; Brown 1978).

Concern about an imbalance between colonizations and extinctions has led ecologists to search for landscape features that promote colonization and minimize extinction. This topic is discussed in more detail in Chapter 10, in which we shall see that island biogeographic theory has been applied to the design of nature reserves in an attempt to maximize the species richness of protected areas.

8.4.4   Population viability analysis: A tool for assessing risk of extinction

Conservation biologists would like to be able to evaluate the risk of a population becoming extinct. Population viability (or vulnerability) analysis (PVA) is a tool for doing this (Gilpin and Soulé 1986). It involves two steps: Estimating the probabilities of various adverse events during a specified time period and estimating the probabilities that if those events occur they will cause a population to become extinct. PVA seeks to estimate the minimum size of a population that will persist for a given amount of time. Many species-specific PVA models have been developed to estimate the vulnerability of small populations to random events. PVA models have also been used as a basis for making recommendations regarding the minimum size of a reserve that will have an acceptable likelihood of maintaining a population for a specified period of time.

PVA models generate testable predictions. As always, their predictions are only valid if their assumptions and reasoning are correct. It is important to keep in mind the distinction between models and data, but this distinction is often obscured. For example, in discussing one of the first attempts to predict the size of a population that would be safe from extinction, Gilpin and Soulé (1986:27) stated that MacArthur and Wilson "found a critical limit of 10 individuals" for asexually reproducing microorganisms. Actually, MacArthur and Wilson *predicted* this on the basis of a mathematical equation.

Preservationist managers have applied insights about the problems faced by small populations to the challenge of preventing extinctions. In the next two chapters we consider techniques for doing this. Chapter 9 considers efforts to preserve populations and species, and Chapter 10 covers efforts to preserve habitats.

References

Allendorf, F. W. and R. F. Leary (1986). Heterozygosity and fitness in natural populations of animals. In *Conservation Biology: The Science of Scarcity and Diversity*, ed. M. E. Soulé, pp. 57–76. Sunderland, MA: Sinauer Associates.

Allison, A. C. (1961). Genetic factors in resistance to malaria. *Annals of the New York Academy of Sciences* **91**:710–729.

Bonnell, M. L. and R. K. Selander (1974). Elephant seals: genetic variation and near extinction. *Science* **184**:908–909.

Brown, J. H. (1971). Mammals on mountaintops: nonequilibrium insular biogeography. *American Naturalist* **105**:467–478.

Brown, J. H. (1978). The theory of insular biogeography and the distribution of boreal birds and mammals. *Great Basin Naturalist Memoirs* **2**:209–227.

Caughley, G. and A. Gunn (1996). *Conservation Biology in Theory and Practice*. Cambridge, MA: Blackwell Science.

Cohn, J. P. (1987). Red wolf in the wilderness. *BioScience* **37**:313–316.

Darwin, C. (1958). *The Origin of Species*, 7th printing. New York, NY: New American Library.

Diamond, J. M. (1969). The Channel Islands of California. *Proceedings of the National Academy of Sciences, U.S.A.* **64**:57–63.

Dowling, T. E., B. D. DeMarais, W. L. Minckley, M. E. Douglas, and P. C. Marsh (1992). Use of genetic characteristics in conservation biology. *Conservation Biology* **6**:7–8.

Ehrlich, P. and A. Ehrlich (1985). *Extinction: The Causes and Consequences of the Disappearance of Species*, 2nd Ballantine printing. New York, NY: Ballantine Books.

Eldredge, N. and S. J. Gould (1972). Punctuated equilibria: an alternative to phyletic gradualism. In *Models in Paleobiology*, ed. T. J. M. Schopf, pp. 82–115. San Francisco, CA: Freeman, Cooper.

Erwin, T. L. (1991). How many species are there? revisited. *Conservation Biology* **5**:330–333.

Frankel, O. H. and M. E. Soulé (1981). *Conservation and Evolution*. Cambridge: Cambridge University Press.

Futuyma, D. J. (1997). *Evolutionary Biology*, 3rd edn. Sunderland, MA: Sinauer Associates.

Gaston, K. J. (1991). The magnitude of global insect species richness. *Conservation Biology* **5**:283–296.

Gilpin, M. E. and M. E. Soulé (1986). Minimum viable populations: processes of species extinction. In *Conservation Biology: The Science of Scarcity and Diversity*, ed. M. E. Soulé, pp. 19–34. Sunderland, MA: Sinauer Associates.

Gittleman, J. L. and S. L. Pimm (1991). Crying wolf in North America. *Nature* **351**:524–525.

Gould, S. J. and N. Eldredge (1977). Punctuated equilibria: the tempo and mode of evolution reconsidered. *Paleobiology* **3**:115–151.

Johnson, N. K. (1975). Controls on number of bird species on montane islands in the Great Basin. *Evolution* **29**:545–574.

Laurenson, K., N. Wielebnowski, and T. M. Caro (1995). Extrinsic factors and juvenile mortality in cheetahs. *Conservation Biology* **9**:1329–1331.

MacArthur, R. H. (1972). *Geographical Ecology: Patterns in the Distribution of Species*. New York, NY: Harper and Row.

MacArthur, R. H. and E. O. Wilson (1967). *The Theory of Island Biogeography*. Princeton, NJ: Princeton University Press.

Martin, P. S. and R. G. Klein (eds) (1984). *Quaternary Extinctions: A Prehistoric Revolution*. Tucson, AZ: University of Arizona Press.

Meffe, G. K. and C. R. Carroll (1997). *Principles of Conservation Biology*, 2nd edn. Sunderland, MA: Sinauer Associates.

Nowak, R. M. (1992). The red wolf is not a hybrid. *Conservation Biology* **6**:593–595.

O'Brien, S. J. and E. Mayr (1991). Bureaucratic mischief: recognizing endangered species and subspecies. *Science* **251**:1187–1188.

O'Brien, S. J., M. E. Roelke, L. Marker, A. Newman, C. A. Winkler, D. Meltzer, L. Colly, J. F. Evermann, M. Bush, and D. E. Wildt (1985). Genetic basis for species vulnerability in the cheetah. *Science* **227**:1428–1434.

Patterson, B. D. (1984). Mammalian extinction and biogeography in the southern Rocky Mountains. In *Extinctions*, ed. M. H. Nitecki, pp. 247–293. Chicago, IL: University of Chicago Press.

Phillips, M. K. and V. G. Henry (1992). Comments on red wolf taxonomy. *Conservation Biology* **6**:596–599.

Primack, R. B. (1998). *Essentials of Conservation Biology*, 2nd edn. Sunderland, MA: Sinauer Associates.

Rojas, M. (1992). The species problem and conservation: what are we protecting? *Conservation Biology* **6**:170–178.

Simberloff, D. S. (1976). Species turnover and equilibrium island biogeography. *Science* **194**:572–578.

Simberloff, D. S. and E. O. Wilson (1969). Experimental zoogeography of islands: the colonization of empty islands. *Ecology* **50**:278–296.

Simberloff, D. S. and E. O. Wilson (1970). Experimental zoogeography of islands: a two-year record of colonization. *Ecology* **51**:934–937.

Thompson, J. N. (1994). *The Coevolutionary Process*. Chicago, IL: University of Chicago Press.

U.S. Army Corps of Engineers (2001). Ecosystem Management and Restoration Information Systems CD. Vicksburg, MS: U.S. Army Corps of Engineers.

Wayne, R. K. and S. M. Jenks (1991). Mitochondrial DNA analysis implying extensive hybridization of the endangered red wolf, *Canis rufus*. *Nature* **351**:565–583.

Weddell, B. J. (1991). Distribution and movements of Columbian ground squirrels: are habitat patches like islands? *Journal of Biogeography* **18**:385–394.

Wilson, E. O. and D. S. Simberloff (1969). Experimental zoogeography of islands: defaunation and monitoring techniques. *Ecology* **50**:267–268.

Wolfheim, J. H. (1976). The perils of primates. *Natural History* **85**(8):90–99.

9

Techniques – protecting and restoring species

We have seen that concern about loss of biodiversity spearheaded an interest in resource management directed at protecting species and habitats. In this chapter we will examine measures that are being taken to slow the earth's extinction rate.

9.1 Overview of options: Strategies for preventing extinctions

Only two basic approaches are available if we want to prevent extinctions – reducing or reversing the factors that trigger species' declines and assisting with recovery. Stemming the tide of species loss involves identifying the proximate causes of species declines and addressing them.

The most direct way to decrease mortality caused by people is to regulate exploitation. A decrease in harvest may also lead to an increase in reproductive rate, if young are taken or if exploitation inhibits reproduction. Of course, limiting harvest will only help populations in trouble if other factors causing mortality or limiting reproduction are also addressed.

This chapter describes direct methods that attempt to protect and restore populations in trouble. In addition, managers often attempt to decrease mortality and increase reproduction indirectly by protecting or restoring habitats. This is covered in Chapter 10.

9.2   Decreasing losses

9.2.1   Regulating exploitation

International agreements

In North America the first international agreement to regulate traffic in wildlife products was the Migratory Bird Treaty between the United States and Canada, which was first signed in 1916 (see Chapter 1). Two more recent examples of international agreements to regulate exploitation of wild organisms are described below.

As you know from Chapter 1, unregulated exploitation of whales led to the decline and near extinction of the larger whale species. In 1931 the Convention for the Regulation of Whaling was signed by 26 nations, and in 1946 the convention established the International Whaling Commission (IWC) to regulate commercial whaling. The objectives of the convention were utilitarian: to protect whales and (or for) whaling. Composed of members of both whaling and nonwhaling nations, the IWC set quotas with the goal of harvesting whale populations at a level designed to permit a maximum sustained yield (see Chapter 4). The convention's quotas were sometimes so high that commercial whaling vessels could not even take enough whales to fill them, so it is doubtful that these "limits" actually decreased the take of whales. In addition, the IWC has never had the authority to enforce its regulations. Any member nation that files a formal objection to an IWC decision is not bound by it.

The Convention on International Trade in Endangered Species of Wild Fauna and Flora (CITES) was formed in 1973 to regulate international trade in species that are put at risk by that trade. CITES assigns at-risk species to one of three categories:

- Appendix I: Species threatened with extinction that are or may be affected by trade. Export or import of the more than 600 species listed in Appendix I is prohibited, except under certain noncommercial circumstances. For example, scientific use may be permitted if both the exporting nation and the importing nation agree that such use is not detrimental to the survival of the species in question.
- Appendix II: Species that may become threatened with extinction unless their trade is regulated. Commerce in species listed in Appendix II is allowed under special conditions regulated by permits. This appendix contains at least 27 000 species, over 90% of which are plants.
- Appendix III: Species that are protected within the treaty nations where they occur and are considered by those countries to need international

control. For example, several species of snakes that are protected in India are listed in Appendix III because the international leather market places them at risk.

Although over 120 nations now belong to CITES, enforcement of its regulations is politically complex, difficult, and sometimes dangerous. The international wildlife trade has several stages, and each stage has different incentives. With the development of a lucrative international trade in wildlife, the enforcement of harvest regulations has become a dangerous and expensive business. In most cases not enough personnel are funded to be able to make a dent in the problem. Because poaching for the international market offers great profits, fines and jail sentences are seldom sufficient to deter offenders.

Legislation protecting endangered species

Many nations have legislation that regulates the taking of game species. In recent decades, many countries have also adopted protection for rare species or have passed legislation specifically aimed at prohibiting the harvest of species in trouble. In the United States, the first federal legislation that explicitly addressed the needs of species facing extinction was the Endangered Species Preservation Act, passed in 1966. This law did not actually protect endangered species, but it authorized the Secretary of the Interior to determine which wildlife species faced extinction and provided for research and habitat acquisition. In 1969 and again in 1973, modifications of the original act were passed, extending protection first to invertebrates (other than insect pests) and later to plants. The 1973 version, termed the Endangered Species Act (ESA), emphasized ecosystem preservation as a means of conserving endangered species. Endangered species were defined as those likely to become extinct throughout all or most of their geographic range. Threatened species were defined as those deemed likely to become endangered. Under the ESA, subspecies and "distinct population segments" also qualify for protection. For instance, the eastern mountain lion or Florida panther, *Felis concolor coryi*, is a subspecies of the cougar, *Felis concolor*. The Florida panther has been reduced to a few dozen individuals in Florida and is legally protected, although in other parts of the United States there are cougar subspecies that are not in trouble and are not covered by the ESA.

In common usage, endangered or threatened species or subspecies are loosely lumped under the term "endangered species." The notorious northern spotted owl (*Strix occidentalis caurina*) is often called an "endangered species," although technically it is neither endangered nor a species. It is listed under the Endangered Species Act as a *threatened subspecies* of the spotted owl (*Strix occidentalis*).

Section 7 of the ESA requires federal agencies to consult with the Secretary of the Interior to ensure that their actions will not jeopardize threatened or endangered species or their habitats. In 1973, this provision led to a controversy over a proposed dam on the Little Tennessee River. The dam triggered Section 7 of the ESA when a zoologist discovered a population of snail darters in water near the dam site. Because of the likelihood that construction of the dam would jeopardize the only known population of this fish, work on the dam was halted.

The media painted a picture of an obscure, trivial fish standing in the way of economic development. The public perceived the incident as a case of endangered species legislation granting political victories to environmental extremists on behalf of insignificant species. Congress eventually passed legislation exempting the Tellico Dam from the ESA, the dam was built, and populations of the snail darter were transplanted elsewhere. (In addition, other populations of snail darters were eventually discovered.) But the controversy set the stage for future conflicts that would be viewed as battles between species protection and economic gains.

Concern that endangered species legislation would grant political victories to environmental extremists led to passage of the Endangered Species Amendments Act in 1978, which set up a review board with the power to grant exemptions to federal agencies in future conflicts. (The board has become popularly known as the "God Squad" because of its power to make decisions about the ultimate fate of species.) Similar concerns about the power of the act to block development, stemming from controversies surrounding the spotted owl and salmon (see Chapter 11), have led to numerous proposed revisions of the ESA.

9.2.2   Minimizing natural sources of mortality

In addition to controlling mortality by regulating exploitation, it is sometimes possible to boost populations of rare species by manipulating natural sources of mortality, such as competitors, parasites, herbivores, or predators. For example, scientists at the University of Nebraska protected young Pitcher's thistle plants from herbivorous insects by spraying them with insecticide and covering them with screen cages. In portions of their habitat where herbivory was high, protected plants had lower juvenile mortality and higher seed production than plants in the control group (Bevill *et al.* 1999). This approach is not likely to be successful, however, unless all important limiting factors are addressed (Box 9.1).

Box 9.1   Efforts to aid endangered Kirtland's warblers by controlling cowbirds

Kirtland's warbler nests in central Michigan and breeds in the Bahama Islands off the coast of Florida. For breeding, this Neotropical migrant requires stands of jack pine that are 1.5–6.0 m tall and 7–20 years in age. Since jack pines can only reproduce after fire, the warblers can nest only on sites that were burned within a very specific time-frame. Because of these highly specific habitat requirements, Kirtland's warbler populations are small and localized; and the species was listed as federally endangered in 1967 (Walkinshaw 1983).

Even in areas where suitable pine stands exist, they are transient. As the trees age, they outgrow their usefulness to Kirtland's warblers. To address this need, federal and state agency personnel regularly set fire to jack pines so that stands of the appropriate age and structure will be available continuously. The breeding warblers face other problems too, however. One of these is brood parasitism from brown-headed cowbirds. Formerly confined to the eastern states, this nest parasite expanded its range westward as forests of the Midwest were cleared. By the middle of the twentieth century, more than 50% of Kirtland's warbler nests contained cowbird eggs. Brood parasitism by cowbirds appears to operate in a density-independent fashion. This is possible because the cowbirds parasitize several different species. If one, such as the Kirtland's warbler, drops to very low levels, the cowbirds can still maintain their population by parasitizing other species, but they continue to lay their eggs in Kirtland's warbler nests whenever the opportunity arises. So the warblers experience no relief from nest predation as their numbers dwindle.

In 1972, the U.S. Fish and Wildlife Service began removing cowbirds from six Michigan counties in the breeding range of the Kirtland's warbler. Between 1972 and 1981, over 33 000 cowbirds were removed from the area by trapping. The results of this program, in terms of the rate of nest parasitism and fledging success, were striking (Figure 9.1). Cowbird parasitism declined markedly, and there was a concomitant increase in the fledging success from fewer than one fledgling per nest between 1931 and 1972 to an average of 2.8 young fledged per nest during the period when cowbirds were controlled. Yet, in spite of these encouraging results, the Kirtland's

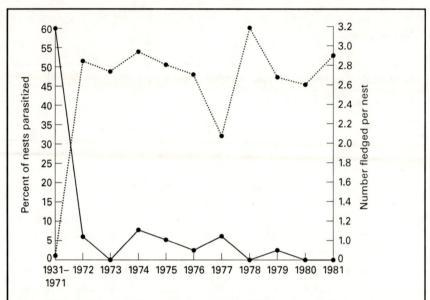

Figure 9.1. Relationship between percentage of Kirtland's warbler nests parasitized by cowbirds (solid line) and fledging success (dashed line) in Michigan from 1931 through 1981. (After Kelly and DeCapita 1982).

warbler population failed to increase markedly. In 1971, the last year before cowbird control was initiated, 201 singing males were counted during the breeding census. During the decade when cowbirds were removed, the average number of singing males counted was 207, just slightly higher than the precontrol number. Evidently, other factors that were not addressed during this period were limiting the Kirtland's warbler population (Kelly and DeCapita 1982).

9.3   Enhancing the size and range of populations

9.3.1   Increasing population productivity

Techniques
Populations of rare species can potentially be increased directly through programs that boost birth rate or survival. One approach to perpetuating rare species involves breeding them in zoos, aquaria, botanical gardens, or research

centers. This is termed *ex situ* (off-site) conservation. Wild organisms in these institutions can be used for research and to educate the public about the plight of declining species, but the greatest conservation benefits of *ex situ* conservation come from returning individuals to the wild, as a means of either augmenting small populations or re-establishing populations in areas from which they have disappeared. Another type of breeding program maintains organisms at a stage of the life cycle that can easily be stored, such as sperm, embryos, or seeds. In this capacity, zoos, aquaria, botanical gardens, and research centers are literally libraries of genetic material.

The survival of young individuals in the wild is low for most populations. Conservation biologist Malcolm Hunter put it succinctly when he stated that "one of the fundamental laws of nature is that little things tend to die quickly. They get eaten by big things. They get outnumbered by big things and then starve or desiccate" (Hunter 1996:315). It would seem logical then that anything that increases juvenile survival could potentially increase population growth. Furthermore, many animals are capable of producing far more young than they can care for, and some species provide no care at all for their young. It stands to reason that we should be able to increase reproductive output by taking seeds, eggs, or young from the wild, caring for the juveniles, and subsequently returning them to the wild.

Several intriguing techniques for doing this have been developed. One method takes advantage of the fact that many species of birds produce a second clutch of eggs if the first one disappears. (This is an adaptation to nest predation.) In double-clutching or double-brooding, biologists remove the first clutch and rear it artificially, hoping that the parents will produce and rear a second clutch. This method has been used to increase the reproductive output of the highly endangered California condor.

In cross-fostering, eggs are removed from a nest and placed in the nest of a closely related species, the adults of which act as surrogate parents. For example, eggs of the black robin, a species native to the Chatham Island archipelago near New Zealand, have been raised successfully in nests of the Chatham Island tit. This technique – along with supplemental feeding, artificial nest boxes, control of competitors and predators, and a variety of other forms of intervention – helped increase the black robin population from a low of five to more than a hundred individuals (Butler and Merton 1992).

Cross-fostering has also been used to increase production of the highly endangered whooping crane. Since whooping cranes typically produce clutches of two eggs but only one survives, the second egg can be removed without decreasing reproduction in the wild (assuming that the disturbance caused by the removal doesn't lower reproductive success). The more abundant sandhill crane has proved to be a good foster parent at Grays Lake, Idaho, but so far the results of cross-fostering with sandhill cranes have been disappointing. Between 1975 and 1988 nearly 300 whooping crane eggs were placed in foster nests, yet in 1991 there were only 13 whooping cranes in the Grays Lake flock, and they had failed to breed. This points to a fundamental problem with cross-fostering – the risk that birds will fail to form pair bonds with members of their own species, and therefore they will not engage in normal reproductive behavior.

Cross-fostering works only with organisms that lay eggs and provide parental care for them, that is, birds. Invertebrates, fishes, amphibians, and reptiles are not candidates for cross-fostering because adults usually do not care for their young. Mammals are not candidates either, because they don't lay eggs. Embryo transfer is a variation of this technique that is used with mammals. It involves transplanting embryos of rare species into closely related species that are common. This method has been used with some relatives of domestic livestock.

Even for species that do not raise their young, however, it is possible to provide care during the vulnerable juvenile stage. Eggs or hatchlings can be removed from the wild and raised in captivity, where they are fed and protected from predation, exploitation, adverse weather, and other mortality factors. This technique, termed head-starting, has been used with sea turtles. The long-standing horticultural practice of growing plants from seeds to a stage where they can be transplanted outdoors is a variation on this theme that is now widely used in habitat restoration projects.

Advantages and pitfalls

Most of the methods described above involve removing individuals from the wild, breeding them to increase their numbers, and returning the descendants to their natural habitat. This approach has both advantages and disadvantages. It is fairly easy to arouse public sympathy for certain species, especially large, appealing animals such as pandas, whooping cranes, and condors. Consequently, in some cases funding is available to breed these species in captivity. Often the necessary techniques have been worked out as well. This is especially likely to be true for organisms that are closely related to domestic plants or animals (e.g., Przewalski's horse; see Box 9.2) or that have been kept as captives for centuries (e.g., falcons). Such projects have enormous public relations value and can be a powerful educational tool for raising the consciousness of the general public about the plight of rare species in particular and environmental issues in general. Finally, for species for which there is virtually no habitat left, *ex situ* conservation is the only hope (Conway 1980). If they are successfully maintained and bred in captivity, then the presence of the captive population can serve as an incentive for habitat protection or restoration (Cohn 1988).

Box 9.2   Genetic changes in zoo populations of Przewalski's horse

The Asian wild horse, or Przewalski's horse, is a heavily built, yellowish brown horse with a short, stiff black mane (Figure 9.2). Although herds once roamed the Gobi Desert of southern Mongolia and northern China (Figure 9.3), by the 1930s herds of wild horses were rare in the region, and by the 1960s it seemed unlikely that a wild population could sustain itself. At that point, conservationists focused their attention on breeding captive Przewalski's horses, with the eventual goal of returning them to the wild. Hunting by local tribes and habitat deterioration caused by a combination of factors – including extensive military activity in the region; periods of unusually icy ground; and increases in grazing by domestic sheep, camels, and goats – may have played a role in the horses' decline (Nowak and Paradiso 1983; Ryder 1993).

Przewalski's horse became extinct in the wild, but as of the early 1990s, there were over 1100 individuals in captivity in 30 institutions. Zoos keep detailed studbook records of matings, and used this information to design pairings that would minimize problems associated with inbreeding (Chapter 8). Fortunately there is a high degree of cooperation and coordination around the world between institutions with populations of Przewalski's horse. For example, in 1982, zoos in the U.S.A. and the U.S.S.R. exchanged individuals in order to increase the genetic diversity within each facility. Yet in spite of the large size of the captive population, the availability of detailed data on parentage, and the care taken to arrange matings, there have been problems from inbreeding (Flesness 1977). This is because all of the captive individuals descended from 13 ancestors. Not only did the number of births per mare decline, but zoo horses began dying at younger ages. Before World War II, 21% of 191 horses lived to the age of 20, subsequently, only 7% (11) of 163 horses lived that long (Bouman 1977). Both of these findings suggest inbreeding depression. Further evidence in support of this interpretation is provided by genetic analyses showing that a horse's inbreeding coefficient (a measure of how closely related its parents are) is closely related to its reproductive output (Flesness 1977).

Figure 9.2. Przewalski's horse. Note the erect mane and lack of a forelock. (Drawn by M. Rockwood.)

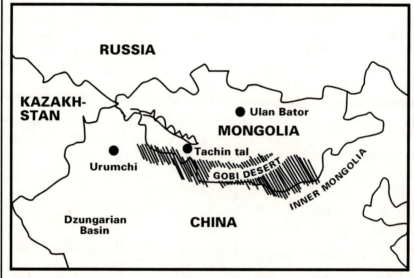

Figure 9.3. Map of the Gobi desert. (After Ryder 1993.)

In addition to these problems, the behavior of the captive horses changed in ways that suggested they were adapting to captivity. In the wild, mares dropped foals only in the spring, but in captivity they produce young at any season. Furthermore, Przewalski's horse breeds at a younger age in captivity. These changes are not detrimental in a captive setting, but in the wild they would be maladaptive. Giving birth throughout the year would result in animals being born when resources were not available. Similarly, animals that reach sexual maturity at a young age in the wild might not yet be capable of rearing young, though this is not a problem in captivity. Both these changes, which are thought to have a genetic basis, resulted from just five to eight generations of captive breeding, which entails a different suite of selective pressures than the Gobi Desert (Frankel and Soulé 1981).

This example illustrates some of the genetic problems that plague breeding programs, even with an organism that is relatively easy to raise in captivity.

The ultimate goal of the techniques described above is to establish self-sustaining populations that do not depend on people. Although such measures seem like a logical way to increase the size and range of dwindling species, they face a variety of problems ranging from technical hurdles to genetic problems.

Breeding programs have some direct negative impacts. First, they cause losses in donor populations. In addition, the removal process itself may disturb the donor population, disrupt breeding, or introduce diseases. Where zoo stocks or seed banks are available, these can be bred or reared with no current adverse effects on wild populations; however, zoo animals may make poor parents of animals destined to be returned to the wild. Handling wild animals and maintaining them in captivity also entails a risk of mortality for the captives, particularly in the early stages of a project, when scientists have much to learn.

Breeding animals in captivity is difficult, especially for species that are not closely related to domestic animals. An analysis of annual census data from British zoos for the period 1962 through 1977 reported that only 22% of 274 species of rare mammals had been bred more than a few times in captivity, and just 9% of those species had self-sustaining populations in captivity (Pinder and Barkham 1978). Most of those were ungulates, carnivores, or primates with domesticated relatives.

It is even more difficult to return captive-bred organisms to the wild, because they are usually unaccustomed to dealing with the rigors of a natural environment. In addition, young animals must acquire the ability to find and obtain food before they can survive on their own. Especially for predatory birds and mammals, hunting skills learned from parents are often critical. These animals are unlikely to acquire such skills in a breeding facility. The same thing is true for learned migration patterns.

A variety of techniques have been developed to assist organisms that are returned to the wild, but these measures are expensive and can introduce a degree of artificiality. For plants this might take the form of watering, weeding, fertilizing, and fencing the transplants to prevent herbivory. With young animals, it is often necessary to have a transition period in which they are gradually weaned from dependence on people. One way of meeting this challenge is with a technique called hacking (Zimmerman 1975). In this method, developed by falconers, a captive-reared falcon is placed in a shed and fed regularly. After the bird becomes accustomed to the feeding schedule, a window or door is opened, so that the fledgling is free to leave. Scheduled feedings continue, however, for as long as the bird returns for food (see Box 9.3). This technique provides nourishment the animal can depend upon until it learns to hunt and becomes self-sufficient. The U.S. Fish and Wildlife Service used a similar technique when it released endangered red wolves in North Carolina (Chapter 8). Plant breeders (and gardeners) commonly try to circumvent this problem by hardening, a process in which plants are gradually acclimated to conditions outside the greenhouse before they are moved out-doors.

Another practical problem is the difficulty of finding appropriate habitat where a population will be able to sustain itself: "'reintroduction' of wildlife is fundamentally limited; if you don't have . . . any place to put it, you can't reintroduce it" (Conway 1988:132). Even if the habitat has all the required resources, if reintroduced individuals are killed by people, either deliberately or accidentally, mortality may limit population growth. This is a problem with plants as well as animals. If the presence of a rare species is perceived as a threat to development, people sometimes kill the offending plants and animals. In addition, it may be necessary to reintroduce scores of individuals simultaneously to establish a viable population. This is an especially serious problem for animals that breed in groups, such as prairie grouse, which mate on communal dancing grounds where males gather and display for females.

Captive breeding exacerbates the genetic problems of already small populations. Breeding programs usually involve species that have gone through a bottleneck. Even if their numbers increase dramatically because of successful breeding projects, as in the case of Przewalski's horse, genetically uniform species may have a reduced chance of long-term survival in the wild because of reduced evolutionary potential (see Chapter 8). Zoos and botanical gardens can take steps to minimize this problem by keeping records on who breeds with whom, breeding individuals from different institutions, and avoiding crosses between closely related individuals. Again, Przewalski's horse is an example. But although these measures may increase the genetic diversity of animals born in zoos over what it would be if animals mated only with individuals from the same facility, the result is a far cry from the genetic results produced by adaptation to local conditions in the wild.

In addition, the initial capture process selects for individuals able to tolerate handling, and breeding programs inevitably produce artificial selection for the ability to survive in captivity (Box 9.2). The characteristics that enhance survival in a greenhouse or zoo are unlikely to be the same traits that will be advantageous in the wild. If this goes on for several generations, adaptations to the wild may be lost, resulting in partial domestication.

Furthermore, breeding projects divert attention and resources from the important issue of what causes species to decline in the first place. At best, only a small number of rare species can be bred in institutions and successfully returned to the wild. These endeavors are extremely expensive and labor-intensive. The money and effort that are spent on breeding are not available for other projects. Some critics argue that we should make habitat preservation, which results in the protection of entire communities, our top priority, instead of using large amounts of money and expertise in attempts to preserve a few species. Taking individuals from the wild may even facilitate habitat destruction, if removal from the wild (without successful transplantation to a different site) removes a major obstacle to a development project.

Finally, critics argue some that *ex situ* conservation involves a high degree of intervention and results in artificial, semi-tame products of human manipulation and technology rather than in free, wild organisms.

9.3.2   Increasing geographic range

We can increase a species' geographic range by taking individuals from the wild and releasing them elsewhere (with or without an intermediate period of breeding in captivity). Utilitarian resource managers have a long history of translocating selected species. Wildlife managers reintroduce game species into areas from which they have been extirpated and introduce them into areas where it is deemed desirable to establish new populations. Foresters reintroduce tree seedlings on logged sites to hasten the process of reforestation, and fisheries managers stock waters with a variety of native and non-native species produced in hatcheries. This kind of translocation seeks to establish valuable species in areas where they can be harvested. In many instances in the past, managers did not care whether the species was native to the area where it was being introduced.

Preservationist managers also translocate species, but with a different purpose. They seek to restore populations in habitats from which they have disappeared or to augment populations where they have declined. Box 9.3 describes the results of a translocation program aimed at returning the red kite to parts of its former range in the British Isles.

Box 9.3   Translocation of the red kite

The red kite is a hawk of European fields, pastures, and woodlands (Figure 9.4). It was once common in the British Isles and throughout much of Europe; but, like other birds of prey, kites were poisoned, trapped, and shot; bounties were paid for their carcasses; and their eggs were collected. As a result, their numbers dwindled, and their distribution shrank. By 1917, the red kite was extinct in England and Scotland, although a small population remained in Wales.

Kites were legally protected in Britain in 1880, but protection did not lead to recovery. A century later, British kites still were confined to Wales, and the productivity of the Welsh kite population remained low. The average number of young fledged per kite pair was 0.5; this is less than half the productivity of red kite nests on the continent. There is indirect evidence that the area they inhabited in Wales, which was on the western edge of the species' former geographic range, provided poor-quality habitat, and that this was the cause of the Welsh kites' low breeding success. Productive agricultural land provided abundant prey for raptors elsewhere, but past mortality from persecution was probably high in those lands. As a result, it seems likely that the areas where kites remained in the 1980s were places where prey is scarce but the level of persecution was low (Davis and Newton 1981).

In 1989 the Joint Nature Conservation Committee and the Royal Society for the Protection of Birds initiated a program to reintroduce red kites to parts of their former range in the British Isles. The Welsh population was too small and its productivity too low for it to donate nestlings for reintroduction elsewhere. (A few nestlings were raised from eggs removed from Welsh nests that were considered vulnerable to egg collectors, however. These were replaced by dummy eggs, some of which were in fact later stolen!) It has been suggested that major genetic differences between

Figure 9.4. The red kite. (Drawn by M. Rockwood.)

the kite populations of Great Britain and continental Europe are unlikely, because they were once part of a large, interconnected continental population. For this reason, the continent was considered a suitable source of young birds. Initially, nestlings were taken from kite populations in Sweden and Spain, because those populations had high enough productivity to compensate for losses. In addition, only nests with large broods were used as donors, so that two nestlings were usually left in each brood. Young were collected from widely scattered donor nests, to maximize the genetic variability of the transplants.

Nestlings were fed initially on finely minced meat and bone, and contact with people was minimized. When the young were 10–12 weeks old, they were fitted with wing tags and radio transmitters and released. Food, in the form of fish, rabbits, or other mammals was provided for several weeks at the release location. Between 1989 and 1994, 93 kites were released in northern Scotland and another 93 in southern England. In 1995 and 1996,

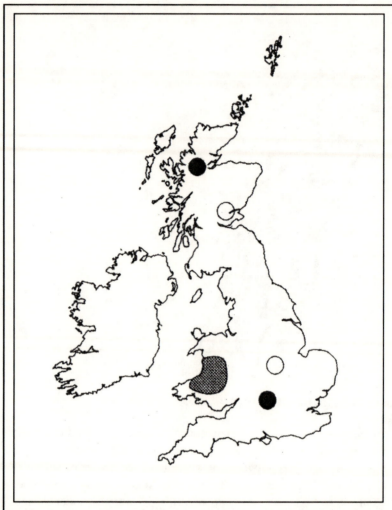

Figure 9.5. Approximate locations of populations of red kites in the British Isles. (Information on the exact locations of release sites is kept confidential.) Cross-hatched area indicates range of the breeding population in Wales, filled circles denote initial release sites, and open circles indicate most recent release areas. (After Evans *et al.* 1997. Reprinted with permission from *British Birds*.)

birds were released at two additional sites, with the goal of eventually linking breeding populations throughout the British Isles (Figure 9.5).

Survival was high in the new populations (76% of the birds released in England survived at least one year). Furthermore, in 1991, some of the released birds engaged in courtship and other breeding behaviors, and two pairs laid eggs. Both these nesting attempts failed, but translocated birds bred successfully in both England and Scotland in 1992. Between 1992 and 1995, 162 young were fledged, and the newly established populations in England and Scotland appeared to be as productive as the donor populations on the continent (Evans *et al.* 1994, 1997).

The fate of the red kite underscores how important it is to address all of the factors contributing to a species' decline. Programs that overlook some limiting factors stand to waste a lot of time and money, because the situation is not likely to improve until each part of the problem is resolved (see Box 9.1). Kite persecution was a major factor contributing to the species' decline, yet protection did little to improve the situation because other factors (perhaps food supply) limited the Welsh population. It was not until kites were reintroduced into other portions of their former range, where food was apparently abundant, that the situation improved. The red kite program appears to have succeeded in increasing both the size and the range of kite populations in Britain, whereas nearly a century of single-factor conservation failed to do so.

Care should be taken in selecting stock for transplants. On the one hand, taking individuals from the wild for augmentation or reintroduction poses a risk to the donor population if it is small and localized. On the other hand, if the donor population is large and widespread, it may be genetically differentiated from a rare or isolated population that is being replaced or augmented. Transplants may fail or do more harm than good if the transplants are not adapted to their new environment. (Prairie-chickens that were introduced to augment the declining heath hen population on Martha's Vineyard in 1902 probably died out because they were not adapted to the local environment.) If transplants do survive and interbreed, this may dilute the gene pool of the declining population and cause an irretrievable loss of critical adaptations to local conditions. Alleles for adaptations to the local environment may be swamped by introduced alleles that evolved in response to a different suite of selective pressures. As a result, the augmented population, though larger, may be less well adapted to its environment than the original population prior to augmentation.

To avoid some of these problems, biologists can use genetic analyses to determine which population is most closely related to the declining population. This was done by scientists who were considering augmentation of a population of about 10 European brown bears remaining in the Pyrenees of southwestern France. The nearest relatives of the Pyrenees bears occur in isolated groups in France, Spain, and Italy, but these, too, are becoming rare. For this reason, the French Ministry of the Environment contemplated introducing bears from larger but more distant populations in Russia and Romania. Analyses of DNA from the various groups of bears indicated that the eastern European bears are not as closely related to the Pyrenees population as bears from less distant localities, however. On the basis of this information, officials rejected the plan to release Russian or Romanian bears in the Pyrenees (Dorozynski 1994).

9.3.3   Guidelines

The following precautions might help to minimize the practical problems and undesirable side-effects of programs that breed and redistribute organisms (Brambell 1977; Evans *et al.* 1997).

Captive breeding
1. Conditions in the breeding facility should replicate those of the species' natural habitat as closely as possible.
2. Care should be taken to prevent individuals from becoming dependent upon people or upon the environment in the breeding facility.
3. Pedigree information should be used to arrange matings that minimize inbreeding and preserve genetic variation.
4. Care should be taken to avoid having individuals imprint on people or become conditioned to the presence of people.

Reintroduction or augmentation
1. There should be good evidence of former natural occurrence, if the taxon is currently not present where captive-bred animals are to be released.
2. The reasons for a taxon's decline should be clearly understood, and the factors causing the decline should have been rectified.
3. There should be suitable habitat available that meets a taxon's requirements; together with item (2), this means that the habitat should be capable of supporting more individuals than it currently supports.
4. The donor population should be closely related genetically to the original native population, and there should not have been admixtures of genes from other taxa.
5. The loss of individuals from the donor population should not place it at risk.

9.4   Setting priorities: Which species should we try to save?

Saving dwindling species from extinction takes enormous amounts of energy and funding. Clearly, it is not practical to assist all species. This necessitates choosing which species will receive aid, in other words, setting priorities. The species we target for protection reflect a host of biases that are not necessarily related to their biological importance. Protection efforts tend to focus on large vertebrates, especially birds and mammals. Groups of organisms that are poorly known or lacking in charisma get neglected.

It has been suggested that in some cases the protection of a single species serves to protect a great many other organisms as well. If such focal species can be identified, then protecting them should serve as a useful shortcut, ensuring that maximum benefits result from species protection programs. There are several ways in which this might work.

Indicator species are species that reflect the fate of other attributes of an ecosystem (Spellerberg 1995). A species should not be used as an indicator of the status of other species unless there is clear evidence that populations of the indicator and the species it serves as an indicator for rise and fall in tandem. Species that are monitored for political or social reasons – for instance, because they are designated as threatened or because they have charismatic appeal – should not automatically be assumed to indicate trends in populations of other species (Landres *et al.* 1988).

Umbrella species are those that require large expanses of habitat, usually because they themselves are large and have vast home ranges. In protecting the habitat of such species, managers hope to protect habitat for many smaller, less well-known species as well. This is especially likely to be the case if the range of the umbrella species encompasses many different types of habitat.

A species that serves as a symbol for conservation, usually because of its charismatic appeal, is termed a flagship species. Large animals, especially mammals, often serve as flagship species. The giant panda, northern spotted owl, koala, and Florida panther are examples of flagship species.

Species that play a pivotal role in ecosystem functioning, out of proportion to their abundance or biomass, are said to be keystone species (Power *et al.* 1996). The beaver, for instance, modifies its environment profoundly by building dams. As a result, many other organisms depend upon the presence of beavers. Protecting keystone species makes sense, because if a keystone species disappears, other species are likely to go too.

A species can play more than one of these roles. The spotted owl is considered an indicator of old-growth forest. It also is an umbrella species because it requires large areas of forest habitat, and it is definitely a flagship for conserving ancient forests. Similarly, the giant panda is an umbrella species and a flagship species. Care should be exercised in designing management strategies around indicator, flagship, umbrella, or keystone species, however. If the species in these categories really do what we hope they do, then protecting them will maximize the gains from species protection efforts. It is important to remember, however, that hypotheses about the relationships between species in these categories and other species have seldom been tested. Managers should not just assume that protecting a species in one of these categories protects other species (Simberloff 1998).

In addition to protecting species, preservationist managers seek to preserve habitats. This is both an end in itself and a means of protecting species that depend upon those habitats. The next chapter looks at ways of doing this.

References

Bevill, R. L., S. M. Louda, and L. M. Stanforth (1999). Protection from natural enemies in managing rare plant species. *Conservation Biology* **13**:1323–1331.

Bouman, J. (1977). The future of Przewalski horses. *International Zoo Yearbook* **50**:62–68.

Brambell, M. R. (1977). Reintroduction. *International Zoo Yearbook* **17**:112–116

Butler, D. and D. Merton (1992). *The Black Robin: Saving the World's Most Endangered Bird*. Auckland: Oxford University Press.

Cohn, J. P. (1988). Captive breeding for conservation. *BioScience* **38**:312–316.

Conway, W. (1980). An overview of captive propagation. In *Conservation Biology: An Evolutionary Ecological Perspective*, ed. M. E. Soulé and B. A. Wilcox, pp. 199–208. Sunderland, MA: Sinauer Associates.

Conway, W. (1988). Editorial. *Conservation Biology* **2**:132–134.

Davis, P. E. and I. Newton (1981). Population and breeding of red kites in Wales over a 30-year period. *Journal of Animal Ecology* **50**:759–772.

Dorozynski, A. (1994). A family tree of European bears. *Science* **263**:175.

Evans, I. M., J. A. Love, C. A. Galbraith, and M. W. Pienkowski (1994). Propagation and range restoration of threatened raptors in the United Kingdom. In *Raptor Conservation Today, Proceedings of the 4th World Conference on Birds of Prey and Owls*, ed. B.-U. Meyberg and R. D. Chancellor, pp. 447–457. Berlin: Pica Press.

Evans, I. M., R. H. Dennis, D. C. Orr-Ewing, N. Kjellén, P.-O. Andersson, M. Sylvén, A. Senosiain, and F. C. Carbo (1997). The re-establishment of red kite breeding populations in Scotland and England. *British Birds* **90**:123–138.

Flesness, N. R. (1977). Gene pool conservation and computer analysis. *International Zoo Yearbook* **17**:77–81.

Frankel, O. H. and M. E. Soulé (1981). *Conservation and Evolution*. Cambridge: Cambridge University Press.

Hunter, M. L., Jr. (1996). *Fundamentals of Conservation Biology*. Cambridge, MA: Blackwell Science.

Kelly, S. T. and M. E. DeCapita (1982). Cowbird control and its effect on Kirtland's warbler reproductive success. *Wilson Bulletin* **94**:363–365.

Landres, P. B., J. Verner, and J. W. Thomas (1988). Ecological use of vertebrate indicator species: a critique. *Conservation Biology* **2**:316–328.

Nowak, R. M. and J. L. Paradiso (1983). *Walker's Mammals of the World*, vol. 2. Baltimore, MD: Johns Hopkins University Press.

Pinder, N. J. and J. P. Barkham (1978). An assessment of the contribution of captive breeding to the conservation of rare mammals. *Biological Conservation* **13**:187–245.

Power, M. E., D. Tilman, J. A. Estes, B. A. Menge, W. J. Bond, L. S. Mills, G. Daily, J. C. Castilla, J. Lubchencko, and R. T. Paine (1996). Challenges in the quest for keystones. *BioScience* **46**:609–620.

Ryder, O. A. (1993). Przewalski's horse: prospects for reintroduction into the wild. *Conservation Biology* **7**:13–15.

Simberloff, D. (1998). Flagships, umbrellas, and keystones: is single-species management passé in the landscape era? *Biological Conservation* **83**:247–257.

Spellerberg, I. F. (1995). *Monitoring Ecological Change*. Cambridge: Cambridge University Press.

Walkinshaw, L. H. (1983). *Kirtland's Warbler: The Natural History of an Endangered Species*. Bloomfield Hills, MI: Cranbrook Institute of Science.

Zimmerman, D. R. (1975). *To Save a Bird in Peril*. New York, NY: Coward, McCann, and Geoghegan.

10

Techniques – protecting and restoring ecosystems

The preceding chapter explored techniques used by preservationist managers to protect organisms facing extinction. Preservationist management also seeks to protect entire communities, for two reasons. First, these assemblages of interacting organisms are an important facet of biodiversity in their own right. Second, habitat protection potentially offers a means of simultaneously helping large numbers of species more efficiently and effectively than single-species conservation efforts. Because of the enormous number of species on earth, many of which have not even been identified, species-based recovery programs cannot, by themselves, save all species. Such programs usually require considerable amounts of effort and funding, and even then success is not guaranteed. Once a population goes into a downward spiral, it is not always possible to save it (for the reasons discussed in Chapter 8).

This chapter describes approaches to protecting and restoring ecosystems. It reviews the history of preserves set aside to protect their natural features, procedures for designing nature reserves to maximize their conservation value, guidelines for identifying sites worthy of protection, and approaches to restoring degraded ecosystems.

10.1   Historical background

The terms "preserve" or "reserve" are used quite loosely. Sometimes government owned lands are considered "protected" areas, even though substantial amounts of resource extraction are allowed on public lands. In the U.S.A. the reasoning is that the regulations on resource use are more stringent on lands owned by the government than on private lands, and that federal land management agencies are required by law to consider the needs of wildlife in general and threatened and endangered species in particular. (See Chapter 5 for information on managing for multiple uses and Chapter 8 for a discussion of endangered species legislation.)

This chapter considers lands set aside specifically to protect their natural features. At the federal level, this means national parks, wilderness areas, and wild and scenic rivers. Here too, however, the terms reserve and preserve do not necessarily imply that use of these lands is prohibited. Under some circumstances, resource extraction (especially fishing) is allowed even in these lands. Further, in most cases, they are managed for nonconsumptive uses, so recreational, scientific, and educational activities, which are not without impacts (see Chapter 7), are permitted in these preserves. Because they are set aside primarily for their natural features, this type of preserve has a cultural significance that is quite different from that of national forests or other lands set aside for utilitarian purposes. Our attitudes toward national parks and wilderness areas are very much tied up with our feelings about human nature and the nonhuman world (see Chapter 11).

Preserves can be designated by any government jurisdiction, including municipal, county, state or provincial, and federal governments, by individuals or private organizations, or by international agreements. Yellowstone National Park, established by an act of the U.S. Congress in 1872, was the world's first national park. (Although Yosemite was set aside in 1864, it did not become a national park for several decades, so Yellowstone has the distinction of being first.) Australia, New Zealand, South Africa, Germany, Switzerland, Sweden, and Russia quickly followed suit, establishing national parks prior to World War I (Allin 1990).

Yellowstone was set aside to preserve scenic and geologic wonders reported by an earlier exploratory expedition. It and America's other early parks encompassed scenic sites at high elevations. In general, these sites were regarded as economically worthless, so as far as Congress was concerned, setting them aside did not entail any economic sacrifice. Senator John Conness advocated setting aside Yosemite, because it was "for all public purposes worthless. . . . It is a matter involving no appropriation whatever" (quoted in Runte 1997:48–49). (Recall from the Introduction, however, that within a few decades Congress would change its mind about the value of the park's resources.)

Beyond a general sense that Yellowstone's lands should be held in trust for the American people and their children, the U.S. Congress had no clear idea of how to manage a national park. A small budget was appropriated, and a superintendent was appointed, but poaching was rampant, and exploitation of the park's resources continued unabated. One administrator even removed and sold large chunks of the mineral formations the park was supposed to preserve! In 1886 the army was sent in to protect the park from market hunters and vandals, but it lacked the authority to press charges.

In spite of these problems, it was not until 1916 that Congress created the National Park Service to administer Yellowstone and other national parks, monuments, recreation areas, historic sites, and seashores. The Park Service was instructed to manage national parks in such a way as to "leave them unimpaired for the enjoyment of future generations." There has always been a tension between the two parts of the Park Service's mission encapsulated in this phrase. If parks are maintained for the enjoyment of people they are not going to remain unimpaired. Even nonconsumptive recreation has substantial impacts (Wright 1992).

In 1964 the Wilderness Act provided for the designation of wilderness areas, a category of public land that was to remain more pristine than national parks, forests, or wildlife refuges. The act stated that "a wilderness, in contrast with those areas where man and his own works dominate the landscape, is hereby recognized as an area where the earth and its community of life are untrammeled by man, where man himself is a visitor who does not remain." Wildernesses are large areas (usually at least 2000 ha) within which buildings, roads, motor vehicles, and commercial activities are prohibited. In 1968, Congress passed the Wild and Scenic Rivers Act in response to concerns that the natural values of the nation's rivers were being degraded. The act stipulated that rivers with outstanding scenic, cultural, geological, historical, biological, or recreational values "shall be protected for the benefit and enjoyment of present and future generations."

Private conservation organizations also set aside and administer natural areas. In the U.S.A., the Audubon Society, the Nature Conservancy, and Ducks Unlimited are examples of private conservation organizations that designate and maintain preserves.

In 1971 an International Convention on Wetlands of International Importance was negotiated in Iran. The purpose of the treaty was to slow the destruction of wetlands, especially those of importance to migratory waterfowl. Member nations are required to designate at least one wetland of international significance that will be managed to maintain its ecological characteristics. Over 80 nations have signed this treaty, known as the Ramsar Convention. (Ramsar is not an acronym; it is the name of the city where this convention was negotiated.)

Most wilderness areas and national parks were set aside for their scenery or because they were not considered suitable for development. Thus, it is not surprising that the borders of these areas do not coincide with biologically meaningful boundaries. Recently, however, conservation biologists have attempted to address this problem by applying insights about the processes of colonization and extinction to the practical problem of designing nature reserves. Their goal is to maximize the probability that reserves will maintain biodiversity.

10.2   Protecting communities

10.2.1   Designing reserves

Theoretical considerations
When faced with decisions about setting land or waters aside, how much is enough? Is a single large reserve better than two or more small ones? What shape should reserves have? How should they be arranged with respect to one another? The answers to the above questions depend on the purposes of the reserve. If, for example, the goal is to create a wildlife refuge for the production of a maximum number of waterfowl, a manager will not make the same choices he or she would make if the goal is the long-term protection of biological diversity. In this chapter, we will be concerned with reserve design where the objective is the perpetuation of a variety of plant and animal populations, that is, the conservation of species richness.

Preservationist managers seek to minimize the chances of species going extinct within reserves. In general, when managing for rare or declining species, it is desirable to minimize sources of mortality (although overpopulation should also be avoided). Under some circumstances, a carefully designed reserve can minimize losses from predators or diseases. Even within large reserves, however, some mortality will of course occur. Therefore, in designing a reserve, it may be prudent to maximize the chances that desirable organisms from outside the reserve will arrive and settle. Arranging reserves so that organisms can move between them can tip the scales in favor of immigration from other nearby sources. On the other hand, there are also some situations where it may be desirable to isolate populations in reserves.

Many species are organized into fairly independent local populations that are connected by occasional movements of individuals between them. These population networks are termed metapopulations. Movements between populations in a metapopulation can play a significant role in the regional dynamics of a species for two reasons. First, if some populations dwindle to very low numbers, from which they are unlikely to recover on their own, they may be "rescued" by immigrants from a nearby population (Brown and Kodric-Brown 1977). Second, if some of these populations do disappear entirely, their habitats may be recolonized by individuals dispersing from other parts of the metapopulation. (The only difference between these two phenomena is that in the first case the dispersers arrive before all members of the original population are gone, whereas in the second instance they do not get there until local extinction has occurred. Of course, sometimes it is difficult for the field biologist to distinguish between these two situations.)

Where habitat quality is variable, reproduction may exceed mortality in some areas but not in others. Productive habitats can serve as sources of individuals that disperse into less productive habitats, which are termed sinks (Pulliam 1988). But in highly developed landscapes, it has become harder for organisms to move between populations within a metapopulation.

The protection of habitat for migratory or nomadic species poses special challenges. Appropriate habitat must be available at essential points throughout the annual cycle, including breeding grounds, overwintering sites, and stopover points in between.

Specific recommendations

On the basis of the theory of island biogeography (see Chapter 8), Diamond (1975) made the following recommendations for the design of nature reserves:

1. Reserves should be as large as possible. This conclusion is based on empirical data showing that larger areas support more species than small areas. This is referred to as the species–area relationship (see Figure 10.1A).
2. All other things being equal, a single large reserve will conserve more species than several small reserves with the same total area. This is controversial. The choice between a large preserve or several small ones is often referred to by the acronym SLOSS, Single Large Or Several Small (see Figure 10.1B).
3. Reserves should be close together. This should maximize chances for movements between reserves (see Figure 10.1C).
4. Reserves should be clustered rather than arranged in a linear fashion. Again, this is to facilitate inter-reserve movements (see Figure 10.1D).
5. Reserves should be connected by corridors. The purpose of this is to make it easier for organisms to move between reserves (see Figure 10.1E). This and rules 3 and 4 are also controversial.
6. Reserves should have a low ratio of edge to interior habitat. This assumes that the goal is to provide habitat for organisms that are sensitive to problems such as predation or brood parasitism, which are greatest at habitat

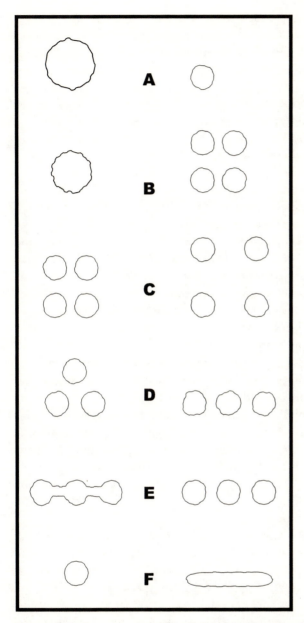

Figure 10.1. Recommendations for the design of nature reserves. Diamond suggested that in each of the paired designs below, the configuration on the left would be likely to have a lower extinction rate than the configuration on the right. (Adapted from *Biological Conservation*, 7, Diamond, J.M., The island dilemma: lessons of modern biogeographic studies for the design of nature reserves, pp. 129–153, 1975, with permission from Elsevier Science.)

edges (Chapter 7). Of course, for species that thrive at the junctions between different habitats, the opposite principle will apply. Most edge species are quite tolerant of disturbance, however, and therefore they do not depend on reserves (see Figures 5.5 and 10.1A and F).

Diamond's second principle, that a single large preserve will provide for more species than several small preserves of equivalent area, has generated considerable controversy. The argument that a single large preserve conserves more species than several small ones of equivalent area is based on inferences from the biotas of land-bridge islands (LBIs), islands that are located near a large land mass (Diamond 1972, 1976; Terborgh 1974, 1976). During the Pleistocene, when sea level was as much as 100 m lower than it is today because massive amounts of water were locked up in glaciers, these areas were connected by "bridges" of land to nearby continents. For instance, until about 10000 years ago a number of islands that now surround New Guinea were connected to it. Before the LBIs became islands, they probably supported the same kinds and numbers of organisms as similar habitats on other parts of the mainland. But when the glaciers melted, they became isolated by rising sea level. At that point extinction rates might have increased and colonization rates decreased, resulting in a decline in the species richness of the islands. New Guinea has 134 species of lowland birds with large area requirements and poor colonizing abilities. Diamond assumed that all of these were present on LBIs before they were isolated and concluded from present distributions that the smaller islands had lost species more rapidly than the larger ones. (The resulting decline in species richness, which occurs when populations become extinct because an island is too small to support viable populations, is termed relaxation.) This scenario, he argued, is what we should expect to happen when development divides large expanses of contiguous habitats into smaller fragments; therefore, to preserve as many species as possible, reserves should be made as large as possible.

Barro Colorado, a 15.7-km^2 island that was created by rising water levels when the Panama Canal was built, provides a more recent example of what happens when a continuous expanse of terrestrial habitat is converted to an island. On this modern-day LBI, a number of bird species apparently became extinct after the hilltop was surrounded by water, so that fewer species now occur in the island's forests than in comparable-sized forest on the nearby mainland (Willis 1974; Wilson and Willis 1975).

If small LBIs lose species more rapidly than large ones, and if the creation of a nature reserve is analogous to the isolation of an LBI, then large reserves would be expected to retain their species longer than small ones. Except for

Barro Colorado, however, the pre-isolation species composition of LBIs (like the evidence for species turnover on the Channel Islands, discussed in Chapter 8) was arrived at on the basis of inference, rather than observation (Simberloff and Abele 1976b). Furthermore, it is very likely that habitats on LBIs have changed over the past 10 000 years (if for no other reason than because of the climate change that caused the sea level to rise and the islands to be created in the first place). On Barro Colorado, where the composition of the original bird fauna is documented, major habitat changes are known to have occurred in the past century. Thus, as in the case of the increasing Kaibab deer herd (Chapter 2), more than one variable changed during the period in question, making it impossible to determine the effects of area on extinction rate.

Regardless of the effect of area on extinction rate, large reserves have a number of potential advantages. Because they have relatively little edge in relation to total area, large reserves contain habitats that are buffered from the negative influences of predators, competitors, or parasites in the surrounding landscape. Large reserves also may meet the area requirements of large carnivores and other wide-ranging species, provide more types of habitats and therefore allow long-term adjustments to environmental change, and contain entire metapopulations, including source populations that can recolonize sinks.

But there are situations in which a single large reserve might not be the best conservation option. If a species has very specific habitat requirements that are found in only a few small, widely spaced patches, that species will derive more benefit from protection of those small, isolated patches than from a reserve that contains vast expanses of unsuitable habitat. Furthermore, when a disease or a catastrophe such as a fire or a flood occurs, it might be better to have several populations in separate reserves, to avoid putting all our eggs in one basket. In a large preserve, an entire population or group of subpopulations can be wiped out from such an event.

For these reasons, critics of the idea that a single large reserve always preserves more species than several smaller reserves of equivalent area argue that the design of nature reserves should be based on detailed knowledge of the ecologies of the species in question, rather than on general theoretical principles (Simberloff and Abele 1976a,b, 1982; Abele and Connor 1979; Gilbert 1980; Williams 1984; Soberón 1992).

Because many organisms are unable to move through the habitats surrounding reserves, conservation biologists have focused a great deal of attention on corridors, connecting strips of habitat that bridge the gaps between reserves (Harris 1984; Noss 1987; Harris and Scheck 1991). (The term "corridor" has actually been used to mean several different things, ranging from linear strips of habitat that physically connect reserves, to "stepping-stone" chains of discrete patches along a migratory bird route, to underpasses and tunnels beneath highways.) It has been suggested that strips of habitat which connect reserves may fulfill several functions, including (1) increasing colonization rates between habitats, thereby increasing the number of species that connected patches can support, (2) increasing the size of patch popula-tions, thereby decreasing the risks to small populations from fluctuations in birth and death rates, (3) minimizing inbreeding, (4) providing habitat within which wide-ranging animals can move, and (5) allowing processes such as fire to move across a landscape. Habitat strips connecting reserves should not be assumed to fulfill these suggested functions, however, unless it can be shown that organisms actually move through them and will not move through the matrix of habitats separating the reserves (Simberloff and Cox 1987; Simberloff *et al.* 1992). Furthermore, corridors can potentially facilitate the spread of unwanted organisms such as exotic species or pathogens (Hess 1994).

Landscape considerations

Principles 3, 4, and 5 relate to the distribution of reserves across a landscape. Although it is convenient to treat an area that supports a particular type of vegetation as if it were a discrete unit, it is important to remember that the earth is not really a collection of separate habitats. We may choose to study a patch of old-growth forest as if it were a separate entity, but in reality it influ-ences and is influenced by the contrasting habitats that border it.

Matter and energy are exchanged between habitats in a variety of ways. Water evaporates from lakes, rivers, and seas to form clouds which drop their moisture some distance away, where it may percolate through the soil and become groundwater or flow across the surface and enter streams. Surface water flows from higher elevations to lower elevations, downcutting stream channels and transporting rocks, silt, minerals, organic matter, and pollutants in the process. Glaciers scour rock surfaces and push along debris as they advance. Soil and rocks move downslope, gradually or in sudden catastrophes. Wind and water erode soils. Climate is influenced by ocean currents and the movements of air masses. Furthermore, currents of air and water move pol-lutants to destinations far removed from their sources. These physical pro-cesses do not stop at ecosystem boundaries (or, for that matter, at political or administrative boundaries).

Similarly, when an aquatic habitat is adjacent to a terrestrial habitat, the two interact in myriad ways. Organisms that inhabit streams depend upon litter that falls or is washed in from the surrounding uplands. Logs that fall into a stream create special microhabitats (Harmon *et al.* 1986). They form small dams that trap debris and allow it to decompose, thereby adding additional nutrients to the stream. The pools that form behind these log dams provide spawning habitat for fishes. Trees along a stream's edge provide shade and moderate water temperature. These are just a few of the ways that different habitats interact across a landscape by exchanging matter and energy.

Like matter and energy, organisms also cross the boundaries between habitats. These movements may involve permanent relocations, occasional excursions, or regular movements. In many species of plants and animals, young individuals disperse, that is, they permanently move away from their parents' home range. This may or may not involve movements to or through different ecosystems. In addition, some animals feed in one type of habitat and rest in another on a daily basis. On a seasonal basis, some undertake predictable round-trip movements, termed migrations, or less predictable nomadic movements. (Recall the passenger pigeon; Chapter 1.) Long-distance migrants are capable of transporting nutrients across hundreds or even thousands of kilometers. For instance, adult salmon ingest nutrients when they feed in the ocean. When they migrate upstream to spawn, terrestrial predators eat them and transport the ocean-derived nutrients into the adjacent forests. In this way inland forests are fertilized with nutrients from marine environments hundreds of kilometers away (Ben-David *et al.* 1998; Hilderbrand *et al.* 1999).

At times, enormous numbers of migrants concentrate at migratory stopover points, termed staging areas. For example, millions of shorebirds funnel into the shores of Chesapeake Bay, on the east coast of North America, during their spring migrations. Another example of a migratory staging area is along the Platte River in Nebraska, where hundreds of thousands of sandhill cranes stop during their spring migration. If the habitats at these staging areas become degraded, the consequences for dependent species reverberate thousands of miles away.

The study of the interactions between habitat patches is termed landscape ecology (Naveh and Lieberman 1984; Forman and Godron 1986). The management of species that move between ecosystems is not a new issue. In fact the Migratory Bird Treaty was one of the first major conservation efforts in North America. But the issue has asserted itself with new urgency as biologists have become increasingly concerned about the effects of habitat fragmentation and isolation.

Edge is one feature of landscape configuration that has received a lot of attention, but landscape pattern also affects community dynamics in less obvious ways. Sites with frequent disturbances provide reservoirs for species typical of early successional stages. For example, riparian zones along rivers, streams, and creeks experience frequent flooding, scouring, and deposition of sediment. These processes remove or damage existing vegetation and create seedbeds for new vegetation; consequently, colonizing species tend to inhabit streams and riverbanks. When riparian zones occur within a matrix of late-successional habitat, which experiences disturbances less frequently, the riparian ecosystem provides reservoirs where these pioneer species are maintained. Then when disturbances occur within a mature forest, sources of colonists are likely to be present in nearby riparian zones; from here these early-successional species can move out to colonize the new forest openings (Agee 1988).

Similarly, when a disturbance occurs in a reserve, we would expect surrounding habitats to provide a source of immigrants to colonize the site. Daniel Janzen (1983) tested this hypothesis by recording vegetation that became established in an opening created when a tree was blown down in Costa Rica's Santa Rosa National Park. Although the forest was considered "pristine," the plants that became established were those of adjacent anthropogenic habitats, such as roads, pastures, fencerows, and fields. Thus, the consequences of a natural disturbance in the forest reserve were profoundly affected by the habitat surrounding the reserve.

Long-term considerations: Coping with environmental changes

Can reserves be designed to increase the chances that their biota will persist for 500 or 1000 years? In the past, if climate changed, organisms could gradually shift their geographic ranges unless they were prevented from doing so by a natural barrier. If the climate became colder, a population of rabbits might evolve adaptations to cope with the cold; but if the population's gene pool lacked the necessary mutations (e.g., for a thicker coat), the animals would be forced to move to lower elevations or farther south, where they would find more suitable conditions. If the population were unable to adapt *and* were prevented from shifting its range, it would eventually go extinct.

Different organisms are adapted to different stages in succession. Large reserves that include vegetation in many different stages of succession are likely to have high species richness. For this reason, Pickett and Thompson (1978) suggested that reserves should be large enough to encompass a minimum dynamic area, the smallest area that will allow natural disturbances to run their course.

Organisms confined to reserves cannot adjust their geographic ranges in response to changing environments if the matrix surrounding the reserves is inhospitable. They must evolve adaptations to new conditions or perish. In these cases, the maintenance of genetic variation is especially important (see Chapter

8). If we wish to maximize their capacity to respond to environmental changes, the protection of several populations will preserve more than one gene pool and will increase the chance that at least one population will, in the long run, contain the genetic raw material necessary to succeed in an altered environment. On the other hand, if we wish to maximize genetic diversity, large populations are likely to have more genetic diversity than small ones. In addition, instead of conserving large expanses of fairly homogeneous habitat, it might be better to protect areas with pronounced environmental gradients, so that organisms can shift their distributions in response to long-term environmental changes without having to move through developed habitats (Hunter *et al.* 1988).

The distributions of organisms shift as environmental conditions change (Risser 1995). This has happened many times in the past when the earth's climate was modified. Climates changed as continents drifted to different latitudes, the tilt of the earth's axis shifted, mountain ranges arose or were eroded, land masses rose from beneath the seas or became submerged, glaciers formed and melted, and sea level rose and fell.

The Great Plains of the U.S.A. are characterized by an east–west climatic gradient. The western plains are relatively dry, and moisture increases to the east. At the eastern margin of the Great Plains, steppe borders deciduous forest. In this transition zone, conditions are marginal for trees. Any microhabitats within the steppe where there is slightly more moisture available may be colonized by invading trees. For instance, trees extend westward into the steppe along watercourses, forming forested projections known as gallery forests. The transition between forest and steppe is dynamic; the ecotone continually shifts in response to changing conditions. Dry conditions or fire inhibit tree regeneration and favor eastward expansion of the steppe. In moist, fire-free years the reverse is true; steppe contracts and forest expands to the west (Daubenmire 1978).

Although the earth's climate has never been static, anthropogenic climate changes may have more serious implications for biodiversity, for two reasons. First, although climates have changed many times in the past, our activities are irreversibly speeding up the pace of such changes. Second, global climate change will occur in a landscape fragmented by agriculture and development. In this landscape, many organisms will have difficulty in shifting their geographic ranges northward or upslope to suitable environments (Peters and Darling 1985; Murphy and Weiss 1992). Organisms that are adapted to a wide range of environmental conditions or with good dispersal abilities will be the least affected by climate change. On the other hand, poor dispersers with narrow tolerance ranges, like the small mammals on mountaintops discussed in Chapter 8, will be placed at risk. One implication of this for conservation is that managers need to think in terms of changes occurring on time scales of hundreds to thousands of years and spatial scales of thousands of kilometers.

Practical considerations

It should be clear from the preceding discussion that preserve managers often have long and unrealistic, and sometime conflicting, wish lists. In practice, theoretical considerations do not usually prevail when preserves are created. Scientists do not just sketch a reserve plan on a blank slate and then carve it out of the landscape. Reserves are superimposed on existing land uses, ownership patterns, and traditions. So the design of nature reserves is never as abstract as this discussion might suggest. It must take into consideration political, economic, and social constraints, and often theoretical debates such as SLOSS become moot in the face of practical realities.

10.2.2   Setting priorities: Which habitats should we save?

General considerations

Clearly some choices must be made about which habitats should be protected in preserves. Often "naturalness" is the criterion for preserve status. But what do we mean by naturalness? In the New World, the goal is often to return to pre-Columbian conditions (Noss 1983). This is based upon the premise that because precontact Native American populations often existed at low densities, had limited technology, and were spiritually connected to the natural world, their impact on the nonhuman world was relatively minor. Although there were and are undoubtedly differences in the environmental impacts of Europeans and Native Americans, this distinction is also somewhat problematic, because it seems to imply that some people (Europeans and their descendants) are not part of the natural world, but others (pre-Columbian Native Americans) were (see Chapter 11 for more discussion of this point). To avoid the problems involved in using pre-Columbian conditions as a benchmark, or standard, for conservation, some conservation biologists advocate substituting "no human influence" as the criterion (Hunter 1996:696).

Filters

In practice, however, decisions about preserve designation usually revolve around identifying specific entities in need of protection. Sometimes preserves are set aside for particular species, but since it is impossible to target all species for conservation efforts, an alternative approach to biodiversity conservation is to try to protect all biological communities. If we have a good classification system that identifies all communities, and if our efforts to protect those communities are effective, then all the components of those communities, that is, all the species they contain, should be protected. This approach has been termed a coarse-filter approach. The filter is the classification system for identifying communities, usually on the basis of vegetation (Nature Conservancy 1982). The goal of a coarse-filter strategy is to protect representative examples of all communities. One advantage of this approach is that it may conserve poorly known species that would otherwise be overlooked. Many species, particularly microorganisms, which perform important ecological functions, are poorly known. They cannot possibly be protected by a species-by-species approach to conservation (Noss and Cooperrider 1994).

Of course, some species may not be addressed by a coarse filter. Therefore, a comprehensive strategy for preservationist management should complement a coarse filter with conservation measures focusing on the needs of specific organisms that are not addressed by the coarse filter. This strategy, which strives to aid species that fall through the cracks of coarse-filter management, is termed a fine filter.

Coarse filters based upon extant vegetation have their limitations, however. Data from research on the distribution of fossil plants suggests that even in relatively undisturbed habitats, the communities we see today are, geologically speaking, fairly young (less than 8000 years old). For this reason, some ecologists consider present-day plant communities to be transitory assemblages that are likely to shift in composition in the future, especially if climate change accelerates. Proponents of this view suggest that coarse filters should be based on characteristics of the physical environment, rather than on the relatively "ephemeral" groupings of plants that we see today (Hunter *et al.* 1988).

Gap analysis
While working in the field of endangered species conservation, ecologist J. Michael Scott realized two troubling things. First, huge amounts of money and effort are poured into programs to save a small number of highly endangered species. Scott reasoned that it would never be feasible to save all endangered species through such programs. It seemed to him that conservationists were trying to stop a massive hemorrhage of biodiversity, but they were focusing their efforts ineffectively – waiting until it was too late to do much good. Second, he noticed that protected lands often failed to include the habitats of species at risk. For example, when Scott mapped the distributions of endangered finches in Hawaii in 1982, he found that the ranges of most of the finches fell outside the ranges of nature preserves. Out of these insights, the

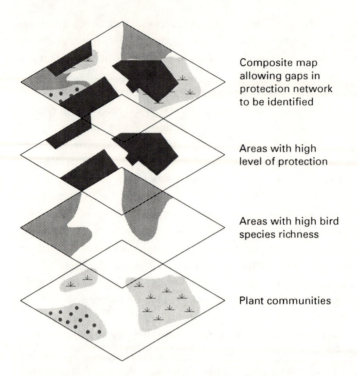

Composite map
allowing gaps in
protection network
to be identified

Areas with high
level of protection

Areas with high bird
species richness

Plant communities

Figure 10.2. Gap analysis. Maps showing the distributions of vegetation, taxa of interest, and land ownership are overlaid to identify gaps – areas where biodiversity concentrations are unprotected. In the top map, areas that are not indicated by black shading are areas that are not currently protected. Unprotected areas that have high species diversity or support important plant communities are logical targets for future protection efforts.

Gap Analysis Program was born. Gap analysis uses the technology of computerized geographic information systems to map the distributions of a variety of taxa in relation to land ownership. These maps can then be overlaid to identify "gaps" in biodiversity protection (Figure 10.2). Areas where biological diversity is concentrated are identified, and the degree to which these lands are protected is ascertained. These may be places where there are many different species, or concentrations of endemic species, or especially high numbers of rare or endangered organisms. Once the locations of such biological hotspots have been identified, their distributions can be compared to maps showing land ownership, and unprotected areas can be identified. Given the fact that most public lands in the United States were set aside because they contained economically valuable resources (national forests) or were of scenic or geological interest (national parks), rather than for their biodiversity, it is not surprising that government lands do not always overlap biodiversity hotspots. For this reason, gap analysis provides a tool for prioritizing and focusing habitat protection efforts, with the goal of reaping the greatest benefits from scarce conservation resources (Scott *et al.* 1993).

Gap analysis and similar geographic approaches to conservation planning provide useful information only if they are developed from sound assumptions and good data, however. Furthermore, gap analysis itself, though it is a technique for identifying priorities, must prioritize indicators of biodiversity. We cannot map the distributions of all organisms, so again we come up against the problem of identifying suitable indicator species or groups of species (Flather *et al.* 1997) (see Chapter 9). Can we assume that patterns in the diversity of butterflies or birds coincide with biodiversity patterns for other taxa? Some other ramifications of the question of how to select biodiversity indicators are discussed in Box 10.1.

Box 10.1   Gap analysis for the state of Washington

The varied environment of the state of Washington supports diverse communities and species. Running roughly along a north–south axis, the Cascade Range divides the state into a wet western portion and a drier eastern section. Most of the area to the west of the Cascades has the potential to support moist coniferous forest, whereas the potential vegetation of most of the area on the east side of the state is steppe or dry temperate forest (see Boxes 13.3 and 13.4).

Cassidy *et al.* (1997) used the techniques of gap analysis to identify priorities for biodiversity in Washington state. To do this, they classified the state's land into four categories. The highest level of protection was assumed to occur on lands maintained primarily in a natural state, such as national parks and wilderness areas. Lands that are maintained in a natural state but where limited extractive uses are permitted (such as national wildlife refuges, state wildlife areas, and national recreation areas) made up the second highest level of protection. Lands having some protection from development but subject to either locally intense resource extraction or broader scale, low-intensity use (such as Bureau of Land Management lands) made up the third category, and lands with little or no legislated protection (such as private lands) comprised the lowest protection category (Figure 10.3).

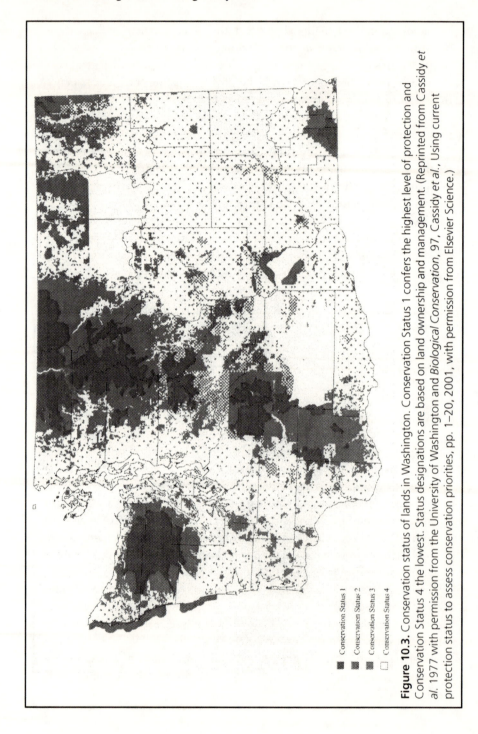

Figure 10.3. Conservation status of lands in Washington. Conservation Status 1 confers the highest level of protection and Conservation Status 4 the lowest. Status designations are based on land ownership and management. (Reprinted from Cassidy et al. 1977 with permission from the University of Washington and *Biological Conservation, 97,* Cassidy et al., Using current protection status to assess conservation priorities, pp. 1–20, 2001, with permission from Elsevier Science.)

Conservation Status 1
Conservation Status 2
Conservation Status 3
Conservation Status 4

The distribution of these levels of protection was then compared with the distributions of vertebrate species and of vegetation (Figure 10.4). The patterns that emerged highlighted lands where large numbers of at-risk species occur on lands with a low level of protection. The most glaring gaps in Washington's biodiversity protection network occur in the steppe zone, most of which has been converted to agriculture. In addition, low-elevation forests in western Washington are poorly represented in protected areas.

As this example illustrates, gap analysis identifies species or communities that are not well represented on protected lands. Thus, it is a proactive approach to conservation that seeks to allocate scarce resources efficiently by identifying high-priority sites that will protect as many species as possible. When Cassidy *et al.* (2001) evaluated the use of current protection status to identify conservation priorities, however, they concluded that gap analysis is not a magic bullet and can, in fact, underestimate the level of protection faced by some species or communities while overestimating the security of others. This problem stems from the fact that gap analysis does not explicitly incorporate historical factors. Focusing on a species' or community's current level of representation on protected lands gives us only a snapshot at a single point in time. Communities that are sensitive to human activities have probably disappeared or nearly disappeared from intensively used lands. Since they now remain only on protected lands, it will appear that they are well protected, but this could be misleading. For example, the marbled murrelet inhabits old-growth forests. Within Washington, it is estimated that about one-quarter of the marbled murrelet's distribution occurs on lands with a high level of protection. This is a relatively high proportion of protected habitat, yet this is a very rare species. To assume that this species was adequately protected because a substantial proportion of its habitat occurred on protected lands would probably be unwise.

On the other hand, communities and species that are well adapted to human activities will be abundant and widespread outside of protected areas and poorly represented within them. They give the impression of being poorly protected, because they are found mostly outside of preserves, yet perhaps they really should not be conservation priorities. In western Washington, the early-successional forests that develop after logging are a case in point. This type of forest is widespread, but not in need of special protection, because it is not likely to become rare in the foreseeable future.

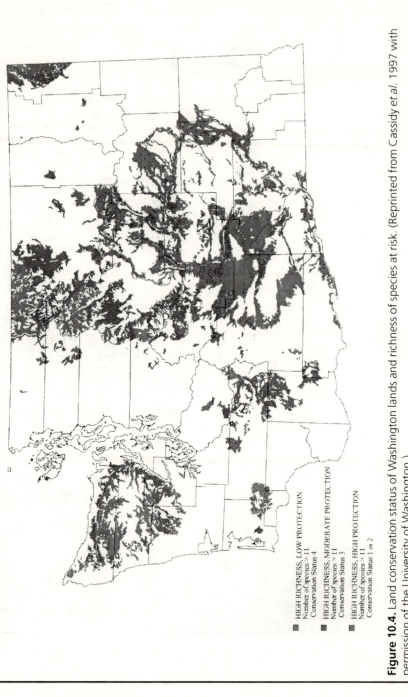

Figure 10.4. Land conservation status of Washington lands and richness of species at risk. (Reprinted from Cassidy *et al.* 1997 with permission of the University of Washington.)

To address this type of problem, care must be taken to interpret the results of gap analysis in their ecological and historical contexts. One way to do this is by including a component in gap analysis that evaluates the vulnerability of communities and species to future human activities (Stoms *et al.* 1998). For example, though a large proportion of Washington's old-growth and late-successional forests have a high degree of protection, these forests are not entirely secure because their economic value generates pressures to liquidate them (Cassidy *et al.* 2001). Prioritization based solely on current status might paint an overly optimistic scenario for this vegetation type, but incorporating an assessment of future risks could correct this.

10.2.3   Managing reserves

After reserves have been set aside, we must decide what to do with them. Reserves are intended as natural areas, but it is not always easy to define "natural." Once a park has been established, what next? Should managers simply build a fence and watch what happens? Is it appropriate to try to modify preserve environments? If so, how much and what kinds of management are called for? If left alone, the environment within a reserve will change. Natural disturbances and processes of succession will alter a park's habitat unless managers deliberately modify those processes (see Chapter 11). In any case, reserves are never truly natural systems, for at least four reasons. First, some elements of the original landscape are invariably missing. Reserve boundaries rarely encompass all the ecological elements and landscape features necessary to preserve an intact and fully functioning system. Even a large national park such as Yellowstone does not include all the requirements of its wildlife. Few large carnivores remain, and the elk and bison that summer in the park do not have enough winter forage to meet their needs within the park's boundaries. Should managers intervene to supply or compensate for these missing elements? Second, most reserves have been invaded by alien plants and animals that have altered ecosystem structure and function. Third, reserve biotas are impacted by regional and global atmospheric changes such as ozone thinning, climate change, radioactive fallout, and acid rain. Fourth, visitors themselves have impacts.

Very few reserves are managed solely for their natural features. In most cases, people make demands on them as well, ranging from recreational use to resource extraction. This means that managing reserves involves managing people: controlling foot traffic so it doesn't destroy sensitive vegetation, cause erosion, or disturb nesting animals; managing vehicular traffic to prevent congestion and air pollution; disposing of human wastes and garbage; and providing interpretive information to educate visitors. In addition, the public often resists attempts to let nature take its course in natural areas. People want fires to be put out, and they don't like animals to starve. They may want injured or diseased animals to be rescued, and they are likely to oppose efforts to control excess animals (Wright 1992). Finally, people usually want to harvest the resources within park boundaries. This problem is especially acute in parts of the developing world, where in many cases people have been evicted from their traditional homelands to create wildlife reserves, a problem we shall return to in the next chapter.

10.2.4   Providing economic incentives to set aside preserves: Debt-for-nature swaps

For economic reasons, people in developing nations are often reluctant to set aside land in preserves. When poor nations accumulate substantial foreign debts, the need to repay them leads them to even more intensive resource exploitation. Any arrangement that allows debtor nations to reduce their debt in exchange for conservation measures thus provides a powerful incentive for conservation. These programs are termed debt-for-nature swaps (Hansen 1989).

This innovative strategy involves agreements between international non-governmental conservation organizations (NGOs), banks, and governments of debt-ridden nations. Debt-for-nature swaps are possible because the probability of debtor nations fully repaying their debts is so low that international banks are sometimes willing to write off part of a nation's debt in exchange for receiving a much smaller amount of money. In a debt swap, an NGO and the government of a debtor country agree on a conservation project such as the protection and management of a reserve. The NGO then negotiates with a bank from the debtor nation to get the bank to agree to reduce the amount it will accept as repayment for the loan. In exchange for paying off the discounted debt, the developing nation implements the agreed upon conservation program.

The first debt-for-nature swap took place in 1987 between Bolivia and the NGO Conservation International. The latter paid $100000, in return for which Bolivia's debt was reduced by $650000, and the Bolivian government established a conservation fund for the Beni Biosphere Reserve. Debt swaps are one tool for ensuring that local people benefit economically from habitat conservation. (Other such tools are discussed in Chapter 14.)

10.3   Restoring communities

10.3.1   The need for ecological restoration

We noted in Chapter 9 that preservationist management is not just interested in preventing the loss of biodiversity; it also seeks to recover some of what has been lost: "At best, preservation can only hold on to what already exists. In a world of change we need more than that. Ultimately, we need a way not only of saving what we have but also of putting the pieces back together when something has been altered, damaged, or even destroyed" (Jordan 1988:311).

Although the protection of communities through the creation of reserves is a central feature of preservationist management, many communities have already been degraded to such an extent that there is not much good-quality habitat left to protect. This fact has given rise to the discipline of restoration ecology, in which scientific knowledge is applied with the goal of recreating ecosystems as they existed at some time in the past. The emphasis is on restoring species assemblages in order to maintain gene pools in areas where they previously flourished.

Ecological restoration is especially useful in situations where existing preserves are inadequate but are surrounded by degraded habitat that is capable of being restored. Successful restoration of neighboring lands can dramatically increase the effectiveness of reserves in such cases.

10.3.2   Methods of restoring ecosystems

Restoration is achieved through alteration of the physical and biotic environment of a degraded ecosystem. Manipulation of the physical environment can involve soil, water, and disturbances such as fire regimes or streamflow dynamics. The restoration ecologist uses many of the same techniques that utilitarian resource managers have used for decades (see Chapter 5). But unlike the utilitarian manager, who seeks to improve habitat for selected species of economic value, the restoration ecologist seeks to restore a functioning natural ecosystem.

Manipulation of the biotic environment for ecological restoration usually involves attempts to establish and promote the reproduction of native species. This can be done by seeding or by transplanting plants and animals to the site or by restoring processes such as seed dispersal and pollination, which depend on biotic interactions. Since most disturbed ecosystems have been invaded by exotic species, the removal or control of these is an essential component of many restoration programs.

Restoration ecology is both a science and an art. Much knowledge of what works has been achieved through trial and error. In the process, restoration ecology provides valuable knowledge about the structure and functioning of the restored ecosystems (see Boxes 10.2 and 10.3).

Box 10.2   Restoring native plant communities in Wisconsin

Aldo Leopold and John Curtis began research on the restoration of native ecosystems at the University of Wisconsin in 1934. Their work led to the restoration of several hundred hectares of land altered by logging, agriculture, grazing, and development, including steppe vegetation as well as deciduous and coniferous forests. This example illustrates what can be accomplished by restoration ecologists and highlights some of the limitations of restoration.

In the 1930s Civilian Conservation Corps crews carried out much of this work; it has been continued by university researchers and staff ever since. As a result of this research, the University of Wisconsin Arboretum now contains native communities in which most of the original plant species are present and non-native plants are largely absent.

Note, however, that in describing the successes of this project, I have addressed only the plant communities. In many cases the restoration of plant species is fairly straightforward. It is usually assumed that the appropriate animals will find the restored habitats and colonize them. Although this is often (but not always) true for birds, which are highly mobile, it is not necessarily true for invertebrates. When some species are missing from the original community assemblage, other species that depend on them will be absent too. In the arboretum's restored maple forest, for example, the distributions of two native forbs, wild ginger and bloodroot, have been affected by the absence of an ant species that normally disperses their seeds. Furthermore, restored patches are usually small – often too small to support populations of important large-bodied, native species. Bison herds require millions of hectares; clearly the absence of this species from all but the largest areas of restored midwestern steppes has important consequences for community structure and function (Jordan 1988).

Box 10.3   Ecological and biocultural restoration of dry tropical forest in Costa Rica

Much of the formerly extensive dry tropical forest of Central America has been converted to savanna or pasture by cattle ranching and farming. The forest's deciduous trees, shrubs, and woody vines have been replaced by grasses, especially tall African pasture grasses. Restoration of this eco- system requires a two-pronged strategy: suppressing fires and promoting seed dispersal (Holden 1986). The first is necessary because the ecosystem's disturbance regime has been altered; the second, because mutualistic inter- actions between plants and their seed dispersers have been disrupted.

Fire suppression is important because fire is not an important part of the natural disturbance regime in dry tropical forest. Natural fires are ignited by lightning during the wet season when they fail to burn large areas. Fires set deliberately during the dry season, on the other hand, inhibit tree reproduction and promote the conversion of forest to savanna. Cattle ranchers set fires annually during the dry season to prevent trees from becoming established in their pastures. This change in disturbance regime favors savanna but is detrimental to the forest (see Box 13.6).

Enhancing seed dispersal is important because the native herbivores that once dispersed seeds in Central America are now extinct. Until the end of the last ice age, a variety of other large herbivores – including horses, giant ground sloths, and relatives of the mastodon – roamed the region. The diversity and biomass of the Central American megafauna at that time rivaled today's African game parks. Many of these herbivores fed upon tropical trees and shrubs that produce large fruits with tough seeds (Janzen and Martin 1982). Some of these seeds survived passage through mam- malian digestive tracts and were dispersed in this way to new locations. About 10000 years ago, however, many of these large herbivores became extinct, for reasons that are not entirely clear (see Chapter 8).

The guanacaste tree, a large member of the legume family, is a conspic- uous element in Costa Rica's dry tropical forest. Today, the role of seed disperser is played by domestic horses and cattle. When domestic horses feed on guanacaste seeds, many of the seeds die before they are defecated. But some seeds retain their hard seed coat and are carried to new locations, where they are deposited in feces. Eventually, the hard seed coat breaks down, and guanacaste seedlings emerge (Janzen 1981). Where these herbi- vores are absent, the guanacaste fruits fall beneath the parent tree and are not dispersed. As part of the forest restoration effort, horses and cattle are fed seeds of the guanacaste tree and other important forest trees, and then allowed to roam through the area to be restored.

Fire suppression and enhanced seed dispersal have effectively allowed dry tropical forest to re-establish itself within Costa Rica's Guanacaste National Park. One reason this project has been so successful is because it identified two critical factors that were limiting tree regeneration – the altered fire regime and the loss of mutualists – and restored them. Another reason is because it has the support and cooperation of local people. Janzen views ecological diversity and cultural diversity as being inextricably linked. He points out that the forest is of enormous cultural and intellectual value to people of the tropics (Janzen 1988). In a process Janzen terms ecological and biocultural restoration, local people participate in many aspects of forest rehabilitation. School children collect guanacaste seeds from horse dung, and local farmers participate in a variety of activities, including planting seeds, fighting fires, and patrolling for poachers (Allen 1988).

Ecosystems differ in their ability to recover after disturbance. Tundra, desert, and steppes dominated by bunchgrasses, are examples of ecosystems that are slow to recover after disturbances (see Chapter 13). In these instances, ecological restoration is especially challenging.

10.3.4   Mitigation

In the U.S.A., the National Environmental Policy Act requires the preparation of an Environmental Impact Statement (EIS) for any activity requiring federal permits. Many activities that alter habitats, such as the construction of dams, airports, and highways, fall into this category. An EIS must be reviewed to determine whether the benefits of a project outweigh the costs. If it is determined that a project will cause a loss or degradation of habitat, mitigation (the alleviation of or compensation for negative effects) may be required. Mitigation can take several forms, including restoration of the affected environment, restoration of an area of comparable habitat, or creation of similar habitat to compensate for loss of habitat at the affected site. State and local regulations may also require mitigation for habitats lost to development.

The mitigation process is often used where wetlands are lost or altered. Substantial losses of wetland ecosystems have occurred as a result of drainage and diversion. Since the late 1970s, however, the filling of wetlands in the U.S.A. has been closely regulated by the federal government under the Clean Water Act. One aspect of this regulation allows developers to substitute a restored or created wetland for a developed one. It is difficult or impossible to re-establish a functioning ecosystem with all its original components, however. Wetland mitigation projects are sometimes designed with scant understanding of ecosystem functions, and the designs themselves are not always followed. As a result, the habitats created to fulfill mitigation requirements bear little resemblance to the wetlands they were intended to replace.

Other problems are more fundamental, however. Even using the most advanced ecological knowledge, it is not easy to create wetlands that function like natural ecosystems. In 1984 as a result of highway construction that damaged a marsh which provided habitat for two endangered birds (the light-footed clapper rail and the California least tern) and an endangered plant (the salt marsh bird's beak), the California Department of Transportation (Caltrans) was required by a federal court to provide habitat for the endangered organisms by restoring a wetland within the Sweetwater National Wildlife Refuge in San Diego Bay. The problems encountered by this mitigation project illustrate the difficulties of wetland restoration. As with terrestrial ecosystems, some plant species usually can be successfully re-established in restored wet-lands without a great deal of difficulty. Although cordgrass, the dominant plant cover, was restored to the Sweetwater site, and the area came to resemble a natural wetland in terms of water level and plant composition, it failed to attract nesting clapper rails. Closer analysis revealed that even several years after restoration was begun, the cordgrass had not achieved its full height. Clapper rails, however, must have tall grass for nesting cover. Extensive research suggested that the short stature of the cordgrass was due to inade-quate nitrogen supplies in the site's sandy soils. This problem was addressed by applying fertilizer, but within a few years the grass was attacked by insects, sug-gesting that some insect predator was missing – one that normally would keep the herbivorous insects in check. Further research identified a missing preda-tory beetle. Thus, after nine years and millions of dollars, researchers had learned a great deal about the functioning of the Sweetwater marsh, but the restored wetland still failed to attract the endangered species for which it was designed. Each time a missing link in the functioning ecosystem was identified and replaced, another missing piece connected to it was discovered.

Although these results must be extremely disappointing for Caltrans, from a scientific standpoint, this experience has been extremely valuable. Each stage in the project was an experiment designed to test hypotheses about ecosystem functioning. The results, though disappointing from the point of view of the desired practical results, suggest additional questions and testable hypotheses (Zedler 1988).

10.3.5   Evaluating restoration and mitigation

It is clear that under some circumstances ecological restoration can enhance the size and quality of natural areas. Furthermore, restoration projects are a valuable research tool that can be used to increase our understanding of community dynamics. On the other hand, some conservationists fear that mitigation actually adds to the likelihood of habitat loss. If a developer is allowed to degrade a habitat in return for a promise to restore or create similar habitat elsewhere, and if mitigation fails to do so, as is often the case, then the result is a net loss of habitat. Thus, in evaluating restoration and mitigation it is prudent to remember that we are still a long way from being able to assemble a fully functioning ecosystem.

It is easy enough to plant, fertilize, and water the desired vegetation on a site. But habitat restoration will not succeed in restoring self-sustaining communities unless managers identify what is missing from degraded ecosystems and understand how to put the missing elements back. The experience of Caltrans at Sweetwater Marsh clearly illustrates this. The success of a restoration effort depends in part upon its goals. If the goal is the creation of mallard habitat, for example, success is likely, because we know a lot about managing habitat for mallards, and mallards do not have highly specialized habitat requirements. But if the goal is to recreate a functioning ecosystem with all its components and processes, success has been more elusive. In Part III we see that novel approaches are being developed for looking at ecosystems more holistically and for restoring processes and contexts.

This discussion of the techniques of preservationist management began with species-oriented conservation (Chapter 9). In many ways, management focused on preserving populations of rare species is quite like utilitarian management, except that the featured species is usually not utilized for economic gain. Both approaches emphasize maximizing populations of selected species, and both utilize a high degree of intervention at times. Preserve management and restoration, the subject of this chapter, can also be focused on particular species. But in many cases the objective of this type of management is to restore or maintain ecological processes, such as hydrological cycles, nutrient cycles, or disturbance regimes, perhaps as a means of conserving biological diversity. This takes us to another type of management, which will be explored in more depth in Chapters 12 through 14. First, however, we will consider the political, economic, social, and theoretical developments that set the stage for that type of resource management.

References

Abele, L. G. and E. F. Connor (1979). Application of island biogeographic theory to refuge design: making the right decision for the wrong reasons. In *Proceedings of the 1st Conference on Scientific Research in the National Parks*, vol. 1, ed. R. M. Linn, pp. 89–94. Washington, DC: U.S. Department of the Interior National Park Service.

Agee, J. K. (1988). Successional dynamics in forest riparian zones. In *Streamside Management: Riparian Wildlife and Forestry Interactions,* ed. K. J. Raedeke, pp. 31–43. Seattle, WA: University of Washington Press.

Allen, W. H. (1988). Biocultural restoration of a tropical forest. *BioScience* **38**:156–161.

Allin, C. W. (1990). *International Handbook of National Parks and Nature Reserves.* New York, NY: Greenwood Press.

Ben-David, M., T. A. Hanley, and D. M. Schell (1998). Fertilization of terrestrial vegetation by spawning Pacific salmon: the role of flooding and predator activity. *Oikos* **83**:47–55.

Brown, J. H. and A. Kodric-Brown (1977). Turnover rates in insular biogeography: effect of immigration on extinction. *Ecology* **58**:445–449.

Cassidy, K. M., M. R. Smith, C. E. Grue, K. M. Dvornich, J. E. Cassady, K. R. McAllister, and R. E. Johnson (1997). *Gap Analysis of Washington State: An Evaluation of the Protection of Biodiversity, Washington State Gap Analysis Final Report*, vol. 5. Seattle, WA: Washington Cooperative Fish and Wildlife Research Unit, University of Washington.

Cassidy, K. M., C. E. Grue, M. R. Smith, R. E. Johnson, K. M. Dvornich, K. R. McAllister, P. W. Mattocks, Jr., J. E. Cassady, and K. B. Aubry (2001). Using current protection status to assess conservation priorities. *Biological Conservation* **97**:1–20.

Daubenmire, R. (1978). *Plant Geography.* New York, NY: Academic Press.

Diamond, J. M. (1972). Biogeographic kinetics: estimation of relaxation times for avifaunas of Southwest Pacific Islands. *Proceedings of the National Academy of Sciences, U.S.A.* **69**:3199–3203.

Diamond, J. M. (1975). The island dilemma: lessons of modern biogeographic studies for the design of natural reserves. *Biological Conservation* **7**:129–153.

Diamond, J. M. (1976). Island biogeography and conservation: strategy and limitations. *Science* **193**:1027–1029.

Flather, C. H., K. R. Wilson, D. J. Dean, and W. C. McComb (1997). Identifying gaps in conservation networks: of indicators and uncertainty in geographic-based analyses. *Ecological Applications* **7**:531–542.

Forman, R. T. T. and M. Godron (1986). *Landscape Ecology*. New York, NY: John Wiley.

Gilbert, F. S. (1980). The equilibrium theory of island biogeography: fact or fiction? *Journal of Biogeography* **7**:209–235.

Hansen, S. (1989). Debt for nature swaps: overview and discussion of key issues. *Ecological Economics* **1**:77–93.

Harmon, M. E., J. F. Franklin, F. J. Swanson, P. Sollins, S. V. Gregory, J. D. Lattin, N. H. Anderson, S. P. Cline, N. G. Aumen, J. R. Sedell, G. W. Lienkaemper, K. Cromack, Jr., and K. W. Cummins (1986). Ecology of coarse woody debris in temperate ecosystems. *Advances in Ecological Research* **15**:133–302.

Harris, L. D. (1984). *The Fragmented Forest: Island Biogeography Theory and the Design of Nature Reserves*. Chicago, IL: University of Chicago Press.

Harris, L. D. and J. Scheck (1991). From implications to applications: the dispersal corridor principle applied to the conservation of biological diversity. In *Nature Conservation, 2: The Role of Corridors*, ed. D. A. Saunders and R. J. Hobbs, pp. 189–220. Chipping Norton, NSW, Australia: Surrey Beatty.

Hess, G. R. (1994). Conservation corridors and contagious disease: a cautionary note. *Conservation Biology* **8**:256–262.

Hilderbrand, G. V., T. A. Hanley, C. T. Robbins, and C. C. Schwartz (1999). Role of brown bears (*Ursus arctos*) in the flow of marine nitrogen into a terrestrial ecosystem. *Oecologia* **121**:546–550.

Holden, C. (1986). Regrowing a dry tropical forest. *Science* **234**:809–810.

Hunter, M. L., Jr. (1996). Benchmarks for managing ecosystems: are human activities natural? *Conservation Biology* **10**:695–697.

Hunter, M. L., Jr., G. L. Jacobson, Jr., and T. Webb iii (1988). Paleoecology and the coarse-filter approach to maintaining biological diversity. *Conservation Biology* **2**:375–385.

Janzen, D. H. (1981). *Enterolobium cyclocarpum* seed passage rate and survival in horses, Costa Rican Pleistocene seed dispersal agents. *Ecology* **62**:593–601.

Janzen, D. H. (1983). No park is an island: increase in interference from outside as park size decreases. *Oikos* **41**:403–410.

Janzen, D. H. (1988). Tropical ecological and biocultural restoration. *Science* **239**:243–244.

Janzen, D. H. and P. S. Martin (1982). Neotropical anachronisms: the fruits the gomphotheres ate. *Science* **215**:19–27.

Jordan, W. R. iii (1988). Ecological restoration: reflections on a half-century of experience at the University of Wisconsin–Madison Arboretum. In *Biodiversity*, ed. E. O. Wilson, pp. 311–316. Washington, DC: National Academy Press.

Murphy, D. D. and S. B. Weiss (1992). Effects of climate change on biological diversity in North America. In *Global Warming and Biological Diversity*, ed. R. L. Peters and T. E. Lovejoy, pp. 355–368. New Haven, CT: Yale University Press.

Nature Conservancy (1982). *Natural Heritage Program Operations Manual*. Arlington, VA: The Nature Conservancy.

Naveh, Z. and A. S. Lieberman (1984). *Landscape Ecology: Theory and Application.* New York, NY: Springer-Verlag.

Noss, R. F. (1983). A regional landscape approach to maintain diversity. *BioScience* **33**:700–706.

Noss, R. F. (1987). Corridors in real landscapes: a reply to Simberloff and Cox. *Conservation Biology* **1**:159–164.

Noss, R. F. and A. Y. Cooperrider (1994). *Saving Nature's Legacy.* Washington, DC: Island Press.

Peters, R. L. and J. D. Darling (1985). The greenhouse effect and nature reserves. *BioScience* **35**:707–717.

Pickett, S. T. A. and J. N. Thompson (1978). Patch dynamics and the design of nature reserves. *Biological Conservation* **13**:23–37.

Pulliam, H. R. (1988). Sources, sinks, and population regulation. *American Naturalist* **132**:652–661.

Risser, P. G. (1995). The status of the science of examining ecotones. *BioScience* **45**:318–331.

Runte, A. (1997). *National Parks: The American Experience,* 3rd edn. Lincoln, NB: University of Nebraska Press.

Scott, J. M., F. Davis, B. Csuti, R. Noss, B. Butterfield, C. Groves, H. Anderson, S. Caicco, F. D'Erchia, T. C. Edwards, Jr., J. Ulliman, and R. G. Wright (1993). Gap analysis: a geographic approach to protection of biological diversity. *Wildlife Monographs* **123**:1–41.

Simberloff, D. S. and L. G. Abele (1976*a*). Island biogeography theory and conservation practice. *Science* **191**:285–286.

Simberloff, D. S. and L. G. Abele (1976*b*). Island biogeography and conservation: strategy and limitations. *Science* **193**:1032.

Simberloff, D. S. and L. G. Abele (1982). Refuge design and island biogeographic theory: effects of fragmentation. *American Naturalist* **120**:41–50.

Simberloff, D. S. and J. Cox (1987). Consequences and costs of conservation corridors. *Conservation Biology* **1**:63–71.

Simberloff, D., J. A. Farr, J. Cox, and D. W. Mehlman (1992). Movement corridors: conservation bargains or poor investments? *Conservation Biology* **6**: 493–504.

Soberón M., J. (1992). Island biogeography and conservation practice. *Conservation Biology* **6**:161.

Stoms, D. M., F. W. Davis, K. L. Driese, K. M. Cassidy, and M. P. Murray (1998). Gap analysis of the vegetation of the intermountain semi-desert ecoregion. *Great Basin Naturalist* **58**:199–216.

Terborgh, J. (1974). Preservation of natural diversity: the problem of extinction-prone species. *BioScience* **24**:715–722.

Terborgh, J. (1976). Island biogeography and conservation: strategy and limitations. *Science* **193**:1029.

Williams, G. R. (1984). Has island biogeography any relevance to the design of biological reserves in New Zealand? *Journal of the Royal Society of New Zealand* **14**:7–10.

Willis, E. O. (1974). Populations and local extinctions of birds on Barro Colorado Island, Panama. *Ecological Monographs* **44**:153–169.

Wilson, E. O. and E. O. Willis (1975). Applied biogeography. In *Ecology and Evolution of Communities*, ed. M. L. Cody and J. M. Diamond, pp. 522–534. Cambridge, MA: Belknap Press of Harvard University.

Wright, R. G. (1992). *Wildlife Research and Management in the National Parks*. Urbana, IL: University of Illinois Press.

Zedler, J. B. (1988). Restoring diversity in salt marshes: can we do it? In *Biodiversity*, ed. E. O. Wilson, pp. 317–325. Washington, DC: National Academy Press.

The figure opposite shows how a landscape managed with a sustainable-ecosystem approach to conservation might look. Because disturbances such as fire have occurred, the forest is heterogeneous. There are trees of different sizes and patches with varying tree densities, and there are many openings. The pasture contains isolated trees, which serve as biological legacies. The woodlot is also heterogeneous; it contains vegetation of different sizes and shapes and openings of various sizes. A substantial area of wetland has been restored. Emergent vegetation is interspersed with channels of open water and floating plants. Traditional use of resources is allowed; a woman collects bulrushes from her canoe. The cultivated field is divided up by brushy fencerows, and more than one crop is being grown

PART THREE

Management to maintain processes and structures – a sustainable-ecosystem approach to conservation

11

Historical context – pressures to move beyond protection of species and reserves

Scientists seeking to maintain biodiversity have made crucial contributions to conservation by developing strategies for protecting and restoring populations and ecosystems, but changing social, economic, and political conditions are presenting new challenges that are not fully addressed by these strategies. These are explored in this chapter.

So far, we have seen that wild organisms can be managed for utilitarian goals or for preservationist goals. These two threads have been woven through the conservation movements of the western world for nearly a century and a half. Recently, however, a third strand has appeared. This third approach strives to maintain ecosystem structure and function as a means of maintaining both biodiversity and productive capacity. Thus, its twin goals, management to produce goods and services and to maintain species and communities, encompass the goals of both utilitarian and preservationist management. (Of course, it may not be possible to do both in every situation.)

I call this approach to resource management the sustainable-ecosystem approach. Like "biodiversity," the terms "ecosystem management" and "sustainable development" have both become popular buzzwords. "Ecosystem management" refers to a specific approach adopted by agencies managing federal lands in the U.S.A. (see below). I use the term "sustainable-ecosystem approach" not to refer to the program of any specific agency or organization, but to denote a type of resource management that seeks to conserve biodiversity and productivity by maintaining heterogeneous structures, components, and functions in interconnected ecosystems.

This alternative mode of resource management developed in response to a set of five kinds of problems faced by preservationist management. First, recovery programs and reserves in and of themselves cannot guarantee that extinctions and ecosystem degradation will not continue. (This is a practical problem.) Second, conservation that focuses on rescuing declining species and degraded habitats involves a huge amount of intervention that may itself have unappreciated ecological consequences (an ecological problem). Third, excluding people from the decision-making process and from using resources they have traditionally made use of creates resentment and erodes support for conservation (a political problem). Fourth, policies that aim to protect bio-diversity have sometimes failed to deal in a fair way with the needs of people who depend upon resources, and these policies have had the effect of under-mining cultural diversity (ethical problems). And fifth, the idea that people are not a part of nature and that conservation should concern itself primarily with protecting nature from people has been challenged (a philosophical problem). These problems are, of course, interrelated. Excluding people from their resource base creates political problems, but it is also problematic on ethical grounds. If we re-evaluate our philosophical position regarding our place in nature, that has ethical implications. But, for the sake of simplicity, these five types of problems are considered separately below.

11.1   Practical considerations

A major impetus in the development of new approaches to conservation in the 1980s came from the recognition of practical constraints. No matter how effective efforts to protect species and habitats are, it is simply not possible to protect more than a small fraction of the earth's biodiversity using the strate-gies described in Chapters 9 and 10 (Franklin 1993). Furthermore, "protec-tion" is no guarantee of success. (Preserves have been called lifeboats of diversity, but even lifeboats can sink.)

This problem is complicated by the fact that most habitats and most species are exploited by people. We depend upon these species for food, clothing, shelter, medicines, and fuel. If we do not find ways of using species without compromising their chances for persisting into the future, then our prospects and theirs are grim indeed.

In addition, preservationist management is inefficient when it focuses on species that are already in dire straits. It uses enormous amounts of money and effort in attempts to prevent "train wrecks," but because it is crisis-driven, it often comes too late to make a difference. Some scientists and managers have questioned whether such management is the best use of scarce assets. In the United States, the crisis-driven response mode stems from legal considerations. The Endangered Species Act is the principal tool for enforcing protection of endangered species, but it does not require action until a species is already in serious trouble.

Another practical matter is the need for local support. The hostility and resentment of people who feel that they were not consulted in decisions about resource use can undermine conservation programs. Not surprisingly, where reserves deny people access to resources they have traditionally utilized, these people fail to support conservation. Rather, "local people see these protected areas as a clear symbol of an anti-people government which wants to throw the poor out and open up nature's bounties to middle and upper class tourists" (Agarwal 1992:297).

11.2   Scientific considerations

In an influential paper entitled "Sea turtle conservation and halfway technology" published in the journal *Conservation Biology*, Nat Frazer (1992) argued that head-starting, hatcheries, and captive breeding are "halfway technologies," a term borrowed from an essay by Lewis Thomas on medical technology. Halfway technologies in medicine are things we do to compensate for a disease or to postpone death, when we don't really understand the disease process. An example would be a heart transplant. It may prolong the life of the patient, but it does not cure the disease. Frazer argued that halfway technologies in conservation fail to address the causes of species' declines:

When we define the impending extinction of a sea turtle species solely in terms of there being too few turtles, we are tempted to think of solutions solely in terms of increasing the numbers of turtles. . . . Programs such as headstarting, captive breeding, and hatcheries may serve only to release more turtles into a degraded environment in which their parents have already demonstrated that they cannot flourish. (Frazer 1992:179)

In addition to suggesting that technological fixes like head-starting treat symptoms but not underlying causes, Frazer objected to such manipulations for a more fundamental reason: they prevent turtles or other species that receive special care from serving important ecological roles in their natural environment. He suggested that the death of the very individuals whose mortality we strive to prevent may fulfill ecological functions that we do not understand or even suspect. Although they are not contributing to the next generation of sea turtles, young turtles that die in the wild might be providing food for parasites or predators; they might limit the extent of beds of seaweed by grazing on it; or they might decay and provide detritus to marine micro-organisms. Successful head-starting could therefore interfere with ecological relationships in unknown ways:

> Do we assume that the millions of little turtles that used to come off our beaches play no important role simply because we rarely see them playing it? We like to see big turtles nesting, and we like to see eggs being laid and little turtles hatching and entering the ocean, and we like to see large juveniles feeding in our seagrass beds and reefs and coastal waters. But we almost never see little turtles doing whatever little turtles do in their natural environment. . . .
>
> The halfway technology of headstarting . . . prevents the turtles, while they are being held in captivity, from performing whatever ecological function they normally serve in the natural environment. (Frazer 1992:181)

Frazer's point is akin to the point Leopold (1966) made in the essay "Thinking like a mountain" half a century ago: our appreciation of the roles that organisms play in ecosystems is often superficial and self-serving (Chapter 6). Leopold was concerned with the implications of utilitarian management, but Frazer points out that even preservationist management often involves manipulating ecosystems to favor a narrow subset of organisms or processes of interest to people, with unintended ecological consequences.

Problems also arise when preserves are managed in a way that seeks to freeze a snapshot of a pristine past. Some scientists question the wisdom of this approach. They argue that because the world is constantly changing, there is no single correct scenario for preservation (see Chapter 12). Furthermore, they challenge the idea that preserves should safeguard the prehuman past and the underlying assumption that human use inevitably degrades nature (Bell 1987; Collett 1987; Seligman and Perevolotsky 1994; Perevolotsky and Seligman 1998).

11.3   Political considerations

11.3.1   Confrontations over the U.S. Endangered Species Act

The northern spotted owl and old-growth forests
In the U.S.A., Section 7 of the Endangered Species Act (ESA) requires that federal agencies consult with the Secretary of the Interior about any activities that might jeopardize a listed taxon. The controversy over the construction of the Tellico Dam (Chapter 9) inaugurated a series of contentious battles over

how to balance the needs of small populations against pressures for economic growth. Two other well-known and heated controversies over the ESA involve the northern spotted owl and salmon in the Pacific Northwest.

In 1990 the northern spotted owl, an inhabitant of old-growth forests in Washington, Oregon, and northern California, was listed as threatened under the Endangered Species Act. This subspecies inhabits coniferous forests, especially old-growth, and has large home range requirements. Conservative estimates suggest that a single pair needs hundreds of hectares in which to breed; consequently, a viable population of northern spotted owls requires thousands of hectares of unfragmented late-successional forest. Most of the remaining habitat that meets the owls' needs is on public lands, areas that also provide high-value timber (Dixon and Juelson 1987; Simberloff 1987; Proctor 1996).

The debate over the fate of the Northwest's ancient forests has been acrimonious, characterized by lawsuits and countersuits and by bitter attacks on both sides. The media once again have portrayed the conflict as a choice between protection of a single species and jobs (Foster 1993). In reality, of course, economic viability and ecological health are intimately intertwined, for if the old-growth forests of the Northwest were to disappear, the economic outlook for the region's timber towns would be bleak indeed. Nevertheless, negative fallout from the conflict remains, and the perception that endangered species protection stands in the way of economic health has become even more entrenched.

Salmon and dams

The annual upstream migration of Pacific salmonids (members of the salmon family, including salmon and anadromous trout) to their spawning grounds in the Columbia River Basin once numbered in the tens of millions. It is estimated that 12 to 16 million salmon returned each year to spawn in the 1880s, but a century later the salmon migration had declined precipitously to about 2.5 million fish. In 1991 the American Fisheries Society issued a report entitled "Pacific salmon at the crossroads," in which 214 populations of salmonids from Oregon, Idaho, Washington, and California were identified as facing a high or moderate risk of extinction (Nehlsen *et al.* 1991).

Many factors have contributed to the decline in Pacific salmon. Foremost among them is habitat loss, which itself has many causes. Salmon and trout require clean beds of gravel for spawning; these are produced by periodic flushes of rapidly moving water. Anything that prevents or reduces these flushing flows has a negative impact on salmon reproduction. That includes dams, because they control flooding, as well as activities such as logging, road-building, and livestock grazing, which increase the amount of silt entering streams. Dams also kill juveniles drawn into the turbines on their way downstream. In addition, the introduction of hatchery fish and overexploitation both played a role in the salmon's demise.

By 1999, the National Marine Fisheries Service had listed dozens of populations of Pacific salmon as threatened or endangered, and more await consideration. Since the U.S. Army Corps of Engineers operates the dams on the Snake and Columbia Rivers, and Section 7 of the Endangered Species Act requires federal agencies to avoid harmful impacts to listed organisms, the fate of salmon has become a pivotal issue in the Columbia Basin. In addition, salmon also play a central role in the Native American cultures of the Northwest and, to a lesser extent, in the region in general. Because of the negative impacts of dams on salmon, breaching the dams to restore salmon populations is seriously being considered. At the same time, the dams on the Columbia and Snake Rivers are important to the region's economy. They provide inexpensive electricity, energy-efficient transportation for crops on their way to the markets of Portland (and from there to the Far East), and a source of water for irrigation. Once again, the failure to manage resources in a sustainable fashion has brought us to the point where utilitarian and preservationist modes of resource management appear to point in opposite directions (Gillis 1995).

11.3.2   Changing directions in natural resource management

In the 1980s, as a result of conflicts between biodiversity conservation and socioeconomic stability, the management philosophy of federal land management agencies in the U.S.A. was modified. The goal of managing for multiple uses (Chapter 1) was replaced by an approach termed ecosystem management, in which the goal is management of biological systems in a manner that maintains ecological integrity while accommodating human use (Agee and Johnson 1988; Slocombe 1993; Grumbine 1994; Kaufmann et al. 1994; Christensen et al. 1996). A crucial characteristic of ecosystem management is its emphasis on integrating ecological considerations with sociopolitical and economic factors. As a result of the spotted owl controversy and similar conflicts, managers came to realize that natural resources cannot be managed in a social vacuum. People are part of ecosystems, and the social context of natural resources must be considered if management is to be effective.

This suggests that resource managers can benefit from listening to the various stakeholders (parties who live or make their living in or near a managed area and institutions with activities that affect the managed area) in a dispute. Having different interest groups participate in the decision-making process can improve communication and understanding. Although the discussions are often heated when people with radically different views sit down at the same table, it is also true that in face-to-face situations people are more likely to respond to each other as individuals. It is easier to demonize an abstract idea than a flesh-and-blood person. Bringing people with different interests together also increases the likelihood (although it does not guarantee) that their different interests will be balanced. In addition, people are more likely to support a management plan if they feel that they participated in its creation.

None of these benefits of bringing stakeholders to a common table is a foregone conclusion. Getting parties with disparate values and interests to cooperate and compromise is extremely challenging, and in many cases some or all participants may come away disappointed. Consensus, shared vision, and mutual respect are not always the outcomes of the process, but when they are, they can accomplish a lot.

Another key characteristic of ecosystem management is that management actions are viewed as hypotheses to be tested and revised if necessary. In the past, policies were often treated as ends in themselves, and studies were conducted to justify them. In ecosystem management, specific objectives should be stated, and these objectives should pertain not just to narrow goals such as resource extraction but to ecosystem structure and function as well. The hypothesis to be tested, then, is that the proposed management action will have the effect of meeting the stated objectives. Studies should be undertaken to test the relevant hypothesis. If the data that are gathered fail to support the hypothesis that the objectives are being met, then the management action should be adjusted and further studies carried out. This approach to management is termed adaptive management. Rather than simply justifying business-as-usual, managers practicing adaptive management should use monitoring as a tool to help them to be flexible and responsive, continually readjusting policies as new information becomes available (Holling 1978; Walters 1986).

11.4   Ethical considerations

The conflicts described above have raised troublesome questions about the relationship between conservation, social justice, and political power. The pertinent questions here are: Do some groups unfairly bear more of the costs of protecting species and habitats than others? Are decisions about resource management imposed upon those groups unfairly? Does biodiversity conservation undermine efforts to distribute resources fairly? Does it serve the interests of some groups at the expense of others? Do the people affected by decisions about conservation participate in making those decisions?

There are several reasons for these concerns. First, biodiversity conservation often means giving up something, cutting back on the level of resource exploitation. Simple rules of fairness demand that those who benefit should be the ones who do the giving up, or conversely, that those who do the giving up should in some way be compensated by those who receive the benefits of their sacrifices. Second, in a fair world those who make the decisions about what is to be given up should be those who are affected by those decisions. Third, we have not contributed equally to resource depletion. Once again, fairness dictates that those who are most responsible for depleting resources should be the ones who bear the primary responsibility for restoring them and should have to make the greatest sacrifices to do so. Some examples that illustrate these problems are discussed below.

11.4.1   Who bears the costs of protection?

Setting aside preserves has obvious economic costs. It can also have social costs as well.

When the creation of nature reserves causes local people to be driven from their lands or prevents them from obtaining products they have traditionally harvested, they are unlikely to support conservation. This is true whether the local people are indigenous hunters prevented from hunting in reserves or loggers enjoined from working because of owls.

This problem is especially acute in parts of the developing world where the establishment of protected areas has been accomplished by moving people off their ancestral lands. The consequences of such policies include increased poverty when people are denied access to resources they need for subsistence, erosion of social institutions, loss of traditional techniques and knowledge, hostility to conservation measures, and poaching from protected areas. Such policies have threatened cultural integrity and caused losses of human life. In addition, protected populations of wildlife that are crowded into inadequate remnants of habitat expand and move into croplands and villages, where they damage fields and attack villagers. In India attacks by tigers on people living near tiger reserves have become a major problem. In some instances police have even fired upon crowds of protesters (Agarwal 1992).

Clearly, any approach that protects the diversity of plants and animals while eroding cultural diversity is objectionable on ethical grounds.

11.4.2   Who makes decisions about access to resources?

When local people are excluded from the decision-making process in conservation planning, resentment rises. Looking at the issues in this way raises some troubling questions. North American schoolchildren can purchase pieces of tropical rainforest to protect it from exploitation. Does this imply that the third-grader living in a temperate climate thousands of miles away is more entitled to or more qualified to make decisions about use of that tropical forest than an adult who has lived there all of his or her life? If so, where does this entitlement stem from, and what are these qualifications? (Surely the greater wealth of the child doesn't necessarily make him or her either wiser or more entitled to power than the indigenous forest-dweller.) Who is benefiting here? Who is sacrificing? (See Box 11.1).

Box 11.1   Access to mountain gorillas

The question of who should control access to resources is dramatically highlighted by the events leading up to the death of the North American biologist Dian Fossey, a well-known authority on the behavior of the mountain gorillas of central Africa. Passionately committed to gorilla protection, Fossey publicized the fate of these primates through books, magazine articles, and TV specials that generated international sympathy for their cause (Fossey 1983). While conducting her research in the Virungas of Rwanda, Fossey found that poachers posed a threat to the wild gorillas. She employed Rwandans as porters and trackers to help her with her work in the physically demanding gorilla habitat, but she never delegated responsibility for conducting scientific studies to Rwandans, and she worried that if gorillas became habituated to dark-skinned people the animals might be less likely to flee from poachers. For similar reasons, Fossey attempted to keep local villagers, including herders, beekeepers, firewood collectors, and farmers, out of gorilla habitat. Thus, she tried to exclude Africans (whom she referred to as "wogs") from all activities that would bring them in contact with gorillas, and especially activities that involved observing them – the very things that she herself found most moving and meaningful.

Other conservationists argued that local people would be more supportive of gorilla conservation efforts if they reaped some benefits from conservation. Tourists from developed nations are willing to spend enormous amounts of money to see large primates in the wild (Chapter 14), but Fossey opposed programs that involved education and tourism because she considered them soft on the issues of poaching and habitat protection.

Convinced of the urgency of the situation, Fossey waged her own war against poaching by cutting traps, intimidating poachers and their families, and campaigning for stiff penalties. In 1985 Dian Fossey was murdered. She has been portrayed in books and movies as a martyr in the cause of gorilla conservation (Mowat 1987), but some feel that her inflexibility, unsympathetic attitude toward local people, and insistence on control of access undermined more effective conservation measures (Adams and McShane 1992).

11.4.3   Who is responsible for causing the problem?

A related issue is the question of whether the parties responsible for depleting resources in the first place are the ones who are being asked to make sacrifices in the name of conservation. In North America, indigenous hunters are required to restrict their traditional harvests of whales. But the depletion of whale stocks to the point of endangerment was brought about by the industrialized whaling nations. Are the sacrifices being borne by the perpetrators here?

Similarly, eagle populations have declined as a result of habitat loss, predator control, and bioconcentration of pesticides. Bald eagle feathers are used for ceremonial purposes by Native Americans, but the U.S. Supreme Court has ruled that the federal Bald and Golden Eagle Protection Act supersedes treaty provisions that give tribal members the right to hunt anything on their reservations, even though exploitation by Native Americans was not the main factor causing the eagle declines.

Furthermore, the protection of resources in the developed world can lead directly to increased exploitation in the developing world. Often the countries that take over production have less stringent environmental regulations than the nations that impose environmental regulations within their own borders. Forest protection in the U.S.A. deflects exploitation pressure onto forests in the tropics (Myers 1979) and Siberia (Dekker-Robertson and Libby 1998). If conservation is not accompanied by a reduction in demand, commodities are simply supplied from elsewhere, an approach that can be considered "analogous to locating a landfill for an affluent city in a neighboring community that needs the money and is willing to put up with the smell" (Dekker-Robertson and Libby 1998:475).

Although the populations of developed nations are not growing as fast as those of developing nations, the average individual in a developed country uses far more resources than a person in a developing country, because rates of consumption are much higher. High rates of resource consumption in the developed world put pressure on natural resources both within the developed world and in less developed nations from which resources are imported to meet the demands of people in richer nations. Yet overpopulation is often presented as a problem of high birth rates in less developed countries. Should people in Africa or Asia or Latin America be expected to forego using resources to meet their immediate needs in order to protect species or habitats of interest to North American or European visitors? Should the seriousness of the sacrifice be a consideration? Are people being asked to give up their livelihood or a luxury?

These are not easy questions to answer. It is certainly difficult to define who benefits from conservation, since all parts of the earth are interconnected, and decisions about resource use have effects far removed in space and time from their source (Peters 1996). We rightly feel that we have a stake in decisions made far from where we live. In a sense, we are all "local." Furthermore, it is not easy to quantify ecosystem values or the negative impacts stemming from their loss, so reckoning who benefits and who loses is a daunting matter. Nevertheless, the issue of the relationship between conservation and social justice is not going to go away.

In some ways, the above discussion has been oversimplified. Conservation is a necessary condition for an equitable distribution of resources, because resource depletion exacerbates inequalities. (When there is not enough to go around, the *haves* get first pick and the *have nots* lose out.) When resources are overexploited, the poor and the powerless are the first to find their resource base eroded. For example, when tropical forests are leveled to provide timber and beef for North American consumers, resources are transferred from poor to rich classes and nations, while the potential for sustainable use of tropical forest ecosystems is reduced.

So the question becomes: How can we conserve natural resources in ways that are fair to all parties involved? We will explore this issue further in Chapter 14.

11.4.4   Biodiversity versus cultural diversity?

All too often, policies designed to protect nonhuman biological diversity have eroded cultural diversity. In Uganda nomadic hunters belonging to the Ik tribe were excluded from their ancestral homeland when a national park was created. They were encouraged to settle in arid lands bordering the park, where they were expected to take up farming. When anthropologist Colin Turnbull visited the Ik in the 1960s, he found them in the midst of a famine. Lacking the experience, technology, and social organization appropriate for farming, and without assistance from the government that had cut them off from their former means of subsistence, many Ik died of starvation (Turnbull 1972).

This story is not unique. Except in extremely remote areas, most nature preserves were formerly inhabited by or at least provided resources for local people. When the Vwaza Marsh Game Reserve in Malawi, the Masai Mara National Reserve, and Amboseli, Nairobi, Tsavo, Tarangire, Serengeti, and Lake Manayara National Parks in Kenya and Tanganyika (now Tanzania) were created, hunting tribes such as the Waliangulu, or Wata, as well as Maasai pastoralists were evicted from their ancestral lands (Parker and Amin 1983; Adams and McShane 1992). Similarly, in India people were "relocated" to create reserves for tigers (Guha 1989; Sutton 1990; Seidensticker 1997). This problem is not limited to Africa. Yosemite Valley was home to the Ahwahneechee Indians before they were driven out by the U.S. Army in 1851, to make way for mining interests. (The National Park Service later created an imitation Indian village within the park's boundaries.) By 1929, there was only one survivor of the Ahwahneechee band that had inhabited the valley (Olwig 1996).

In these cases, and many others, traditional patterns of resource use were prohibited and defined as inimical to conservation. Ironically, though, often the role of people in maintaining the featured ecosystems was not appreciated. Peter Raven, Director of the Missouri Botanical Garden, notes that "the removal of people not only creates enemies, but disrupts the central ecological dynamics" (Raven 1998:1510).

Although poaching for profit is a serious problem, many instances of "poaching" have nothing to do with the international trade in wildlife parts. Rather, they involve people trying to make a living the way they have always made a living, in the places where their ancestors made a living for centuries (Parker and Amin 1983; Adams and McShane 1992). (Recent studies suggest that in parts of India and Africa the cessation of traditional land uses, particularly grazing, has actually led to ecosystem degradation; see Chapter 12.)

Even if modern substitutes are available, the traditions associated with obtaining, sharing, and preparing wild plants and animals may have considerable cultural importance. For example, a report to the International Whaling Commission on subsistence whaling found that because the hunting of bowhead whales is central to the culture of Alaskan Inuit, nutritionally adequate substitute foods are not regarded as culturally satisfactory. The report concluded that:

whaling is a focal point of Eskimo culture in which values are expressed and actualized, individual achievement is fulfilled, and social integration is manifested to its highest degree. . . .

The north Alaskan Eskimos place an extremely high cultural value on their customary diet. Most people express a conviction that meals are incomplete without native food, and emphasize that they cannot remain strong, healthy, and satisfied when they rely on imported foods. (International Whaling Commission 1982:41)

The cultural value of traditional practices is often overlooked or dismissed. On May 17, 1999, Makah hunters in the state of Washington killed their first gray whale in 70 years, an event that was celebrated by native peoples but lamented by animal rights groups. While representatives from as far away as Fiji and Tanzania traveled to join the Makah in celebrating the survival of their traditional culture, the Seattle *Post-Intelligencer* reported that the overwhelming majority of comments they received were critical of the Makah. Many expressed bitterly anti-Indian sentiments, such as "Save a whale; harpoon a Makah."

Tropical ecologist Daniel Janzen (Chapter 10) disputes the notion that we have to choose between biological diversity and cultural diversity:

The natural world is by far the most diverse and evocative intellectual stimulation known to humans. Tropical humans are experiencing nearly total loss of this integral part of their mental lives. It is as though they are losing their color vision and most of their hearing. . . . The level of human intellectual deprivation represented by the upcoming obliteration of tropical wildlands is the terminal step in what has been many collective generations of gradual biocultural loss to tropical humanity.

We have long been misled by the view that tropical peoples care only about their stomachs, not the 'luxury' of baubles like national parks. (Janzen 1988:244)

11.5   Philosophical considerations

11.5.1   Attitudes about people and the natural world

The dominant view of nature in western culture has shifted within the last two centuries. Prior to that time, wilderness was considered threatening, barren, and desolate. When Europeans arrived in eastern North America, they reacted to the landscape with fear, considering the forests dark, foreboding, and full of fierce beasts. Similarly, colonists believed that the other wild continents must be tamed. Sir Charles Eliot wrote that "large parts of South America and Africa" should be made to submit to the "influence and discipline" of man: "marshes must be drained, forests skillfully thinned, rivers be taught to run in ordered course and not to afflict the land with drought or flood at their caprice" (quoted in Collett 1987:139).

But in the late nineteenth century, Euroamerican ideas about wild places began to change radically. This was partly a result of romantic disenchantment with modern life. The cultural symbolism associated with wilderness shifted. Instead of seeing wild places as hostile, romantics saw wild places as sublime refuges from civilization and places to renew contact with God. To American romantic writers such as Henry David Thoreau and John Muir, wilderness was a sacred temple.

Wilderness advocates found some kinds of natural places more sublime than others. Cronon (1996:73) notes that "God was on the mountaintop," but evidently not in swamps or steppe. (Everglades National Park was not established until the 1940s, and there are still no national parks in U.S. grasslands.)

The Euroamerican experience of expanding across the North American continent fostered a specific set of attitudes toward nature that involved conquest. This was consistent with the idea that people should dominate nature, a familiar theme in Judeo-Christian traditions. In the United States, this orientation toward nature was expressed as manifest destiny, the idea that it was the responsibility of Euroamericans to expand westward across North America, taming the frontier as they went (see Figure 1.1). As noted above, this mindset presupposed an uninhabited, "virgin" continent, so the original human inhabitants of North America were "rounded up and moved onto reservations . . ., their earlier uses of the land redefined as inappropriate or even illegal" (Cronon 1996:69).

These two views of nonhuman nature, as something to be subdued and as a source of divine inspiration, suggest contrasting attitudes toward human nature. If God appointed people as stewards of the natural world, this seems to imply that people are worthy of this responsibility. This view predominated in western thought until recently. In fact, "from the Renaissance down to the middle of the nineteenth century, European thinkers had generally agreed that to be human was to be something special and splendid." But the romantic revolt challenged this idea. The romantics did not consider civilization something splendid. Rather, they saw the natural world as something good and harmonious from which people were excluded and alienated. Anthropologist

Matt Cartmill suggests that "the Victorian era witnessed the birth of the doctrine that what is nonhuman is sacred and sane, whereas people and their works are inherently cruel, disordered, and sick"(Cartmill 1983:70–71). The atrocities of the twentieth century have done nothing to contradict this view.

A corollary of this is the idea that "wilderness stands as the last remaining place where civilization, that all too human disease, has not fully infected the earth" (Cronon 1996:69). Again, this idea is consistent with Judeo-Christian concepts and symbols: people have fallen from grace and been cast out or separated from a pure and beautiful natural world (Merchant 1996; Slater 1996).

How we feel about the nonhuman world is very much tied up with how we feel about ourselves, about what it means to be human. This is especially true if human nature is seen as contrasting with the natural world, as tends to be the case in western culture. In a dualistic mind-set, if we are the opposite of natural, then if we are bad, nature is good.

The view that people in the modern world are morally tainted goes along with a tendency to romanticize people of nonindustrial cultures, who are sometimes portrayed as living in harmony with nature. In one version of this viewpoint, the effects of native people on their environment are benign, because they lack the technology and systems of property ownership that Euroamericans associate with environmental transformation (Marshall 1999). This view of "the ecologically noble savage" can be patronizing and condescending, however (Redford 1991) if it fails to credit indigenous people with the ability to purposefully and effectively affect the natural world. In its extreme form, it implies that "whites are . . . the only beings who make a difference" (White 1996:175). (This point of view is closely tied to equilibrium theories about change in the natural world, as we shall see in Chapter 12.) Because of the great number and diversity of hunter–gatherer cultures, it is unwise to generalize about the conservation measures typical of these societies or their ecological impacts in different times and places (Nabhan 1995).

In developed parts of the modern world, it is easy to overlook our dependence on resources. As high-technology jobs come to dominate our economy, fewer and fewer people do work that requires them to use natural resources directly. Historian Richard White points out that this has led to polarized attitudes about work in western culture (White 1996). On the one hand, environmentalists tend to condemn forms of work such as logging and farming that directly exploit resources. (Certain forms of archaic labor, such as peasant farming, are excepted.) On the other hand, the wise-use movement equates work with private property rights. White contends that the latter confuses respect for rural labor with property rights, while the condemnation of work in nature is equally problematic because it contributes to an illusion that the work of white-collar workers and intellectuals does not have environmental impacts. Paradoxically, this attitude toward work goes along with a preference for leisure activities that imitate work, such as backpacking, skiing, canoeing, or climbing, and the idea that natural places should be used for recreation only, not for work. (It also goes along with a preference for certain high-impact life-styles, such as commuting to work from a home out in the country where there is no public transportation.)

This failure to acknowledge "the implications of our labor and our bodies in the natural world," argues White, leaves environmentalists in the position of "patrolling the borders" of the natural world (White 1996:185). It also con-tributes to a perception that environmentalists are elitist, concerned only with the interests of people of leisure. This perception is shared by poor people and people of color who have struggled with a different set of concerns: toxic waste dumps in their neighborhoods; asbestos, lead, and polluted water in their homes; and exposure to toxic chemicals in the workplace. Until recently, environmental organizations did not recognize these problems as being "environmental" issues, because mainstream environmentalist concerns were largely limited to the nonhuman environment.

In 1987 the Commission for Racial Justice of the United Church of Christ issued a report on *Toxic Waste and Race in the United States* (United Church of Christ, Commission for Racial Justice 1987). The report concluded that people of color in the United States are disproportionately exposed to haz-ardous wastes. Reaction to this situation crystallized in a new form of envir-onmental activism, termed the environmental justice movement. To environmental justice advocates, environmentalists' preoccupation with the places from which people are absent reflects an insensitivity to the concerns of poor people and minorities (Di Chiro 1996).

11.5.2   Defining biodiversity

The idea that people are outside of nature, and that anything people do in the natural world is bad, is reflected even in some definitions of biodiversity. As noted in Chapter 7, the term biological diversity originally referred to species richness, but it has been expanded to encompass structures and functions at all levels of organization. While there are certainly valid reasons for valuing diversity at different levels of organization, the broadest definition of bio-diversity reflects and contributes to the view that people are apart from nature. It is easy to envision cataloguing structures (genes, species, or communities) to keep track of them. (This is not necessarily easy to do, but at least the concept is straightforward.) But if we include ecological processes as part of biodiversity, a problem arises. It is true that processes are important and some processes are threatened, but how do we catalogue and keep track of biological diversity at the process level? In order to track whether "biodiversity" at this level is being maintained, it is necessary to have some criteria for when a process is functioning adequately and when it is not. The simplest way to approach this is to use "naturalness" as the standard, and to define natural as the absence of all human impacts (Hunter 1996). If this thinking is carried to its logical conclusion, any impact or change of any sort caused by people is tantamount to a loss of biodiversity. In this way of thinking, if people impact a process, then the process is not operating naturally and biodiversity has been lost. So the presence of people implies a loss of biodiversity. A paper entitled "The limits to caring: sustainable living and the loss of biodiversity," published in the journal *Conservation Biology* in 1993, exemplifies this trend. Ecologist John Robinson (1993:24) argues that "*Any* exploitation of a species will remove part of a biological community, with concomitant effects on community dynamics and ecosystem functioning. . . . *Any* use of a species . . . is likely to encourage the overall loss of biological diversity." And again, "*Any* use of a biological community will ultimately involve a loss of biological diversity" (emphasis added). Although Robinson himself recognizes a connection between use and conservation (Robinson and Redford 1991), framing the issue in this way contributes to the perception that there is a dichotomy between people and nature.

11.5.3   An alternative view

The tendency to divorce people from nature is not restricted to preservationists. I pointed out in the Introduction that the preservationist and the utilitarian approaches to natural resource management are grounded in similar views of our relationship to the natural world. The preservationist perspective is that nature will take care of itself and will in fact be better off if people leave it alone. The utilitarian view is that people should dominate nature by manipulating it for their own use. The perception that people are separate from nature underlies both these positions. In one view, people degrade nature; in the other, people improve upon nature, but in either case people are not part of nature.

Aldo Leopold (1966:240) suggested a third alternative: that of "plain member and citizen" of the "land-community," a role that implies respect for other species and for "the community as such." Recall from Chapter 6 that

Leopold himself changed from a killer of wolves to an opponent of predator control because he came to respect wolves and appreciate their ecological role. Writing in the 1940s, Leopold was exceptional in his appreciation of the role of people in nature. The idea that people are part of nature is gaining acceptance, however, as the limitations of both the utilitarian and the preservationist perspective in certain contexts become apparent.

Of course, saying that we regard people as part of nature does not, by any means, imply that everything that people do or have done is ecologically benign. It is still necessary to evaluate our actions in terms of their impacts on populations, communities, and ecosystems and to set standards for how much impact is acceptable.

11.6   Diagnosing the problem

Clearly, management of the earth's living natural resources in the twenty-first century is inextricably bound up with and complicated by cultural, social, economic, ethical, and political matters. The philosophical underpinnings of preservationist management have been criticized for failing to address these complexities, however. For example, Murray Bookchin, one of the first outspoken critics of pollution (Chapter 7), charges that neither mainstream environmentalism nor deep ecology is an adequate response to current ecological problems. The former makes too many compromises and tradeoffs and fails to challenge the *status quo*, according to Bookchin, whereas deep ecology, though it purports to be a radical philosophy, ignores the social dimensions of ecological problems. Both fail, in Bookchin's estimation, to understand the connection between exploitation of the natural world and exploitation of people (Bookchin 1989). (For a more detailed discussion of the tensions between Bookchin and deep ecologists see Ellis (1996).)

The American version of biocentrist management has also been criticized for exporting inappropriate strategies to the developing world. The Indian ecologist Ramachandra Guha contends that the American version of deep ecology, as articulated by Devall and Sessions (1985), ignores the problems of overconsumption and militarization, which Guha considers the two fundamental ecological threats facing the modern world. Instead, he argues, by focusing on the protection of wilderness areas where people are not allowed to live, American deep ecology promotes a strategy that is inappropriate for places like India, which are densely settled by peasants and tribal people who must meet their material needs by harvesting local resources. According to Guha, American environmentalists' narrow focus on wilderness protection has led them to disregard Naess's original concerns about social equity and human diversity (Chapter 7); rather than being universal, American radical environmentalism is born "out of a unique social and environmental history" (Guha 1989:79).

11.7   The response to the problem: The rise of sustainable-ecosystem management

Faced with this cluster of interrelated problems at the end of the twentieth century, scientists and resource managers began to rethink their assumptions about how the natural world operates and how it should be managed. This led to a different perspective on conservation. The underlying concepts and techniques of this approach, as well as some of the challenges it faces, are described in the next three chapters.

Questions about whether people are, or should be considered, part of nature belong in the domains of ethics and philosophy, and questions about what social policies are needed to remedy ecological problems belong to the domain of politics. Science, however, can provide relevant information to help us make these decisions. The next chapter describes how scientists' ideas about the natural world and our place in it are changing and how new perspectives can point the way to strategies for resource management that address the challenges we have been discussing.

References

Adams, J. S. and T. O. McShane (1992). *The Myth of Wild Africa: Conservation Without Illusion*. New York, NY: W. W. Norton.

Agarwal, A. (1992). Sociological and political constraints to biodiversity conservation: a case study from India. In *Conservation of Biodiversity for Sustainable Development*, ed. O. T. Sunderlund, K. Hindar, and A. H. D. Brown, pp. 293–302. Oslo: Scandinavian University Press.

Agee, J. K. and D. R. Johnson (eds) (1988). *Ecosystem Management for Parks and Wilderness*. Seattle, WA: University of Washington Press.

Bell, R. (1987). Conservation with a human face: conflict and reconciliation in African land use planning. In *Conservation in Africa: People, Policies, and Practice*, ed. D. Anderson and R. Grove, pp. 79–101. Cambridge: Cambridge University Press.

Bookchin, M. (1989). *Remaking Society*. Montreal: Black Rose Books.

Cartmill, M. (1983). "Four legs good, two legs bad": Man's place (if any) in nature. *Natural History* **92**(11):64–79.

Christensen, N. L., A. M. Bartuska, J. H. Brown, S. Carpenter, C. D'Antonio, R. Francis, J. F. Franklin, J. A. MacMahon, R. F. Noss, D. J. Parsons, C. H. Peterson, M. G. Turner, and R. G. Woodmansee (1996). The report of the Ecological Society of America committee on the scientific basis for ecosystem management. *Ecological Applications* **6**:665–691.

Collett, D. (1987). Pastoralists and wildlife: image and reality in Kenya Maasailand. In *Conservation in Africa: People, Policies, and Practice*, ed. D. Anderson and R. Grove, pp. 129–148. Cambridge: Cambridge University Press.

Cronon, W. (1996). The trouble with wilderness. In *Uncommon Ground: Rethinking the Human Place in Nature*, 2nd edn, ed. W. Cronon, pp. 60–90. New York, NY: W. W. Norton.

Dekker-Robertson, D. L. and W. J. Libby (1998). American forest policy: global ethical tradeoffs. *BioScience* **48**:471–477.

Devall, B. and G. Sessions (1985). *Deep Ecology: Living as if Nature Mattered*. Salt Lake City, UT: G. M. Smith.

Di Chiro, G. (1996). Nature as community: the convergence of environment and social justice. In *Uncommon Ground: Rethinking the Human Place in Nature*, 2nd edn, ed. W. Cronon, pp. 298–320. New York, NY: W. W. Norton.

Dixon, K. R. and T. C. Juelson (1987). The political economy of the spotted owl. *Ecology* **68**:772–776.

Ellis, J. C. (1996). On the search for a root cause: essentialist tendencies in environmental discourse. In *Uncommon Ground: Rethinking the Human Place in Nature*, 2nd edn, ed. W. Cronon, pp. 256–268. New York, NY: W. W. Norton.

Fossey, D. (1983). *Gorillas in the Mist*. Boston, MA: Houghton Mifflin.

Foster, J. B. (1993). The limits of environmentalism without class: lessons from the ancient forest struggle of the Pacific Northwest. *Capitalism, Nature, Socialism* **4**:11–41.

Franklin, J. F. (1993). Preserving biodiversity: species, ecosystems, or landscapes? *Ecological Applications* **3**:202–205.

Frazer, N. (1992). Sea turtle conservation and halfway technology. *Conservation Biology* **6**:179–184.

Gillis, A. M. (1995). What's at stake in the Pacific Northwest salmon debate. *BioScience* **45**:125–132.

Grumbine, R. E. (1994). What is ecosystem management? *Conservation Biology* **8**:27–38.

Guha, R. (1989). Radical American environmentalism: a third world critique. *Environmental Ethics* **11**:71–83.

Holling, C. S. (1978). *Adaptive Environmental Assessment and Management*. Chichester: John Wiley.

Hunter, M., Jr. (1996). Benchmarks for defining ecosystems: are human activities natural? *Conservation Biology* **10**:695–697.

International Whaling Commission (1982). *Aboriginal/Subsistence Whaling (with Special Reference to the Alaska and Greenland Fisheries)*. Reports of the International Whaling Commission, Special Issue no. 4, Cambridge.

Janzen, D. H. (1988). Tropical ecological and biocultural restoration. *Science* **239**:243–244.

Kaufmann, M. R., R. T. Graham, D. A. Boyce, Jr., W. H. Moir, L. Perry, R. T. Reynolds, R. Bassett, P. Mehlhop, C. B. Edminster, W. M. Block, and P. S. Corn (1994). *An Ecological Basis for Ecosystem Management*. U.S. Department of Agriculture Forest Service, General Technical Report no. RM-246, Washington, DC.

Leopold, A. (1966). *A Sand County Almanac*, 7th edn. New York, NY: Ballantine Books.

Marshall, A. G. (1999). Unusual gardens: the Nez Perce and wild horticulture on the eastern Columbia Plateau. In *Northwest Lands, Northwest Peoples: Readings in Environmental History*, ed. D. D. Goble and P. W. Hirt, pp. 173–187. Seattle, WA: University of Washington Press.

Merchant, C. (1996). Reinventing Eden: western culture as a recovery narrative. In *Uncommon Ground: Rethinking the Human Place in Nature*, 2nd edn, ed. W. Cronon, pp. 132–159. New York, NY: W. W. Norton.

Mowat, F. (1987). *Woman in the Mists: The Story Of Dian Fossey and the Mountain Gorillas of Africa*. New York, NY: Warner Books.

Myers, N. (1979). *The Sinking Ark*. Oxford: Pergamon Press.

Nabhan, G. P. (1995). Cultural parallax in viewing North American habitats. In *Reinventing Nature? Responses to Postmodern Deconstruction*, ed. M. E. Soulé and G. Lease, pp. 87–101. Washington, DC: Island Press.

Nehlsen, W., J. E. Williams, and J. A. Lichatowich (1991). Pacific salmon at the crossroads: stocks at risk from California, Oregon, Idaho, and Washington. *Fisheries* **16**(2):4–21.

Olwig, K. (1996). Reinventing common nature: Yosemite and Mount Rushmore – a meandering tale of a double nature. In *Uncommon Ground: Rethinking the Human Place in Nature*, 2nd edn, ed. W. Cronon, pp. 379–408. New York, NY: W. W. Norton.

Parker, I. and M. Amin (1983). *Ivory Crisis*. London: Chatto and Windus.

Perevolotsky, A. and N. G. Seligman (1998). Role of grazing in Mediterranean rangeland ecosystems. *BioScience* **48**:1007–1018.

Peters, P. (1996). "Who's local here?" *Cultural Survival* **20**(3):22–25.

Proctor, J. D. (1996). Whose nature? The contested moral terrain of ancient forests. In *Uncommon Ground: Rethinking the Human Place in Nature*, 2nd edn, ed. W. Cronon, pp. 269–297. New York, NY: W. W. Norton.

Raven, P. H. (1998). Wildlife conservation in Kenya. *Science* **280**:1510–1511.

Redford, K. H. (1991). The ecologically noble savage. *Cultural Survival* **15**(1):46–48.

Robinson, J. G. (1993). The limits to caring: sustainable living and the loss of biodiversity. *Conservation Biology* **7**:20–28.

Robinson, J. G. and K. H. Redford (eds) (1991). *Neotropical Wildlife Use and Conservation*. Chicago, IL: University of Chicago Press.

Seidensticker, J. (1997). Saving the tiger. *Wildlife Society Bulletin* **25**:6–17.

Seligman, N. G. and A. Perevolotsky (1994). Has intensive grazing by domestic livestock degraded Mediterranean Basin rangelands? In *Plant–Animal Interactions in*

Mediterranean-Type Ecosystems, ed. M. Arianoutsou and R. H. Groves, pp. 93–102. Dordrecht: Kluwer Academic Publishers.

Simberloff, D. S. (1987). The spotted owl fracas: mixing academic, applied, and political ecology. *Ecology* **68**:766–772.

Slater, C. (1996). Amazonia as Edenic narrative. In *Uncommon Ground: Rethinking the Human Place in Nature*, 2nd edn, ed. W. Cronon, pp. 114–131. New York, NY: W. W. Norton.

Slocombe, D. S. (1993). Implementing ecosystem-based management. *BioScience* **43**:612–622.

Sutton, D. B. (1990). From the Taj to the tiger. *Cultural Survival* **14**(2):15–19.

Turnbull, C. M. (1972). *The Mountain People*. New York, NY: Simon and Schuster.

United Church of Christ, Commission for Racial Justice (1987). *Toxic Waste and Race in the United States: A National Report on the Racial and Socioeconomic Characteristics of Communities with Hazardous Waste Sites*. New York, NY: United Church of Christ.

Walters, C. J. (1986). *Adaptive Management of Renewable Resources*. New York, NY: Macmillan Press.

White, R. (1996). "Are you an environmentalist or do you work for a living?": work and nature. In *Uncommon Ground: Rethinking the Human Place in Nature*, 2nd edn, ed. W. Cronon, pp. 171–185. New York, NY: W. W. Norton.

12

Central concepts – the flux of nature

Wherever we seek to find constancy, we discover change. . . . The old ideas of a static landscape, like a single musical chord sounded forever, must be abandoned, for such a landscape never existed except in our imagination. Nature undisturbed by human influence seems more like a symphony whose harmonies arise from variation and change over every interval of time. We see a landscape that is always in flux, changing over many scales of time and space, changing with individual births and deaths, local disruptions and recoveries, larger scale responses to climate from one glacial age to another, and to the slower alterations of soils, and yet larger variations between glacial ages.

(Botkin 1990:62)

We have seen that utilitarian and preservationist resource managers tend to envision nature as moving toward equilibrium. In this view, ecological systems are seen as closed and self-regulating. Like a pendulum, they return to their original state if they are altered. This balance-of-nature view generated many important insights in at least three areas of inquiry. First, when biologists looked at populations from this perspective, they focused on those that were in equilibrium with their resources, that is, populations near carrying capacity (Chapter 2). Intraspecific competition and density-dependent population growth were elucidated in this context. Second, community ecologists viewed communities as proceeding toward a stable climax and focused their attention on disturbances that set back succession. This allowed ecologists to distinguish between the existing plant community on a site and the potential vegetation at that location. Third, the theory of island biogeography turned the attention of conservation biologists toward the equilibrium that occurs if island extinction rates and colonization rates are in balance. Insights from the theory of island biogeography were used to examine the dynamics of extinction and colonization among populations in fragmented habitats.

Balance, stability, equilibrium – this theme recurs over and over again in the writings of resource managers and conservation biologists. But recently, scientists have begun questioning the view that most of nature is in equilibrium or on its way to equilibrium most of the time. This does not mean that the insights generated by equilibrium theories are incorrect or insignificant. Rather, it suggests that important nonequilibrium phenomena in the natural world were overlooked because scientists were not interested in them or regarded them as unimportant.

Before considering an alternative to the equilibrium view, let us re-examine some ideas about equilibrium.

12.1   Revisiting equilibrium theories

12.1.1   Competition and density-dependent population regulation

Reassessing the role of density dependence
In many cases, the idea that populations are regulated by density-dependent processes works well as a guide for managing harvests of game and commercially exploited species. But not always. Until the 1950s, the northern fur seal harvest appeared to be a well-managed and successful hunt based upon conventional harvest theory (Chapter 4). By 1950 the rookeries even began to show signs of overcrowding. Pups were crushed by bulls in the rookeries, and diseases and parasite loads increased. Managers believed that these were density-dependent problems that would be corrected when the harvest was increased and population density was reduced. This was accomplished by killing females (because the killing of polygynous males has little effect on reproductive rate). These measures succeeded in bringing the numbers of fur seals down, but the population did not stabilize at the desired level. Surprisingly, except for a brief increase in the early 1970s, the downward trend in fur seal numbers continued (Gentry 1998). Clearly, there was something going on that conventional thinking had failed to grasp. Between 1800 and 1950, northern fur seals twice recovered from precipitous declines, yet in spite of the fact that for more than a century fur seal populations exhibited classic, density-dependent resilience, the fur seal population continued to decline after 1950.

The reasons for this decline are still not well understood. Apparently the seals' ecology is more complex and their environment is more variable than managers had assumed. Evidently the marine environment changed in ways that led to increased mortality or decreased reproduction, or both, and these effects were independent of population density. One possibility is that the prey base in the North Pacific has changed (Gentry 1998). The simultaneous decline of seabird populations as well as fur seal populations on the Pribilof Islands and on Robben Island supports this interpretation. Thus, in spite of over a century of successful management and several decades of apparently sustainable harvest (see Chapters 1 and 4), equilibrium models of seal population dynamics ultimately proved inadequate. Rather than being a textbook case of density-dependent population dynamics, it now appears that northern fur seals can be regulated by both density-dependent and density-independent processes.

In fact, considerable evidence now suggests that this situation is not unusual. Population regulation is not necessarily a matter of density-dependent *or* density-independent processes; rather, the two types of phenomena can act simultaneously in the same population (Coulson *et al.* 2000). For example, in Morongoro, Tanzania, field studies combined with models of population dynamics in the multimammate rat suggest that the timing of reproduction in this pest species is affected by rainfall, which operates in a density-independent fashion, but density-dependent responses prevent the population from becoming very low or very high. These factors interact in complex ways that are still not thoroughly understood (Leirs *et al.* 1997).

The concept of competition also influenced ideas about human population growth and its relationships to environmental problems. We saw in Chapter 7 that influential scientists like Garrett Hardin and Paul Ehrlich see population growth as the inevitable outcome of self-interested exploitation of commonly held resources and hold that such growth was the root cause of the post-World War II environmental crisis. Their interpretations have been challenged, however.

Revisiting Malthus
The Malthusian explanation of current environmental problems attributes poverty in poor nations to population growth (although, as we have seen, some ecologists such as Paul Ehrlich recognize that high levels of resource consumption and inequities in the distribution of resources play a role too). In the Malthusian view, resource use intensifies as the human population expands, and environmental degradation is the inevitable consequence.

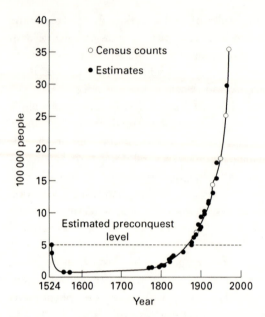

Figure 12.1. Population growth in El Salvador, 1524–1971. (After Durham 1979.)

William Durham examined the evidence for a Malthusian interpretation of deforestation in El Salvador, one of Central America's most densely populated nations (Durham 1979). The growth curve for El Salvador's human population after 1800 certainly appears to demonstrate exponential growth (Figure 12.1). Furthermore, the rise in population from the beginning of the nineteenth century to the mid-twentieth century was accompanied by a rapid rise in deforestation. Estimated forest cover fell from 60%–70% in 1807 to just 8% by 1946, and most of the deforested areas, even steep, highly erodable slopes, were converted to growing food. Between 1770 and 1892, the human population of the forested northern districts of Chalatenango and Cabañas grew by 1.90% and 2.49% per year, respectively, although annual growth in the national population was only 1.38%. Clearly, people had moved into forested regions, and this movement was accompanied by deforestation. But was the colonization of the forests caused by population growth?

According to Durham's calculations, in 1892 farmers owned an average of 7.4 ha. El Salvador's farming population increased by a factor of 3.8 between 1892 and 1971, but changes in access to and use of resources occurred during this period as well. The amount of land dedicated to export crops, particularly coffee, increased, and the amount of land devoted to food production decreased. By 1971, 50.8% of the agriculturally active population was either landless or owned less than 1.0 ha of land. For this group, average land holdings had declined by a factor of 19.5. In other words, for the poorest farmers, the decline in available resources (land) was far greater than the average increase in population. Even if we do not focus on the poorest segment of the rural population, a similar pattern emerges: per capita land decreased more than population increased. This indicates that the pressure on El Salvador's forests was not simply a result of increasing population.

Remember that correlation does not equal causation. The fact that forest exploitation and the density of El Salvador's rural population increased simultaneously does not prove that forest exploitation was caused by or caused solely by population growth. Durham's analysis suggests that change in land use from growing subsistence crops to growing crops for export, and the resulting changes in land ownership, played an important role as well, by creating conditions that led people to migrate into and cultivate the forests of this densely populated Central American nation. Rather than being the sole cause of deforestation, the growth of the human population was linked in complex ways to changes in land use and land tenure, which affected the conversion of forests to fields.

Revisiting the commons

We noted in Chapter 7 that the argument articulated by Garrett Hardin in "The tragedy of the commons" has been interpreted to mean that communally owned lands must inevitably be overused and degraded and that this interpretation stems partly from confusions between open-access resources and communally owned resources. In *Ecology and Equity: The Use and Abuse of Nature in Contemporary India*, Madhav Gadgil and Ramachandra Guha describe the fate of India's common lands under the British Empire. The soils of tropical India are relatively infertile. For centuries agricultural peasants made a living from these poor soils by allowing their cattle to graze on the surrounding communal lands and then fertilize their fields. These communally held lands also provided fuelwood for cooking as well as materials for constructing dwellings and implements. They were managed cooperatively; users were required to contribute to maintenance, and there were restraints on individual use. Some specific areas were also set aside as sacred groves, ponds, or pools. After the British arrived in India, colonial administrators allowed some agricultural lands and irrigation ponds to remain in communal ownership, but grazing lands and forests were taken over by the state, and community control of these lands ended. They were converted to open-access lands, which were often overutilized, or they were set aside as forest reserves dedicated to timber production. Diverse mixed forests were converted to monocultures of commercially valuable trees, such as teak. Traditional land uses were prohibited (Gadgil and Guha 1995).

Rather than tragic degradation of a commonly held resource, with privatization or state regulation coming to the rescue, the reverse seems to have been the case in Indian agricultural villages. The intensity of exploitation increased under private or state control. Similarly, throughout much of Africa traditional institutions regulating communal property have been eroded. Many natural resources that were once community property are now open-access resources (Kiss 1990).

In the scenario Hardin described, access to "the commons" is always unrestricted, users are selfish and their behavior is not influenced by social norms or taboos, and users have perfect information and try to maximize their short-term gains on the basis of that information. Research on societies that hold resources in common indicates that this scenario is by no means universal. The outcome of communal resource ownership depends upon the specific social, historical, ecological, and institutional context in which communal property occurs.

Marine fisheries are prone to overexploitation, in part because they are open-access resources. Yet off the coast of Maine, generations of lobster fishers have harvested lobsters without depleting the resource. Although Maine's Department of Marine Resources has the formal authority to regulate the lobster harvest, in practice, local traditions limit who can fish and where, and the state almost always follows the recommendations of the people who are most familiar with the resource: the lobster fishers themselves (Acheson 1987; Jensen 2000). To understand how communal property will affect resource conservation we need to know things like: Are there rules about using common resources? Can the behavior of group members be monitored or controlled? How do people understand their obligations to other group members and to nature? What kinds of relationships does the group have with neighboring groups (McCay and Acheson 1987)?

Hardin's implication that controlled resource use depends upon restricting access, either through privatization or regulation, underestimates the ability of local groups to cooperatively manage resources for the common good under some circumstances. It also minimizes the problems of private ownership and state regulation. Privatization can lead to commodification and overexploitation, as we saw in Chapter 1, and state control does not guarantee prudent resource use either, for a variety of reasons. Resource depletion can occur under state control because of bureaucratic inefficiency, because administrators are vulnerable to political pressures from special interest groups, because local people do not trust the state and do not comply with regulations, because resource conservation is not as high a priority as military or industrial growth, or because the state does not take advantage of the expertise of local people (McCay and Acheson 1987).

This is not meant to imply that communal ownership always leads to resource conservation. Clearly, under some circumstances tragedies of the commons can occur. The challenge is to understand what those circumstances are and what circumstances are conducive to sustainable use of common resources (Jensen 2000).

12.1.2   The stable climax

The influential ecologist Frederic Clements and other early proponents of the theory of succession considered communities to be tightly-knit groups of species that respond to environmental changes as a unit, much like a single organism (see Chapter 3). They also believed that succession reaches a relatively stable endpoint. If Clements and Tansley were right in their characterization of communities, then we would expect communities to reach a point at which each species would perpetuate itself on a site indefinitely unless there were a disturbance. In this way, the species composition of a climax community would be stable until a disturbance changed conditions on the site and started the process of succession again. But a large body of research (reviewed by Pickett *et al.* 1987) indicates that this is not the case. Each species has unique physiological requirements, dispersal abilities, modes of reproduction, responses to environmental stress, competitive abilities, responses to herbivory, and so on. So each species responds differently to the environment. In other words, most plant ecologists today agree with Gleason's individualistic concept of the community, rather than with Clements' organismic view (see Chapter 3). Community composition is constantly in a state of flux; species are not locked together in tightly coordinated units (Brubaker 1988).

Clements argued that plant communities follow one another in predictable and inevitable sequences that lead to stable climax communities. But it seems that stable climaxes actually occur rarely, if ever, in nature. By the 1970s most ecologists recognized the importance of natural disturbances in a variety of communities, including northern boreal forests, chaparral, tropical forests, and eastern and western temperate forests. In these communities, "natural disturbance is so common that it keeps the system from ever reaching a stable state, so it is unrealistic to assume that climax is the 'normal' condition for ecosystems to be in" (Sprugel 1991:3).

Other work suggests that in addition to being common, disturbances are necessary to maintain certain community attributes that are generally considered desirable. Equilibrium models of community dynamics predict that communities with high species diversity, such as moist tropical forests and coral reefs, are stable, with low rates of disturbance. But data on the population dynamics of corals and trees suggest that the species diversity of coral reefs and moist tropical forests actually decreases in the absence of disturbances (Connell 1978). This has important implications for management, because by excluding disturbances from managed areas we may inadvertently decrease community complexity and diversity (Box 12.1).

Box 12.1   Mettler's Woods

Mettler's Woods, a 26-ha remnant of old-growth oak forest located in central New Jersey, has not experienced a major disturbance for centuries, but the absence of disturbance did not lead to the expected result. The trees in Mettler's Woods were old when they were purchased in 1701 by a Dutch settler, and they have not been burned or logged since that time. In 1749 a botanist described the woods as consisting of oaks, hickories, and chestnuts with very little underbrush. In the 1950s, Rutgers University acquired the woodlot. Since it was thought to be in a climax state, all that was deemed necessary to preserve the forest community was to exclude people and disturbances. But, surprisingly, the existing community is not perpetuating itself, even though it has been rigorously protected from disturbances that would set back succession. Large acorn crops are produced periodically, yet virtually no oak seedlings survive. Old oaks die, but the young saplings beneath the old oaks are sugar maples, not oaks. If these trends continue into the future, the oak forest will give way to a forest dominated by maples. Furthermore, the forest floor is no longer clear of underbrush, as it had been in the eighteenth century. Instead, dense shrubs and saplings have grown up. These changes were unexpected, since by definition a climax community is capable of perpetuating itself under the conditions it creates.

When scientists from Rutgers examined trunk sections from trees that had died during a hurricane, they found that before 1701 the forest had burned about once every 10 years. Evidently, oak regeneration requires frequent ground fires. The oak forest persisted in the face of frequent fires because oaks are more resistant to fire than maples. The exclusion of fire from Mettler's Woods actually degraded the community that was considered "old growth" (Botkin 1990; Pickett et al. 1992).

Even though disturbances are so commonplace that a given patch is unlikely to reach a stable equilibrium, plant communities could still be at equilibrium when viewed at a larger spatial scale. Consider a forested region with a moderately high frequency of fires. The total amount of old growth in the landscape might remain fairly constant over time. Succession is set back at some sites (for instance on dry slopes that burn often), and late-successional communities develop in locations that burn rarely. While the composition of a given patch may change constantly, the total amount of old-growth forest, viewed at the scale of the landscape, can be in a stable equilibrium in which the maturation of some patches to late-successional stages balances the creation of early-successional patches.

Another modification of equilibrium theory proposes that there can be more than one equilibrium state in an ecosystem (Holling 1973, 1986). For example, Dublin *et al.* (1990) tested several alternative hypotheses regarding stable states in the vegetation of the Serengeti–Mara ecosystem of East Africa by combining data on the causes of tree seedling mortality with predictions from mathematical models. Their analyses indicated that there are two stable states in this ecosystem, and a combination of fire and herbivory can shift the system from one state to another. Fire initially kills trees and converts the vegetation dominated by trees to grassland. The new stable state is then maintained by elephants, which prevent trees from regenerating in the grassland (Dublin *et al.* 1990; Sinclair 1995).

Multiple stable state models and equilibrium at the level of the landscape mosaic are refinements of more simplistic equilibrium models that pictured communities as inexorably moving toward a single steady state. But even these more dynamic models do not account for two kinds of forces that can prevent a community from reaching a climax with a predictable composition.

First, the environment is not static. Continents drift, mountains are uplifted and eroded, sea level rises and falls, and climates change. These alterations affect the development of vegetation, so that the potential natural vegetation on a site changes. Communities are "trajectories of change from the past, through the present, and into the future" (Tausch 1996:99). People tend to overlook these changes, because on the time-scale of a human life span they are difficult to perceive. Even when "we recognize succession as a dynamic play, [we] regard the stage on which it is played out as static and unchanging" (Sprugel 1991:6).

Second, each site has its own singular history. No two sites are exactly the same. Even if they have similar climate, geology, and soils, each has a unique history of disturbance, and each past disturbance is unique in its size, timing, and severity. So sites will differ in the availability of resources, seeds, spores, and so on. This variability is another reason why we cannot predict the precise composition of the vegetation that will eventually dominate a site. A single, large-scale, unique event can have long-lasting effects on the proportions of a landscape in different successional stages. For example, about 4800 years ago, the eastern hemlocks in eastern North America were attacked by a novel pathogen. This single event caused hemlock populations to plummet throughout the region. Recovery from this crash took nearly 2000 years (Davis 1981). The landscape certainly could not be considered in equilibrium during that time. The Serengeti savanna is another striking example of vegetation produced by unique historical events (Box 12.2).

Box 12.2   People behind the scenes: Anthropogenic landscapes

The Yosemite Valley: Nature's park?

When Totuya, the last surviving Ahwahneechee Indian, returned to her homeland in the Yosemite Valley in 1929, she remarked that it was "Too dirty, too much bushy." Lafayette Bunnell, the diarist of the military expedition that routed her people from the valley, wrote a similar account of the valley's altered vegetation. He wrote in 1894 that when he had visited the valley in 1855 "there was no undergrowth of young trees to obstruct the clear open views in any part of the valley," but by 1894 the area of "clear open meadow land" had dwindled to one-fourth its former size (Olwig 1996:396). These anecdotal accounts suggest that the parklike quality which so impressed the first whites who visited Yosemite resulted from Native American burning and required regular ignition to be maintained. After the park was established, fires were suppressed, however. In spite of the fact that the famous landscape architect Frederick Law Olmsted was hired to design improvements that would preserve Yosemite's parklike quality (Spirn 1996), changes in the park's disturbance regime led to unanticipated changes in its legendary scenery.

The Serengeti savanna: Vaccines and acacias

Most of us are familiar with images of the Serengeti savanna from television nature shows. Mention of the Serengeti brings to mind grassy plains with widely spaced acacia trees. But recent work suggests that rather than being the climax vegetation of the region, the savanna may be the product of unique historical circumstances and human actions. Prior to the late nineteenth century, a combination of grazing by large herbivores and burning had inhibited tree reproduction throughout most of the region. In 1889 an Italian military expedition introduced rinderpest, a viral disease of cattle and their relatives, to Somalia (Sinclair 1977). The exotic disease spread rapidly throughout East Africa, devastating populations of cattle, wildebeest, and buffalo. The human population of the region also declined because the nomadic tribes starved when their cattle died. During this period, elephant hunting for the ivory trade increased. The absence of cattle, native ungulates, and elephants allowed trees to become established, thereby setting the stage for the development of savanna vegetation (Dublin *et al.* 1990; Sinclair 1995).

Eventually, a vaccine for rinderpest was developed, and native herbivores developed some resistance to the disease as well. Cattle and native ungulates again inhabit the plains. Mature acacia trees are now dying, but young saplings are not taking their place. The landscape we are familiar with is not reproducing itself. Thus, the acacias that we think of as the hallmark of the "natural" Serengeti landscape may be "due to an extraordinary outbreak of trees precipitated by the rinderpest epizootic late in the nineteenth century" (Sinclair 1995:109). What we see as typical is actually unique and idiosyncratic.

Since the tree-dotted savanna is a huge tourist attraction, land managers are in a difficult position. As Sprugel put it, "This has created extraordinary problems for the park and reserve managers of the region; visitors to East Africa (who provide the money that justifies the parks' existence) expect to see big, umbrella-shaped trees, and are disappointed if they do not. Regardless of the fact that the plains without trees may be more 'natural' than plains with trees, our image of the savannas has become fixed in a static early 20th century mold" (Sprugel 1991:8).

Short-turf grasslands in southern England: Interactions between livestock, ants, and butterflies

Populations of the large blue butterfly began to decline in the grasslands of southern England in the 1880s. The habitat of this butterfly was short grass with thyme, the food plant of the large blue caterpillar. The butterfly decline was attributed to collecting, and in the 1930s a reserve was created. Collectors were excluded from the reserve, as well as livestock. With the cessation of grazing, the vegetation in the reserve grew taller and more dense. The butterflies disappeared from the reserve as well, in spite of the fact that plenty of wild thyme remained. It turns out that an increase in vegetation of just 2 cm was enough to create a cooler microclimate at ground level, which caused one species of red ant to replace the species that had been present in the shorter turf. In addition to feeding on thyme, the large blue caterpillars must enter ant nests and parasitize larval ants, but survival is much higher in the nests of the short grass ant species than in the species that predominates in slightly taller grass (Thomas 1980). Thus the cessation of grazing brought about subtle changes in habitat that were unfavorable to the large blue butterfly. In 1979, the last known colony of the large blue butterfly disappeared from England. After short turf was re-established through burning and grazing, large blues from northern Europe were successfully reintroduced to southern England.

The local extinction of the large blue butterfly illustrates again how important it is to understand the ecological requirements and the complex interspecific interactions of declining species. The decline of the large blue butterfly was recognized over a century ago, and considerable effort was expended in trying to save the population. But because the cause of its problems was not correctly diagnosed, these efforts were to no avail. This example also shows how a well-intentioned management program that fails to understand the role of anthropogenic activities can be counter-productive.

Rajasthan's Keoladeo National Park: Surprises from a manmade wetland

Even within a constructed landscape, the factors affecting plant community structure and composition are not always appreciated. Until recently, Keoladeo, a shallow wetland created in the eighteenth century in India's Rajasthan, provided habitat for impressive concentrations of birds. Thousands of waterfowl visited the area in winter, and during the monsoon season it served as a breeding ground for herons, storks, ibises, and egrets. Prior to independence, Keoladeo was controlled by the maharaja of Bharatpur. Resource utilization of the site was at times intensive: during the colonial period, British aristocrats were invited for waterfowl hunts, during which tens of thousands of birds were reportedly killed in a day. Local villagers were also allowed to graze their buffalo and cattle in the wetland. In the early 1980s, grazing was banned. Police fired on local villagers who protested the ban, and several villagers were killed. According to Gadgil and Guha, "The results have been disastrous for Keoladeo as a bird habitat. In the absence of buffalo grazing, *Paspalum* grass has overgrown the wetland, however, choking out the shallow bodies of water, rendering this a far worse habitat, especially for wintering geese, ducks and teals, than it ever had been before" (Gadgil and Guha 1995:93; Sarkar 1999).

If succession leads to stable, undisturbed, predictable communities, then it is fairly simple to decide what a natural ecosystem is, and a hands-off approach seems an appropriate management strategy. But if succession does not operate in that way, the questions "What is natural?" and "How do we go about managing natural vegetation?" are not so easy to answer (Christensen 1988; Sprugel 1991). Often managers avoid confronting this question and just manage for "some past magical time of supposed ecosystem perfection when things were still 'natural'" (Tausch 1996:99). But there is no single scenario that we can objectively identify as the only natural possibility. This means we have to make choices about what to conserve, and those choices inevitably involve values.

12.1.3   The equilibrium theory of island biogeography

MacArthur and Wilson's equilibrium theory of island biogeography states that the biotas of islands reach a state of equilibrium at which colonization and extinction balance each other out. Conservation biologists have applied this model to small populations in fragmented habitats, seeking to understand how community dynamics in this unnatural situation, where the activities of people have carved up habitats into isolated refuges, compare to the equilibrium that theoretically would occur in an unfragmented landscape.

But do colonization and extinction often reach equilibrium on real islands? Simberloff and Wilson's study of arthropods on mangrove islands was designed to test the MW model, and it did provide experimental evidence of the sort of local colonization, extinction, and turnover the model predicts. Empirical evidence in support of the MW model is limited to that study, however. Nevertheless, the model has been widely used as a basis for management recommendations. One textbook on conservation biology states that "the island biogeography model has been empirically validated to the point where it is now accepted by most biologists" (Primack 1993:87). Simberloff himself, however, maintains that in spite of the evidence for turnover that he and Wilson amassed for one group of organisms in one setting, "we must be aware of the possibility that, when the field observations are made, they will show that certain taxa (or ecosystems) conform more closely than others to the theory." He laments the fact that the MW model has become "a theory so widely accepted as an accurate description of nature that failure of an experiment to yield the result deduced from the theory leads *not* to rejection of the theory but rather to attempts to fault the deductive logic or experimental procedure, or simply to willful suspension of belief in the experimental result" (Simberloff 1976:578; emphasis in original).

Diamond offered indirect evidence in support of the MW model in the form of repeated censuses of the Channel Islands of California (Diamond 1969) (see Chapter 10). This evidence is far from a conclusive proof that turnover occurred, however. He did not observe colonization and extinction directly; they were inferred from census data. The censuses might have missed birds that were really present and breeding on the islands. Species that were overlooked initially but found subsequently would be considered colonists, when they had actually been there all along. Similarly, species that were found originally but missed later would incorrectly be scored as extinctions. This "pseudoturnover" would make it appear that colonizations and extinctions had occurred more often than they actually had (Lynch and Johnson 1974). Furthermore, changes in the avifauna of the islands might be due to environmental changes, rather than to the kind of random population fluctuations predicted by the MW model. In addition to these criticisms, the evidence for extinctions on islands that were created relatively recently, such as land-bridge islands or Barro Colorado Island in the Panama Canal, has also been challenged, on similar grounds. (See Chapter 10 for a discussion of the controversy surrounding the implications of those studies for the design of nature reserves.)

Thus it is still not clear whether an equilibrium between colonization and extinction is a common phenomenon in nature. The MW model is useful because it focuses ecologists' attention on the processes of local extinction and colonization in isolated habitats. In particular, the idea that the relationship between the two processes will determine the long-term fates of populations has proved to be a very valuable insight. In this regard, the legacy of the MW model remains, regardless of whether its specific predictions about equilibrium turn out to be true for varied organisms and settings. On the other hand, the specific management recommendations derived from the model are so general as to be of limited use. The argument for larger reserves, with its "all other things being equal" caveat is not very helpful, because all the other relevant factors never are equal. There is no substitute for detailed knowledge about the specifics of each situation, including the habitat requirements and dispersal abilities of the species in question, their reproductive potential and sources of mortality (including predators and parasites) in different habitats, the amounts and sizes of habitat patches in a proposed reserve, and so on. For this reason, the debate over whether a single large reserve will preserve more species than several small ones of equivalent area (Chapter 10) has subsided. More and more ecologists are turning their attention to other pursuits, such as studying the dynamics of metapopulations in habitats of varying quality, as a way of getting the information they need to make recommendations for managers (Caughley and Gunn 1996; Simberloff 1997).

12.2   A new perspective: The flux of nature

12.2.1   Background

George Perkins Marsh complained that wherever "man . . . plants his foot, the harmonies of nature are turned to discords" (Marsh 1874:34). I noted in the Introduction to this volume that to Marsh, people were "everywhere a disturbing agent," upsetting the natural balance of nature. But, although the metaphor of nature in balance remains entrenched in the popular imagination, many ecologists today see nature in a different light, as the quote that opened this chapter indicates. Northern fur seals of the Pribilof Islands, oaks in Mettler's Woods, acacias in the Serengeti – equilibrium models do not satisfactorily explain observed phenomena in these and a good many other cases. In addition to these accumulated observations from a variety of organisms in a variety of ecosystems, equilibrium theories face another problem. Many ecologists have an uneasy sense that all too often their colleagues have molded their interpretations to fit inappropriate equilibrium models, willfully suspending disbelief (in Simberloff's words). In a paper on "The shifting paradigm in ecology," Pickett and Ostfeld suggested that by

idealizing and simplifying the ecological world, the tenets of the classical paradigm have blinded ecologists and managers to critical factors and events that govern ecosystems. The assumptions have also caused scientists and managers to neglect important dynamical pathways and states, and to disregard important connections among different systems. (Pickett and Ostfeld 1995:265)

These misgivings gave rise to an alternative perspective on nature: the idea that the natural world is often not near, or even approaching, equilibrium. This viewpoint does not deny that the phenomena of competition, succession, and turnover occur, but it does suggest that disturbances are so widespread and frequent that stable climax communities occur rarely if ever and many populations are kept below the level at which density-dependent interactions play a pivotal role. Equilibrium states can be found in nature, but they are not the rule. All models are simplifications, and their appropriateness depends on the circumstances and the scale being investigated. Equilibrium models are useful in certain situations, but it is important to recognize their limitations.

Sometimes when a process that seems to be at equilibrium is viewed at a different scale, nonequilibrium dynamics appear. For example, the composition of a mature forest may appear to be stable when viewed on a time-scale of years to decades, but when it is viewed on a scale of thousands of years it may be changing in response to long-term climate change. As early as 1965, geologists S. A. Schumm and R. W. Lichty made a similar point with respect to physical phenomena. They pointed out that although a landform may appear to be in equilibrium when it is viewed over a short period of time and a small area, when larger time-spans and larger areas are taken into consideration, dynamic change becomes evident. For example, the amount of water and sediment entering and leaving a small stream reach could briefly balance each other out, but over a longer time-span and a larger area it would be evident that erosional processes were resculpting the hills (Schumm and Lichty 1965).

Proponents of this new view suggest that "flux of nature" is a better metaphor than "balance of nature" for describing how the natural world operates (Pickett *et al.* 1992; Pickett and Ostfeld 1995). This view is sometimes referred to as the nonequilibrium paradigm or the flux-of-nature paradigm. (A paradigm is a central concept that organizes a body of knowledge; a paradigm shift occurs when a prevailing theory is toppled by the accumulated weight of evidence it can not explain, bringing about a scientific revolution (Kuhn 1970). There is some debate about whether the "flux-of-nature" viewpoint is a revolutionary paradigm or simply a useful metaphor, but clearly it represents a new way of looking at natural phenomena.)

12.2.2   Key points

Six key points that follow from a nonequilibrium perspective (Fiedler *et al.* 1997; Meyer 1997; Ostfeld *et al.* 1997) are discussed below. Techniques for putting these principles into practice are described in Chapters 13 and 14.

Equilibrium is not the usual state for nature

In the flux-of-nature perspective, equilibrium phenomena are the exception rather than the rule in nature. The evidence leading up to this conclusion is summarized above.

Disturbances are widespread and common

Until recently, many American ecologists, led by Clements, tended to minimize the importance of disturbances, seeing them "only as a mechanism resetting the inexorable march toward equilibrium" (Pickett and White 1985:372). Biotic disturbances such as grazing, seed predation, digging, wallowing, and trampling were often overlooked, and abiotic disturbances such as fires, volcanic eruptions, hurricanes, and floods were seen only as catastrophic agents that set back succession. But by the 1980s considerable evidence had accumulated that disturbance occurs in virtually all types of ecosystems (although ecosystems differ in how and how often they are disturbed) and that it affects resource availability and a host of other aspects of community structure and function (Bazzazz 1983; Pickett and White 1985). Whereas, formerly "managers viewed disturbance as having mostly negative impacts, . . . currently, evidence suggests nearly the opposite: preservation of natural disturbance regimes is essential to promote healthy, dynamic ecosystems" (Rogers 1996:13).

Ecosystems are open and interconnected across a landscape

The movements of organisms across ecosystem boundaries make it necessary to manage landscapes rather than single ecosystems (see Chapter 10). Matter, energy, and organisms don't stop at political boundaries; they move between states and nations, through international waters, and between private and public lands. The landscapes we must consider therefore become very large. This is not a new principle; it dates back at least to Leopold's edge effect. Preservationist conservation also considers the configuration of reserves in a landscape. What is new in the sustainable-ecosystem approach is the emphasis on ecological processes in the matrix between reserves.

Heterogeneity has a pivotal influence on ecosystems

Many species are more abundant or have higher survival rates in structurally complex habitats than in homogeneous environments. Temporal heterogeneity is as important as spatial variation. In streams, for example, some species thrive in wet years, while others need dry years, so overall biological diversity benefits from year-to-year variations in stream flow (Poff *et al.* 1997). Unfortunately, much past management has greatly simplified landscapes, by building dams, cultivating fields, straightening stream channels, suppressing fires, planting tree plantations, and so forth (Holling and Meffe 1996), but new ways of incorporating temporal and spatial heterogeneity into ecosystems are being devised. Box 12.3 describes a specific case in which environmental heterogeneity enhances survival of a rare species.

Box 12.3   Habitat complexity and survival of the Bay checkerspot butterfly

For four decades Paul Ehrlich and his colleagues at Stanford University have studied the population dynamics of the Bay checkerspot butterfly, which occurs only on a particular, patchily distributed soil type found on grassy hillsides near San Francisco. The area has a Mediterranean climate, with mild, wet winters and dry summers. In this climatic regime, herbaceous annual plants begin growth after the onset of the fall rains, grow throughout the winter, and then dry up, or senesce, when summer arrives. Female checkerspots lay their eggs in spring, and when the larvae hatch they must develop to the point where they can enter summer diapause (dormancy) before their host plants complete their annual cycle. Since the larvae can move only short distances, they are completely at the mercy of the microclimate in the patch of habitat where they hatch. During unusually dry summers, the host plants senesce early on south-facing slopes, which are warmer and dry out sooner than slopes with other aspects. Many larvae die in this microhabitat because their food dries before the larvae have completed their development, but some survive on the cooler slopes facing north, northeast, or northwest. In wetter years the reverse is true: hatching is delayed on the cooler slopes, and consequently few larvae have time to complete their development there. Survival is better on the warmer, south-facing slopes in those years, however, because the larvae get a head start on development (Murphy and Weiss 1992).

If the Bay checkerspot's habitat did not provide this kind of topographic heterogeneity, populations would not be able to cope with year-to-year climatic variation. Variation in topography produces microclimatic variation, which in turn provides refugia that allow checkerspots to cope with climatic variation. (Their ability to persist in the face of long-term, global climate change is questionable, however; in fact, the Stanford research group has witnessed the extinction of a number of local checkerspot populations associated with extreme weather, such as prolonged droughts or unusually wet winters.) Even a population in a relatively large habitat patch is not secure if the patch fails to provide a variety of micro-climates. For example, the Coyote Reservoir site, which consists mainly of an east-facing slope, supported a dense colony of butterflies in 1971, but by 1976 the population was gone (Ehrlich and Murphy 1987).

Many states previously considered natural are influenced by the activities of people

In the balance-of-nature view, natural systems are closed and self-contained. Since people are not part of natural systems, management should strive to protect or control these systems. Research grounded in this perspective has often overlooked or failed to appreciate the effects of people on ecosystems (Box 12.2). In the flux-of-nature view, however, the role of people in ecosystems is explicitly addressed. People have always affected the ecosystems in which they live in a variety of ways, both intentional and unintentional. Thus, social context should be considered as well as geographic context. This principle grew out of two insights. First, there is no way that people can live without consuming resources. Like other animals we consume food, water, and oxygen and produce wastes; in addition we set or put out fires, move plant and animal populations around, remove vegetation, and move water. Second, even relatively isolated ecosystems are no longer exempt from the impacts of human populations; there are no pristine places. The activities of humanity now affect every centimeter of the globe, even where there are no people.

Not only is it impractical to try to manage natural resources without considering the impacts of people, it is also unwise. Ecologist Judy Meyer puts it this way: "Conservation is essentially management of human activity in the landscape, so to ignore the societal context for conservation efforts, is to invite failure" (Meyer 1997:141). Because the effects of people are widespread, we cannot avoid making decisions about how to respond to or manage those effects. In the previous chapter, we saw that pressures to consider the social context of conservation have been mounting. In Chapter 14 we will consider some cases where this is being done.

The idea that people should be considered as part of ecosystems has generated a heated controversy (see for example Callicott *et al.* 1999, 2000; Hunter 2000; Willers 2000). Much of the argument centers on whether activities of people should be considered "natural" or uniquely different from the rest of the natural world. This seems unproductive. Saying that people are part of nature does not mean that there should be no limits to resource exploitation or that no areas should be managed as wilderness. As Pickett *et al.* point out, "the flux-of-nature paradigm does not excuse human excesses relative to bio-diversity, ecosystem function, and natural heterogeneity" (Pickett *et al.* 1997:362). It does, however, explicitly state that we need to understand present and past human impacts on ecosystems. With this understanding and an understanding of the limits of ecosystems, we can make informed decisions about which states are desirable and which uses ought to be allowed. This should increase the likelihood that our management strategies will produce the results we desire.

12.2.3   Minimum conditions for maintaining ecosystem functions

Advocates of a sustainable-ecosystem approach to resource management often direct their attention to the conservation of functions and structures, under the assumption that doing so will conserve species (Franklin 1993). Although this is an area where a lot of research needs to be done, some general guidelines are clear. Using natural resources in ways that maintain temporal and spatial variability and that maximize the potential for return to predisturbance conditions increases the probability that species richness and genetic diversity will be maintained. At least five interrelated things are involved in conserving functioning ecosystems: (1) productivity, (2) soil structure and fertility, (3) disturbance regimes, (4) the ability of affected species to survive and reproduce, and (5) biotic interactions.

Since all other trophic levels depend upon an ecosystem's producers, it is essential that the ability of plants to trap solar energy and produce biomass be maintained (Franklin 1995). Ecosystem productivity varies dramatically between ecosystems, of course, and there are certain natural constraints on productivity set by climate and geology (see Chapter 13). But within these constraints, the choices we make about resource use affect whether the productive capacity of an ecosystem is sustained or degraded. For plant productivity to be sustained, soil fertility and structure must be preserved. Soils or soil nutrients must not be removed from a site faster than they are replenished, and soil structure must not be drastically altered, for instance by compaction or by irreversible alteration of the microbial community in and on the soil. Otherwise soils may lose their ability to absorb water, and runoff and erosion may increase. Appropriate amounts of water must be available at appropriate times; that is hydrological regimes should be maintained. This is not to imply that none of these parameters may be changed, but they should not be altered beyond the point from which an ecosystem can adjust or recover.

Interactions between species should be maintained at levels that allow coevolved species to persist and ecosystem functions to be fulfilled. This includes antagonistic interactions between species, such as predation, parasitism, and herbivory, as well as mutualistic interactions, such as seed dispersal, nitrogen fixation, and pollination. Decomposers and symbionts, especially microscopic and belowground forms, are often overlooked because they are poorly known, although numerically they are responsible for much of the species richness in many ecosystems. These invertebrates, bacteria, cyanobacteria, and fungi recycle nutrients, and some enhance the availability of nutrients to vascular plants. The interactions of herbivores and pathogens with their target species also play essential roles in ecosystems, even though from a narrow perspective their effects might be considered "negative" because they kill plants. By killing vegetation, they create openings for colonization of early successional species, and provide important habitat features such as coarse woody debris and cavities in standing trees.

To conserve the productive capacity of an ecosystem, environmental contaminants should be kept below the level at which they alter community composition by affecting survival and reproduction. Furthermore, harvests of featured species should not compromise the ability of the harvested species to reproduce or regenerate after harvest. Therefore, harvests should not alter habitat and population structure to a point where normal reproductive behavior is inhibited, and the harvest of young individuals must not be so severe as to interfere with recruitment into a population.

To summarize, "natural systems . . . have many states or 'ways to be' and many ways to arrive at those states" (Pickett *et al.* 1992:70). These many possibilities stem from the fact that disturbances are frequent and ubiquitous, operate at many scales, and are influenced by the unique history and chance events of each population and ecosystem. In this view, ecosystems are not closed, self-regulating systems. Rather, they are open and interconnected. Consequently, to be understood they must be viewed in the context of their surroundings.

12.3   Implications of the flux-of-nature viewpoint for conservation strategies

The flux-of-nature viewpoint suggests novel ways of managing renewable natural resources. The application of insights from the balance-of-nature perspective to resource management has been successful in many cases, but in other instances strategies based upon the balance of nature backfired. For example, in the Clementsian equilibrium view there was an implied tendency to view disturbances in a negative light. One consequence of this was fire suppression, which had many unintended and far-reaching ecological consequences. In dry climates with high biomass production, the suppression of forest fires allowed flammable fuels to build up and led to the destruction of timber – an outcome that was exactly the reverse of what was intended by utilitarian managers. Similarly, the protection of Mettler's Woods resulted in a decline in the old-growth oak community, the opposite of the intended preservationist outcome. In these instances, utilitarian and preservationist goals were not well served by management strategies based on an equilibrium perspective. Both approaches led to management actions that overlooked important ecological interactions and processes.

The flux-of-nature perspective is useful because it sheds light on why these policies failed to achieve their objectives. In addition, as we saw in the previous chapter, many resource managers are now refocusing their efforts and redefining their goals because of recent social, economic, and political developments as well as for ethical and practical reasons. In particular, the role of people in ecosystems and the social context of resource use are receiving more attention. The flux-of-nature approach suggests three principles for management strategies: (1) conserve processes not just species or parcels of land, (2) conserve the geographic context in which processes occur, (3) include people as elements in the landscape.

12.4   Conclusions

Although there is a long-standing tension between those who favor preservation of natural places and those who stress regulated use of natural resources, both groups share a similar world view – the balance-of-nature or equilibrium perspective – and the idea that people are outsiders. Now an alternative point of view has been articulated. The flux-of-nature perspective suggests a new theoretical basis for conservation and new strategies for the challenging task of conserving natural resources in the twenty-first century. In addition, it suggests a useful way of regarding the role of people in nature – as participants in the natural world rather than as outsiders. By reminding us that the past is not static and that people have been part of that past, the flux-of-nature viewpoint compels us to explicitly identify the value judgments and cultural contexts that underlie decisions about how society should manage natural resources. In the next two chapters we will consider examples of how these ideas can be put into practice.

References

Acheson, J. M. (1987). The lobster fiefs revisited: economic and ecological effects of territoriality in Maine lobster fishing. In *The Question of the Commons: The Culture and Ecology of Communal Resources*, ed. B. J. McCay and J. M. Acheson, pp. 37–65. Tucson, AZ: University of Arizona Press.

Bazzazz, F. A. (1983). Characteristics of populations in relation to disturbance in natural and man-modified ecosystems. In *Disturbance and Ecosystems: Components of Response*, ed. H. A. Mooney and M. Godron, pp. 259–275. Berlin: Springer-Verlag.

Botkin, D. B. (1990). *Discordant Harmonies: A New Ecology for the Twenty-First Century.* New York, NY: Oxford University Press.

Brubaker, L. B. (1988). Vegetation history and anticipating future vegetation change. In *Ecosystem Management for Parks and Wilderness*, ed. J. K. Agee and D. R. Johnson, pp. 41–61. Seattle, WA: University of Washington Press.

Callicott, J. B., L. B. Crowder, and K. Mumford (1999). Current normative concepts in conservation. *Conservation Biology* **13**:22–35.

Callicott, J. B., L. B. Crowder, and K. Mumford (2000). Normative concepts in conservation biology: reply to Willers and Hunter. *Conservation Biology* **14**:575–578.

Caughley, G. and A. Gunn (1996). *Conservation Biology in Theory and Practice.* Cambridge, MA: Blackwell Science.

Christensen, N. L. (1988). Succession and natural disturbance: paradigms, problems, and preservation of natural ecosystems. In *Ecosystem Management for Parks and Wilderness*, ed. J. K. Agee and D. R. Johnson, pp. 62–86. Seattle, WA: University of Washington Press.

Connell, J. H. (1978). Diversity in tropical rain forests and coral reefs. *Science* **199**:1302–1310.

Coulson T., E. J. Milner-Gulland, and T. Clutton-Brock (2000). The relative roles of density and climatic variation on population dynamics and fecundity rates in three contrasting ungulate species. *Proceedings of the Royal Society of London, Biological Sciences* **267**:1771–1779.

Davis, M. B. (1981). Outbreaks of forest pathogens in Quaternary history. *Proceedings of the 4th International Palynology Conference* B:216–217. Lucknow, India.

Diamond, J. M. (1969). The Channel Islands of California. *Proceedings of the National Academy of Sciences, U.S.A.* **64**:57–63.

Dublin, H. T., A. R. E. Sinclair, and J. McGlade (1990). Elephants and fire as causes of multiple stable states in the Serengeti–Mara woodlands. *Journal of Animal Ecology* **59**:1147–1164.

Durham, W. H. (1979). *Scarcity and Survival in Central America: Ecological Origins of the Soccer War.* Stanford, CA: Stanford University Press.

Ehrlich, P. R. and D. D. Murphy (1987). Conservation lessons from long-term studies of checkerspot butterflies. *Conservation Biology* **1**:122–131.

Fiedler, P. L., P. S. White, and R. L. Leidy (1997). The paradigm shift in ecology and its implications for conservation. In *The Ecological Basis of Conservation: Heterogeneity, Ecosystems, and Biodiversity*, ed. S. T. A. Pickett, R. S. Ostfeld, M. Shachak, and G. E. Likens, pp. 83–92. New York, NY: Chapman and Hall.

Franklin, J. F. (1993). Preserving biodiversity: species, ecosystems, or landscapes? *Ecological Applications* 3:202–205.

Franklin, J. F. (1995). Sustainability of managed temperate forest ecosystems. In *Defining and Measuring Sustainability: The Biogeophysical Foundations*, ed. M. Munasinghe and W. Shearer, pp. 355–385. Washington, DC: The World Bank.

Gadgil, M. and R. Guha (1995). *Ecology and Equity: The Use and Abuse of Nature in Contemporary India.* New Delhi: Penguin Books India.

Gentry, R. L. (1998). *Behavior and Ecology of the Northern Fur Seal.* Princeton, NJ: Princeton University Press.

Holling, C. S. (1973). Resilience and stability of ecological systems. *Annual Review of Ecology and Systematics* 4:1–23.

Holling, C. S. (1986). The resilience of terrestrial ecosystems: local surprise and global change. In *Sustainable Development of the Biosphere*, ed. W. C. Clark and R. E. Munn, pp. 292–317. Cambridge: Cambridge University Press.

Holling, C. S. and G. K. Meffe (1996). Command and control and the pathology of natural resource management. *Conservation Biology* 10:328–337.

Hunter, M. L., Jr. (2000). Refining normative concepts in conservation. *Conservation Biology* 14:573–578.

Jensen, M. N. (2000). Common sense and common pool resources. *BioScience* 50:638–644.

Kiss, A. (ed.) (1990). *Living with Wildlife: Wildlife Resource Management with Local Participation in Africa.* Africa Technical Department Series, World Bank Technical Paper no. 130, Washington, DC.

Kuhn, T. S. (1970). *The Structure of Scientific Revolutions*, 2nd edn. Chicago, IL: University of Chicago Press.

Leirs, H., N. C. Stenseth, J. D. Nichols, J. E. Hines, R. Verhagen, and W. Verheyen (1997). Stochastic seasonality and nonlinear density-dependent factors regulate population size in an African rodent. *Nature* 389:176–180.

Lynch, J. G. and N. K. Johnson (1974). Turnover and equilibrium in insular avifaunas, with special reference to the California Channel Islands. *Condor* 76:370–384.

Marsh, G. P. (1874). *The Earth as Modified by Human Action.* New York, NY: Scribner, Armstrong.

McCay, B. J. and J. M. Acheson (eds) (1987). *The Question of the Commons: The Culture and Ecology of Communal Resources.* Tucson, AZ: University of Arizona Press.

Meyer, J. L. (1997). Conserving ecosystem function. In *The Ecological Basis of Conservation: Heterogeneity, Ecosystems, and Biodiversity*, ed. S. T. A. Pickett, R. S. Ostfeld, M. Shachak, and G. E. Likens, pp. 136–145. New York, NY: Chapman and Hall.

Murphy, D. D. and S. B. Weiss (1992). Effects of climate change on biological diversity in North America. In *Global Warming and Biological Diversity*, ed. R. L. Peters and T. E. Lovejoy, pp. 355–368. New Haven, CT: Yale University Press.

Olwig, K. (1996). Reinventing common nature: Yosemite and Mount Rushmore – a meandering tale of a double nature. In *Uncommon Ground: Rethinking the Human Place in Nature*, 2nd edn, ed. W. Cronon, pp. 379–408. New York, NY: W. W. Norton.

Ostfeld, R. S., S. T. A. Pickett, M. Shachak, and G. E. Likens (1997). Defining the scientific issues. In *The Ecological Basis of Conservation: Heterogeneity, Ecosystems, and Biodiversity*, ed. S. T. A. Pickett, R. S. Ostfeld, M. Shachak, and G. E. Likens, pp. 3–10. New York, NY: Chapman and Hall.

Pickett, S. T. A. and R. S. Ostfeld (1995). The shifting paradigm in ecology. In *A New Century for Natural Resources Management*, ed. R. L. Knight and S. F. Bates, pp. 261–279. Washington, DC: Island Press.

Pickett, S. T. A. and P. S. White (1985). Patch dynamics: a synthesis. In *The Ecology of Natural Disturbance and Patch Dynamics*, ed. S. T. A. Pickett and P. S. White, pp. 371–384. Orlando, FL: Academic Press.

Pickett, S. T. A., S. L. Collins, and J. J. Armesto (1987). Models, mechanisms, and pathways of succession. *Botanical Review* **53**:335–371.

Pickett, S. T. A., V. T. Parker, and P. Fiedler (1992). The new paradigm in ecology: implications for conservation biology above the species level. In *Conservation Biology: The Theory and Practice of Nature Conservation, Preservation, and Management*, ed. P. L. Fiedler and S. K. Jain, pp. 65–88. New York, NY: Chapman and Hall.

Pickett, S. T. A., R. S. Ostfeld, M. Shachak, and G. E. Likens (1997). Themes. In *The Ecological Basis of Conservation: Heterogeneity, Ecosystems, and Biodiversity*, ed. S. T. A. Pickett, R. S. Ostfeld, M. Shachak, and G. E. Likens, pp. 361–362. New York, NY: Chapman and Hall.

Poff, N. L., J. D. Allan, M. B. Bain, J. R. Karr, B. D. Richter, R. E. Sparks, and J. C. Stromberg (1997). The natural flow regime. *BioScience* **47**:769–784.

Primack, R. B. (1993). *Essentials of Conservation Biology*. Sunderland, MA: Sinauer Associates.

Rogers, P. (1996). Disturbance ecology and forest management: a review of the literature. U.S. Department of Agriculture Forest Service, General Technical Report INT-GTR-336, Ogden, UT.

Sarkar, S. (1999). Wilderness preservation and biodiversity conservation: keeping divergent goals distinct. *BioScience* **49**:405–412.

Schumm, S. A. and R. W. Lichty (1965). Time, space, and causality in geomorphology. *American Journal of Science* **263**:110–199.

Simberloff, D. S. (1976). Species turnover and equilibrium island biogeography. *Science* **194**:572–578.

Simberloff, D. S. (1997). Biogeographic approaches and the new conservation biology. In *The Ecological Basis of Conservation: Heterogeneity, Ecosystems, and Biodiversity*, ed. S. T. A. Pickett, R. S. Ostfeld, M. Shachak, and G. E. Likens, pp. 274–284. New York, NY: Chapman and Hall.

Sinclair, A. R. E. (1977). *The African Buffalo: A Study of Resource Limitation of Populations*. Chicago, IL: University of Chicago Press.

Sinclair, A. R. E. (1995). Equilibria in plant–herbivore interactions. In *Serengeti II: Dynamics, Management, and Conservation of an Ecosystem*, ed. A. R. E. Sinclair and P. Arcese, pp. 91–113. Chicago, IL: University of Chicago Press.

Spirn, A. (1996). Constructing nature: the legacy of Frederick Law Olmsted. In *Uncommon Ground: Rethinking the Human Place in Nature*, 2nd edn, ed. W. Cronon, pp. 91–113. New York, NY: W. W. Norton.

Sprugel, D. G. (1991). Disturbance, equilibrium, and environmental variability: what is "natural" vegetation in a changing environment? *Biological Conservation* **58**:1–18.

Tausch, R. J. (1996). Past changes, present and future impacts, and the assessment of community or ecosystem condition. In *Proceedings: Shrubland Ecosystem Dynamics in a Changing Environment*, pp. 97–101, U.S. Department of Agriculture Forest Service, General Technical Report INT-GTR-338, Ogden, UT.

Thomas, J. (1980). Why did the large blue become extinct in Britain? *Oryx* **15**:243–247.

Willers, B. (2000). A response to "Current normative concepts in conservation" by Callicott *et al. Conservation Biology* **14**:570–574.

13

Techniques – conserving processes and contexts

In the last chapter we saw how ecologists have come to understand that ecosystem processes operate at multiple spatial and temporal scales, that patchiness is a crucial aspect of ecosystem structure and function, and that people have complex and often unappreciated effects on their resource base. In this and the following chapter, we will examine some applications of these concepts to management.

Ecologists' expanding understanding of ecosystems has focused attention on managing ecosystems in ways that maintain or restore critical processes and structures. The underlying hypothesis is that if critical functions and structures are maintained, ecosystems will continue to supply services and products that society needs, and this will allow for the long-term persistence of the species found in those ecosystems as well. The emphasis is less on producing specific products or protecting particular habitats and species and more on sustaining functioning ecosystems as a means of protecting both biodiversity and the resource base that fulfills people's needs. The ecosystems that result from this type of management are not simple; instead they are spatially heterogeneous and temporally variable. The resulting heterogeneity allows for greater species diversity. This type of management does not necessarily replace management that focuses on the needs of individual species, but it complements it.

The flux-of-nature perspective suggests that ecosystems should be viewed as dynamic and open and that by understanding how key components operate, managers can devise novel methods of manipulating ecosystems to protect biological diversity while meeting human needs. This involves developing a model of how resources, organisms, and structures are interrelated. The

Table 13.1. *Examples of ecosystem processes*

Processes resulting from cycles of matter and energy
Production of biomass (primary production)
Decomposition and mineralization of organic matter
Nitrogen fixation and denitrification
Soil formation
CO_2 absorption

Processes resulting from interactions between species
Predation
Competition
Parasitism
Herbivory
Mutualism

Processes resulting from the operation of abiotic ecosystem components
Fires
Floods
Storms
Droughts

---

predictions of the model can then be tested, and specific management actions suggested by the model can be tried. The results of these experiments are used to refine the model and revise management measures if necessary. The bottom line is that ecosystem complexity should be maintained. This includes spatial complexity (habitat heterogeneity), temporal complexity (disturbance regimes), and complexity in community structure (species diversity).

13.1   Conserving processes

The underlying principle here is the idea that by conserving ecosystem processes, managers can increase the likelihood that ecosystem structure and function will be maintained and, therefore, that viable populations of species will persist. We saw in Chapter 7 that biodiversity declines in simplified ecosystems, whether they are agricultural monocultures, tree plantations, or dammed rivers. Some examples of ecosystem processes are listed in Table 13.1, and an example of a key ecosystem process is described in Box 13.1.

Box 13.1   Dynamic sheetflow – a key ecosystem process in South Florida

South Florida is well known for the spectacular natural features of the Everglades. It is also a region of intensive agriculture (principally sugar production) and rapid urban growth. Prior to development of the region, a huge, shallow sheet of water moved slowly over much of the land surface for several months during most years. (This is referred to as dynamic sheetflow.) Its unusual hydrology is one of the key features of the South Florida landscape. Because the topography of South Florida is relatively flat, water flows quite slowly across it. Slight differences in topography provide habitat diversity. Elevated hummocks dry out sooner than the surrounding area and support distinctive vegetation. Conversely, depressions, such as those made by alligators, hold water after it has receded from the rest of the landscape. During the dry season and in dry years, these wet spots provide important refugia for many species of invertebrates and other animals that feed upon them.

As a result of these distinctive features, the landscape of South Florida historically supported a mosaic of different habitats. This heterogeneous landscape, with resources distributed in widely dispersed patches, was utilized by a variety of large-bodied animals, including the Florida panther, alligators, and wading birds. Major disturbances affected this landscape periodically. In years of low rainfall, vegetation dried out sooner than usual and was subject to fires. Hurricanes with extreme storm tides and high winds, occasional freezes, and droughts also periodically damaged local patches of vegetation.

Surface water in South Florida is now regulated by an intricate system of canals, levees, and dikes. Although these measures were successful in terms of their objectives – clearing land for agriculture and development and controlling floods – they disrupted the region's dynamic sheetflow and fragmented and reduced habitat for wildlife. In addition to the problems of altered hydrology and habitat loss, water quality has worsened as a result of pollution. Historically, Everglades water was quite low in nutrients, but agricultural runoff has increased nutrient levels substantially, resulting in a proliferation of cattails and a decline in vegetative diversity (Harwell 1997). A massive restoration effort is currently under way to restore more natural hydrological dynamics to this region.

Studies of the flows of matter and energy through ecosystems provide information that can be used to evaluate and predict the impacts of resource use. In studies in the White Mountains of New England, long-term measurements of hydrological and chemical cycles followed by measurements of the same parameters after experimental manipulation have provided a detailed picture of changes following major perturbations. These studies took place at the Hubbard Brook Experimental Forest (HBEF), a 3076-ha mature hardwood forest ecosystem. In 1955, ecologists began collecting baseline data on inputs and outputs of water and nutrients in six watersheds of the HBEF (Likens *et al.* 1967, 1969; Likens and Bormann 1972). A decade later, investigators removed all of the vegetation on one of the experimental watersheds. Continued monitoring of nutrient levels showed that the forest's nutrient cycles were altered substantially by the experimental treatment (deforestation). Prior to deforestation, nitrogen in the form of compounds such as ammonium ions was conserved because it was taken up directly by plants. As a result, stream waters contained only low levels of these compounds, and nitrogen inputs and outputs to the ecosystem were approximately equal. After vegetation was removed, however, ammonium ions could not be taken up, so they were oxidized to other compounds that were then flushed from the ecosystem in stream water. Likens and Bormann estimated that it would take 43 years for nitrogen inputs from precipitation to replace the lost nitrogen, assuming that all imported nitrogen was retained. Large amounts of hydrogen ions were produced in the oxidation of ammonium ions, which caused a decline in pH. This acidification, in turn, allowed substantial amounts of calcium, magnesium, sodium, and potassium ions to dissolve and be rapidly leached from the system. By blocking the uptake of nutrients, the removal of vegetation perturbed a cycle in which nutrient gains and losses had been approximately balanced, leading to "greatly accelerated export of the nutrient capital of the ecosystem" (Likens and Bormann 1972:63).

Although this experiment indicated that the complete removal of vegetation dramatically changed the HBEF and that recovery would take decades, it also demonstrated that ecosystem studies can be the basis for designing management strategies with less serious consequences. "Man's manipulations may cause serious imbalances in the ecological function of natural ecosystems," concluded Gene Likens and F. Herbert Bormann,

however, with a clear understanding of the functional interrelationships of the ecosystem he may be able to substitute for the natural ecosystem's ability to conserve nutrients, while still extracting products desirable to himself. As a simple example, since bark is relatively rich in nutrients, lumbering operations that strip the bark from logs within the ecosystem rather than at some distant processing plant may act to conserve nutrients within the ecosystem. (Likens and Bormann 1972:62)

This example shows that disturbances, that is, abiotic or biotic events that remove vegetation, affect flows of energy and matter and have pivotal influences on ecosystem structure and function. When conserving disturbance regimes and other processes, it is important to retain or restore variability in process rates and intensities (Swanson *et al.* 1993; Morgan *et al.* 1994; Holling and Meffe 1996; Poff *et al.* 1997; Landres *et al.* 1999), because this variability creates heterogeneous environments that support diverse species. Some organisms have adaptations that allow them to cope with large, severe, hot fires, whereas others prosper in the wake of small, cool fires of moderate intensity but are killed outright by hot fires. A management plan that allowed fires of only one intensity would eventually eliminate some organisms and result in a loss of biodiversity.

13.2   Recognizing limits and assessing vulnerability

"The wise management of our natural resources," wrote Gene Likens and F. Herbert Bormann,

depends upon a sound understanding of the structure and function of ecological systems. To date, the narrow approaches such as specialized agricultural and industrial strategies, designed to maximize production of food, power, and other products, have dominated our management of natural resources and invariably have led to imbalances and instabilities in ecosystems. When resource management is based on an understanding of the ecosystem's interconnections and interactions, the "hidden costs of narrow management strategies" become part of the overall accounting and further environmental deterioration . . . can be avoided.

In the past, when natural resources were plentiful in relation to man's wants and abilities for utilization, the function of natural ecosystems was considered much less important than their structure. Forests were to be cleared, rivers to be tamed, and wild animals to be conquered. Now in the face of man's exploding population and dwindling resource base, his very survival may depend on an accurate knowledge of ecosystem function, i.e., maintaining the continuous flow of ecological systems and life itself. (Likens and Bormann 1972:25)

There are processes that, if we understand them, predict where erosion will be high and where it will not. . . . We learn when we can be effective in controlling erosion and when we cannot, when our actions will be productive and when they will not be. . . . The change in our understanding and management of erosion illustrates how our approach to land use is changing and should continue to change in the future. (Botkin 1990:199)

Managers using an ecosystem approach to conservation need to be able to predict the effects of resource uses on the structures, functions, and components of ecosystems and to recognize limits beyond which resource use will have irreversible negative consequences. "Human-generated changes must be constrained because nature has functional, historical, and evolutionary limits. Nature has a range of ways to be, but there are limits to those ways, and therefore, human changes must be within those limits" (Pickett *et al.* 1992:82).

Organisms become adapted to the disturbance regimes (frequency, intensity, and duration of disturbances) that they experience during their evolution. If they do not adapt to the disturbance regime they are subjected to, eventually they become extinct. For this reason, profound changes in an ecosystem's disturbance regime may cause local extinctions. If we can understand how different ecosystems respond to disturbances, we can predict how our actions will alter ecosystems, and we can design management strategies that take advantage of, or at least do not undermine, an ecosystem's potential for recovery.

It is especially important to understand what situations are likely to change ecosystems irreversibly. Irreversible changes may occur when ecosystem functions and structures are altered so dramatically that conditions cannot return to a predisturbance or pre-exploitation state. Sometimes this happens as a result of natural perturbations, but more often such irreversible changes result from the activities of people. From a manager's point of view, in most cases it is clearly undesirable to alter ecosystems irreversibly.

Some types of ecosystems are more seriously disrupted by natural disturbances and exploitation than others. For example, the grasses of the American Great Plains can tolerate grazing and trampling, but the bunchgrasses of the Intermountain West (located between the Pacific coast ranges and the Rocky Mountains) are very susceptible to such disturbances. After cultivation or heavy grazing, the structure and function of intermountain steppes change in ways that are virtually irreversible (see Box 13.11).

Invasion by alien species is a special type of ecological modification. The establishment of an exotic species may or may not be followed by major disruptions of ecosystem functioning. Sometimes exotic species become integrated into a pre-existing community without a major upheaval in community composition, but in other instances the establishment of exotics is followed by the disappearance of native species and substantial alterations in ecosystem functioning. Some of the same conditions that make certain ecosystems sensitive to fire, logging, and grazing predispose those ecosystems to disruption by introduced species.

Ecosystems that do not recover readily from a particular type of disturbance can be considered sensitive, or vulnerable, to that disturbance. They are sometimes said to be less resilient than ecosystems that return to predisturbance conditions fairly quickly. At least three factors affect an ecosystem's sensitivity to disturbance, to invasion, and to exploitation. The first involves low productivity, which is usually a result of physical constraints such as those imposed by climate and nutrient availability. A second but related factor is low soil fertility. Finally, a third factor relates to evolutionary adaptations to disturbance. These factors are complex and interrelated, and there are exceptions to these generalizations, so they cannot be applied in a cookbook fashion to assess ecosystem sensitivity. They do, however, suggest some things that managers should consider when attempting to predict the effects of their programs. As usual, general insights cannot substitute for in-depth understanding of a specific case.

In addition to sensitivity, it is important to consider the biological value of an ecosystem's components. When we are dealing with an ecosystem that contains many endemic species or unique and unusual species assemblages, the consequences of losing these elements are great. Often, ecosystems that are sensitive also have high biological value.

Boxes 13.2–13.9 describe the biological value of the earth's major types of ecosystems as well as the threats they face and how they respond to resource use. Additional information on the ecosystems described below can be found in Whittaker (1975), Daubenmire (1978), and Brown and Gibson (1983).

Box 13.2   Tundra

Tundra (from a Russian word referring to the marshy plains of northern Eurasia) occurs where there is too little heat to support the growth of trees. This condition is met at high latitudes (the arctic and antarctic regions of the northern and southern hemispheres respectively) and at high altitudes (above the treeline on mountain slopes in temperate and tropical zones). Arctic tundra forms a circumpolar belt across the northern hemisphere, extending through North America and Eurasia. In the southern hemisphere, tundra occurs in parts of Antarctica and on a few islands.

There is relatively little dead organic matter in the tundra, and what there is decays slowly. Thus, tundra soils are shallow and infertile. As a result of their infertile soils and short growing season, tundras are characterized by very low productivity, in spite of the fact that water is abundant (see Table 13.2). These characteristics cause them to recover slowly after they are disturbed. In addition, a third feature, unique to arctic tundra, adds to its

Table 13.2. *Estimated net primary production of major ecosystem types. See Boxes below for more information about ecosystem types*

| | Net primary production (dry matter) | |
Ecosystem type	Normal range (g/m²/yr)	Mean (g/m²/yr)
Terrestrial		
Tropical forest		
Moist tropical forest	1000–3000	2200
Dry tropical forest	1000–2500	1600
Temperate forest		
Evergreen temperate forest	600–2500	1300
Deciduous temperate forest	600–2500	1200
Boreal forest	400–2000	800
Woodland and shrubland	250–1200	700
Savanna	200–2000	900
Temperate grassland	200–1500	600
Tundra (arctic and alpine)	10–400	140
Desert and semidesert scrub	10–250	90
Extreme desert – rock, sand, ice	0–10	3
Cultivated land	100–4000	650
Aquatic and wetland		
Swamp and marsh	800–6000	3000
Lake and stream	100–1500	400
Open ocean	2–400	125
Upwelling zones	400–1000	500
Continental shelf	200–600	360
Algal beds and reefs	500–4000	2500
Estuaries (excluding marsh)	200–4000	1500

Source: Based on Whittaker and Likens (1975).

vulnerability. In inland arctic areas, where the climate is not moderated by proximity to an ocean, only the surface soil thaws during spring and summer; the subsurface soil remains permanently frozen. This phenomenon is known as permafrost. In addition, the annual cycle of freezing and thawing in the tundra causes the soil to expand and contract repeatedly, a phenomenon termed frost churning. Tundra vegetation provides insulation, so when it is removed, the soil thaws more deeply, frost churning is accentuated, and as a result productivity may be destroyed indefinitely.

When an area of tundra is denuded by activities related to oil exploration and pipeline construction, it may remain bare for centuries.

Arctic tundra is of great importance to the millions of waterfowl and shorebirds that breed there.

Alpine tundra occurs on mountaintops. Like arctic tundra, alpine ecosystems do not recover readily after they have been disturbed. Because of the extreme temperatures, short growing season, thin soils, and high winds characteristic of high altitudes, alpine vegetation grows slowly. In addition, freeze–thaw cycles create surface cracks that disrupt root growth. This is why the alpine vegetation of the Olympic Mountains is slow to recover from goat wallowing and trampling (see Box 13.10). In addition, because it occurs on the tops of mountains, alpine tundra is distributed in isolated patches surrounded by low-elevation habitats. This isolation has two important consequences. First, alpine habitats contain many endemic subspecies, species, and even genera because they evolved in isolation. Second, for some groups, extinction rates exceed colonization rates (see Figure 8.5).

Box 13.3   Temperate forest

As we move toward the equator from the poles or downslope from mountain peaks in the temperate zone, we encounter a belt of coniferous forest. This is termed subarctic or boreal forest if it is adjacent to arctic tundra or subalpine forest where it abuts alpine tundra. The Russian term taiga is used to denote subarctic forests in Eurasia. These communities are dominated by conifers, trees that bear evergreen, needlelike leaves. Fire spreads easily through conifer foliage, which contains high levels of flammable oils and resins and decays slowly. Many plant species characteristic of coniferous forests have evolutionary adaptations that allow them to survive fires or to reproduce after fires. Thus, coniferous forest communities can cope with burning if the frequency, size, and intensity of fires is within the range of conditions these forests experienced during their evolutionary history. Because of the cold climate at high latitudes and altitudes, however, subarctic and subalpine forests grow slowly, and they may take many years to recover after a fire.

As we continue our progression equatorward or downslope, we encounter areas of temperate forest. These can be dry or moist. Temperate rainforests occur along the west coasts of New Zealand, Chile, southern

Australia, and North America. They have maritime climates with high rainfall throughout the year (or, in the case of the California coast, summer fogs which compensate for low summer rainfall). These forests are dominated by eucalypts in Australia; southern beech and conifers in New Zealand and Chile; Douglas-fir in Washington, Oregon, and British Columbia; sitka spruce in Alaska; and redwoods in California. In southwestern Oregon, and northern California, the moist temperate forest is dominated by two endemic giant conifers, giant sequoia and redwood. The temperate rainforests of the Pacific coast and Australia are the tallest forests in the world. Redwoods attain heights of 115 m; giant sequoias have trunks up to 8 m in diameter at the base.

Some moist temperate forests burn readily and contain species that persist only where fires are frequent. These include lodgepole pine, Jeffrey pine, and giant sequoia. Although the giant sequoia can live for thousands of years, its presence in the moist temperate forest actually depends on fire, because this species does not reproduce itself in mature moist coniferous forest. Sequoias persist only because their extremely thick, fire-resistant bark allows them to survive fires that eliminate the less fire-tolerant conifers such as Douglas-fir.

When the first European settlers arrived on the eastern seaboard of North America, they did not find an unbroken expanse of mature forest. Instead, they found a mosaic of stands of different ages; in many places they found grassy openings and parklike groves. These conditions resulted from widespread and frequent disturbances. Fires, both naturally started and those set by Native Americans, had been an important disturbance agent in this landscape. Settlers cleared large tracts of eastern deciduous forest. Initially this favored species of early-successional habitats, which took advantage of the shrubs that invaded the cleared areas, and the fields associated with agriculture benefited species that utilized clearings. Not all wildlife profited from the clearing of the eastern forests, however. The passenger pigeon, for example, became extinct in part because of loss of its oak forest habitat (see Chapter 1). Many sites that were originally cleared to create cropland were subsequently abandoned. If sufficient seed sources for trees remained in the vicinity, cleared sites gradually returned to forest through a process termed old-field succession.

It is well known that the principal economic use of temperate forests is timber harvest. Many temperate forests are also used for livestock grazing or cleared for agriculture. Because they are productive and evolved to cope with frequent fires, these forests can often regenerate fairly readily after natural disturbances, such as fires or volcanic eruptions, or those that mimic natural disturbance regimes. Nevertheless, the capacity of moist coniferous forests to recover is not unlimited. Overuse and poor management can exceed a forest's capacity to respond, especially if the vegetative cover is removed from steep slopes. In addition, wide-ranging species and species with specialized habitat requirements found only in old-growth forest decline when logging fragments late-successional and old-growth forests. The extreme fragmentation of temperate forest has been detrimental for many species that require forest interiors, such as Neotropical migrant songbirds (see Chapter 7).

Dry temperate forest is the most drought-tolerant forest type of the temperate zone. This type of vegetation occurs in areas with dry summers, such as parts of California, the Mediterranean region, and southern Australia, as well as in parts of Southeast Asia. In the American West, this type of forest is dominated by two types of conifers – pines and junipers. Oak is also common in some western dry forests. The trees in dry temperate forests are generally not very tall and are widely spaced. Fires are frequent in this dry environment, where they inhibit tree regeneration. Adult trees of some species, like ponderosa pine, can withstand light fires because of their fire-resistant bark, but fires kill seedlings and young pines. For this reason, repeated burning creates stands with widely spaced, mature trees and few saplings. The major land uses in this ecosystem are livestock grazing and agriculture.

Box 13.4   Chaparral and steppe

Chaparral

In temperate zones, vegetation dominated by shrubs with thick, evergreen leaves is termed chaparral. This type of vegetation occurs along the west coasts of all continents. Some chaparral communities have exceptionally high plant diversity. In fact, greater diversity has been described only for tropical rainforests.

The leaves of chaparral shrubs are high in flammable oils, and they decay slowly. These two characteristics allow extremely flammable litter to build up. Fires are common during the hot, dry summers characteristic of chaparral regions. Many chaparral shrubs sprout readily after a fire, and many have seeds that germinate only after being exposed to heat. It is not surprising, therefore, that chaparral communities usually recover quickly after being burned. In some cases it takes less than 30 years to return to preburn conditions. In fact, if the interval between fires is much longer than 30 years, productivity starts to decline.

Some areas of chaparral, such as southern California, have undergone intensive residential development; others are managed for agriculture, livestock grazing, or wildlife habitat. On steep slopes where chaparral vegetation has been removed, increased erosion and frequent landslides have become serious problems. Another grave problem has been created by disruption of the normal disturbance regime. As a result of decades of fire suppression in developed areas, highly flammable litter from chaparral vegetation has accumulated. Fires have become harder to control and more destructive because of this fuel buildup.

Steppe

Just as the distribution of trees is limited by temperature at high latitudes and altitudes, it is limited by available moisture in warm, dry areas at low elevations and latitudes. Areas where conditions are too dry for trees and the vegetation is dominated by perennial grasses are technically termed steppe. (A number of other terms including "grassland" and "prairie" are also used to describe this type of community.) Steppes occur on all continents. The best-known steppes are those of central North America, central Eurasia, the pampas of Argentina, and the veld of South Africa. In parts of western North America, steppe grasses are accompanied by an overstory of shrubs, particularly big sagebrush, a type of community termed shrub steppe. This is the familiar landscape of many western movies. Steppes that occur in climates that are almost moist enough to support trees are sometimes accompanied by abundant perennial forbs. This type of community is termed meadow steppe.

Steppes (referred to as "temperate grassland" in Table 13.2) have high primary productivity. This stems from the fact that they have rich, well-drained soils that are high in organic matter. These soils form from the extensive roots of decaying grasses (Figure 13.1). (Meadow steppe soils

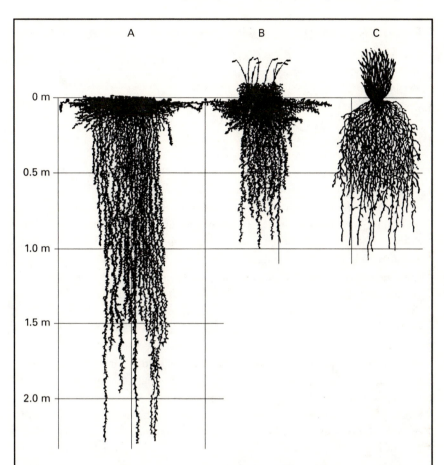

Figure 13.1. Roots of three grasses characteristic of steppes in the midwestern U.S.A. (A) Buffalo grass, (B) blue grama, (C) wire grass. The large amount of underground biomass is one reason why steppes have such deep, rich soils. (After Weaver and Clements 1938.)

were the principal reason for designating Russia's Central Chernozem State Biosphere Reserve, which is described in Box 14.3.)

Agriculture and livestock grazing are the major uses of steppe. Some of the world's most productive croplands are in steppe soils. If excessive amounts of biomass are removed by grazers, however, changes in plant community composition result, and less palatable species increase in abundance (Chapter 3).

Box 13.5   Desert

Where the climate is too hot and dry to support the perennial grasses of steppes, shrubs dominate, and if grasses are present they are annuals (plants that complete their life cycle within a single growing season) rather than perennials (plants that live for several years). This type of ecosystem is termed desert. (Some types of steppe vegetation are also sometimes popularly known as "desert.")

Water is scarce is deserts, and water loss is high (Evenari *et al.* 1971). Rain falls infrequently and unpredictably. When rain does come, it may occur as a downpour falling so suddenly that much water runs downslope instead of soaking into the soil. It is not surprising that productivity is very low in deserts (Table 13.2.). Consequently, desert soils are low in organic matter and relatively infertile. Furthermore, because evaporation rates are high, salts and other chemicals, such as carbonates, accumulate in the soil instead of leaching out of it. For this reason, desert soils tend to be both saline (high in salts) and alkaline (high in pH). When deserts are irrigated, the evaporation of irrigation waters exacerbates these conditions.

The plants and animals that inhabit deserts cope with the rigors of this environment in a variety of ways. Many desert animals are active at night, when the air is relatively cool and moist. This behavioral adaptation is accompanied by physiological and morphological adaptations that facilitate nighttime activity. The eyes of nocturnal animals such as owls and many rodents respond to low levels of light, and sensory modalities that do not require light are also well developed. Bats find food with the aid of ultrasonic sonar, snakes locate their prey by means of heat-sensitive receptors, and many desert animals have excellent hearing and keen smell. Because of their acute hearing, however, desert animals may experience hearing loss as a result of unusually loud, low-frequency noises, such as those caused by off-road vehicles (Berry 1980). It has been suggested that such hearing loss could cause inappropriate behaviors that increase mortality.

As a consequence of their low productivity and limited water supplies, desert ecosystems are very sensitive to disturbance from anything that removes excessive plant biomass or uses unusual quantities of water. The deserts of the American Southwest have experienced serious consequences as a result of introduced plants and animals that are more efficient than native species at using desert resources (see Box 13.10).

Deserts are used for grazing livestock and for agriculture. Farming is possible only near oases or where irrigation water is supplied. Agriculture in ancient Mesopotamia, the fertile "cradle of civilization," depended upon irrigation. Eventually, however, the region became unsuitable for agriculture. The accumulation of salts in Mesopotamia's irrigated soils may have been one cause of this environmental degradation.

Box 13.6   Tropical ecosystems

The geographic region where lowland areas remain free of frost through-out the year is known as the tropics. This occurs at low latitudes, usually south of the Tropic of Cancer and north of the Tropic of Capricorn. As we move toward the equator from the temperate zones, we pass through tropical habitats known as woodland, savanna, and forest. Woodland is the driest of these habitats; tropical forest (which itself is differentiated into dry tropical forest and moist tropical forest) is the wettest. Tropical savanna and forest are described in more detail below.

Savanna

Tropical savanna is characterized by perennial grasses with an interrupted overstory of scattered trees and shrubs. The so-called "Kalahari desert" of southwestern Africa is actually a savanna. Unlike the soils of steppe regions, savanna soils are not very fertile. Similarly, the nutritional value of savanna grasses is relatively low in comparison to that of steppe grasses.

Savanna vegetation burns easily, and fires can convert tropical forest to savanna. Many areas that are dominated by savanna vegetation today were probably created when fire burned dry tropical forests. Herbivores also affect the balance between trees and grasses in this environment. In spite of the low quality of its forage, savanna vegetation is grazed by both live-stock and wild herbivores, including the enormous herds of wildebeests and other ungulates on the Serengeti Plain (see Box 12.2).

Tropical forest

Dry tropical forest, or seasonal tropical forest, is characterized by pro-nounced wet and dry seasons. As noted above, burning can convert this type of forest to savanna. The "monsoon forests" of southeastern Asia are actually classified as dry tropical forest, because even though they experience heavy rains during the wet season, there is also a pronounced dry season.

The soils of dry tropical forests are relatively fertile compared to most tropical soils. The major land uses of this ecosystem are timber harvest, grazing, and agriculture. Like the better-known moist tropical forest, dry tropical forest has been reduced to a fraction of its original area; less than 2% of Central America's dry tropical forest remains intact (Janzen 1988). Fire, which retards the regeneration of tree species, grazing, and clearing for agriculture are largely responsible for this decline. In addition, teak and some other trees of dry tropical forest are highly valued for their attractive wood. Efforts to restore Costa Rica's dry tropical forest through ecological and biocultural restoration are discussed in Chapter 10.

Moist tropical forest has received a great deal of publicity in recent years. Unlike dry tropical forests, moist tropical forests do not experience a pronounced seasonal drought. This type of tropical forest has very high species diversity and productivity (see Table 13.2). Many species found in tropical moist forests have not even been described by scientists yet. Typically, large numbers of species coexist in a moist tropical forest, but they exist at low population densities. In addition, many of the species present in moist tropical forests occur in only a small area. This combination of small population size and limited geographic range puts tropical moist forest species at high risk of extinction when their habitat is profoundly altered by deforestation.

Moist tropical forests are used for subsistence farming, timber harvest, and extraction of other forest products. Timber harvest yields a number of valuable products from moist tropical forests, including mahogany and other fine woods. Other forest products can be obtained from moist tropical forests, as well, such as Brazil nuts, palm fronds, chicle for chewing gum, and rubber. Because these products can often be obtained without killing the tree that produces them, there is a considerable amount of interest in developing systems for extracting nontimber forest products on a basis that will be sustainable into the future. These systems are discussed in Chapter 14.

In general, moist tropical forests recover slowly following large-scale disturbances; therefore, the harvest of tropical hardwoods and softwoods often has serious negative impacts. One reason for this is the fact that most of a tropical forest's biomass is located above the ground, so nutrients from its lush vegetation do not get incorporated into the soil. When timber harvest removes this aboveground biomass, the soils that are left behind are low in organic matter. In addition, some tropical rainforests occur on poor soils that harden into a bricklike substance after they are cleared.

These characteristics also make moist tropical forest sensitive to disturbance from shifting cultivation. When a forest plot is burned, the nutrients released from the ash are rapidly washed away. Additional nutrients are lost when biomass is removed by weeding. Because the soil of moist tropical forests is not very fertile to begin with, a site's productivity declines markedly when its aboveground nutrients are removed by leaching and weeding associated with shifting cultivation. Small clearings in moist tropical forest can probably return to predisturbance conditions in about 100 years, while the projected recovery time for a disturbance of 10 hectares is on the order of thousands of years (Uhl 1983). In a study in Costa Rica, soil fertility declined dramatically in a plot that was kept bare (by hand weeding) for five years. At the end of the study, very few plants had colonized this plot, even though there was a large input of seeds from the surrounding vegetation (Ewel *et al.* 1991). These findings underscore the serious impacts of massive biomass removal in moist tropical forests.

Box 13.7   Marine ecosystems

The major types of marine ecosystems are described below. The principal threats facing these ecosystems are pollution, overfishing, habitat destruction, invasions of exotic species, and global climate change (Carlton 1998).

Open ocean

The upper layer of the ocean is termed the photic zone (*phot*, light). Sunlight penetrates this zone, allowing photosynthesis to take place there. The depth of the photic zone depends on the amount of material suspended in the water. The productivity of the earth's oceans is limited by the amount of photosynthesis that can take place in marine waters and thus by the depth to which sunlight can penetrate. Almost all life in the ocean depends upon photosynthesis carried out in the photic zone. (Chemosynthetic exceptions are discussed below.) Organisms below the

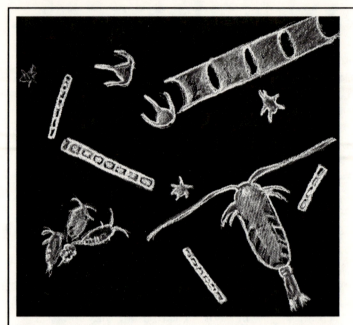

Figure 13.2. Examples of marine plankton. (After Whittaker 1975.)

photic zone consume organic matter, such as feces and dead organisms, that is produced in the photic zone and sinks below it.

The polar regions of the oceans are especially productive because in those regions upwellings bring to the surface nutrients that sink to the ocean floor. These nutrient-rich waters support the growth of plankton (Figure 13.2), which are fed upon by small crustaceans known as krill. The krill in turn are fed upon by baleen whales (Figure 1.4). Thus, plankton nourished by nutrient upwellings form the basis of an important marine food chain.

Ocean floor

Bottom-dwelling, or benthic, organisms live on the ocean floor. This group includes seaweeds that are attached to the substrate, such as kelp, as well as corals, many fishes, crabs, lobsters, clams, snails, octopi, and diverse other invertebrates.

Recent explorations using submersible devices have allowed oceanographers to explore benthic communities at great depths. During these expeditions, unique communities were discovered below the photic zone. In these communities chemosynthetic bacteria (Chapter 3) obtain energy by oxidizing the hydrogen sulfide emitted by submarine hot springs. These autotrophic bacteria are the primary producers on which the food chain of these unusual communities is based. Many animals previously unknown to science have been discovered in this environment.

Bottom trawling and other fishing methods that employ mobile gear can severely impact communities of the ocean floor. By burying, crushing, and exposing benthic structures and animals, mobile fishing gear dramatically reduces the structural diversity of the marine substrate. Ocean floor sites recover slowly from this type of disturbance, and often they are repeatedly disrupted before they have had a chance to recover. In regions such as the outer continental shelf and slope, where there is normally little storm-wave damage, organisms are not adapted to frequent natural disturbances, and growth rates are slow. These areas are the most vulnerable to disturbance from mobile gear. Technological advances have made fishing gear even more efficient at harvesting benthic resources, so that they can now reach previously inaccessible sites that had functioned as refuges until recently (Watling and Norse 1998).

Coral reefs

Corals live in warm, undiluted seawater where light can penetrate, that is in the photic zone of tropical oceans. Because of these requirements, they are excluded from areas with upwellings of cold water and from the vicinity of river mouths where fresh water entering the ocean causes high turbidity. A coral reef is a structure formed from calcium carbonate deposited by plants and animals over thousands of years. The most important of these organisms are the reef-building corals, animals that are related to sea anemones and jellyfishes. A type of red algae termed coralline algae also contributes significantly to reef building.

Reef-building corals depend upon symbiotic algae that live in their tissues. These algae produce food for the corals through photosynthesis, thereby recycling wastes containing phosphorus and nitrogen, and they also enhance their hosts' ability to deposit calcium carbonate. The dependence of reef-building corals upon these photosynthetic mutualists explains why coral reefs are restricted to the photic zone and are unable to tolerate suspended material that impedes the penetration of light.

Coral reefs are among the most productive and diverse ecosystems in the world (Table 13.2). Many reef taxa are not well known, however. Because many organisms that inhabit coral reefs have small geographic ranges, they are vulnerable to extinctions caused by unusual disturbances (Reaka-Kudla 1997). They are, however, adapted to frequent natural disturbances from exceptionally low tides and from storms that stir up sediments. After a disturbance kills surface corals, a reef is normally repopulated by free-swimming coral larvae.

A number of hazards associated with human activities threaten reef ecosystems around the world. These include dynamiting of coral to obtain cement, dredging and associated sediment deposition, discharge of sewage effluent, pollution, and collection of coral for sale as souvenirs. Such disturbances can result in long-term damage to coral reefs.

Fluctuations in the warm-water environment of coral reefs are associated with a phenomenon known as bleaching, in which corals expel their symbiotic algae. Bleached corals sometimes reacquire algae and recover, but if they fail to do so, the corals die and the reef becomes vulnerable to erosion. Bleaching occurs in response to a variety of environmental stresses, including unusually high or low temperatures, high or low salinity, variation in either ultraviolet light or visible light, high sedimentation, and pollution. In 1982 and 1983 the waters of the tropical eastern Pacific were unusually warm as a result of climatic fluctuations associated with El Niño. These fluctuations were linked to a massive bleaching episode in which 85% of the corals on Panamanian reefs died, and two species of coral are thought to have become extinct during that event (Glynn and de Weerdt 1991).

Although bleaching is a fairly common phenomenon, in recent years scientists have reported a dramatic increase in the frequency, severity, and distribution of bleaching episodes. The reefs of Southeast Asia are considered especially vulnerable because in this region high rainfall results in freshwater runoff and sedimentation and because of the dense human population and associated environmental problems. Some scientists suspect that increases in the availability of nutrients resulting from pollution might lower the temperature threshold for bleaching, so that temperatures that would not have posed a problem a few decades ago will now cause corals to expel their algae (Roberts 1991). In addition, pollution may interfere with the reproductive ability of corals and thus retard a reef's recovery after widespread mortality has occurred.

Intertidal zones

The intertidal zone, the area that is exposed to rising and falling tides, is the shallowest part of the ocean. Organisms that inhabit intertidal zones must be adapted to environmental conditions that fluctuate often. They must be able to withstand both exposure to air and submersion in water as well as changes in temperature and salinity, and continual wave action. Mortality is high (Menge 1979), and relatively few organisms are adapted to these conditions, so the species diversity of intertidal ecosystems is not impressive.

Although intertidal environments present special challenges to organisms that inhabit them, these aquatic ecosystems are extremely productive. Rivers bring a rich supply of nutrients washed from terrestrial ecosystems, and tides circulate them. These conditions allow coastal ecosystems to support high densities of individuals and to produce large amounts of biomass.

Box 13.8   Inland aquatic ecosystems

Inland aquatic habitats usually contain fresh water, but there are some exceptions in arid environments, where salinity is high because of high rates of evaporation. Under these conditions, undrained basins accumulate dissolved salts. (The Great Salt Lake and the Dead Sea are well-known examples.)

Flowing freshwater aquatic ecosystems such as rivers, streams, creeks, and springs are inhabited by diverse aquatic invertebrates and fishes, many of which have not been thoroughly studied. Inland bodies of water differ from the earth's oceans in that they are isolated from each other by terrestrial barriers. Although inland waters such as lakes or ponds may be interconnected by branching rivers, these drainages themselves are isolated from one another. Freshwater organisms cannot easily disperse over land to reach a new habitat. As a consequence of this isolation by land barriers, isolated freshwater drainages tend to develop genetically distinct populations and subspecies and to have high numbers of endemic species (Allan and Flecker 1993). Unfortunately, however, many of these endemics have been displaced by introduced species. For instance, in North America numerous extinctions of native fish species have occurred because of introductions of game fishes and accidental introductions of exotic aquarium species.

Riparian zones occur adjacent to flowing fresh water. Typically, they too are productive and diverse. The ecological importance of riparian zones is far greater than their area would suggest. Riparian habitats, particularly those surrounded by arid or semiarid uplands, are used by wildlife for feeding, breeding, escaping, hiding, resting, and traveling. Food resources that are abundant in riparian zones include aquatic invertebrates and vertebrates, aquatic plants, and insects. Because of this, many vertebrates, such as raccoons, salamanders, flycatchers, swallows, shrews, and bats forage preferentially in or over riparian areas. In steppe or desert regions, deciduous trees and shrubs and standing dead trees – both of which provide nesting habitat for a wide variety of birds and small mammals – may be virtually confined to riparian zones. Migrant songbirds make extensive use of strips of woody vegetation along rivers and streams during migration; wading birds concentrate at riparian stopover points during their migrations; and ungulates such as deer and elk utilize riparian vegetation when they migrate between their high-elevation summer ranges and low-elevation wintering grounds (Brinson *et al.* 1981; O'Connell *et al.* 1993).

Under natural circumstances, riparian zones experience frequent floods. During floods vegetation is scoured by moving water and battered by ice and debris; consequently it is killed, uprooted, and washed away. In addition, banks collapse, and sediments and debris are deposited during floods. The species of plants, animals, and microorganisms that live in riparian areas are adapted to these events and can survive and reproduce under these conditions. A pronounced change in this disturbance regime is likely to lead to reduced biodiversity.

Flowing waters include some of the most threatened ecosystems on earth. Throughout history people have settled next to and modified flowing water. Recently, however, the degradation of flowing-water habitats has increased. The major threats to freshwater ecosystems are habitat alteration (from water diversion, impoundment, and channelization), introductions of exotic species (including deliberate introductions of hatchery fish and accidental introductions), and pollution. Some effects of these perturbations are considered in Chapter 7.

Box 13.9   Wetlands: The terrestrial–aquatic interface

Wetlands, areas that are saturated or flooded for long enough during the growing season to support vegetation adapted to life in saturated soil, occur at the ecotone between aquatic and terrestrial communities. Marshes, swamps, and mangroves are examples of wetlands. They perform important functions such as storing floodwaters, filtering out pollutants, and reducing erosion. In addition, wetlands provide extremely valuable habitat for a variety of breeding, wintering, and migrating species of invertebrates, fish, and wildlife (Mitsch and Gosselink 2000).

Wetlands are usually characterized by fertile soils, and since they are located near water they are also in demand for urban, industrial, and recreational development. In the past, wetlands have been drained for agriculture and development, and this demand continues (see Chapter 7).

Marshes and swamps are dominated by emergent plants. In a marsh, grasses and herbs form the dominant vegetation; in swamps, woody plants dominate. Estuaries are areas where fresh water flows into a marine ecosystem. They are characterized by brackish water, which is intermediate in salinity between ocean water and fresh water. Estuaries typically contain salt marsh vegetation.

Swamps and marshes are highly productive (Table 13.2). In addition, salt marshes provide critical habitat for many transient animals, such as fishes and crustaceans that spend part of their life in salt water and part in fresh water. Salt marshes also provide staging areas for migrating waterfowl and shorebirds and valuable nesting habitat for these groups as well.

Several million years ago, glaciers advanced across the Great Plains region of North America. As they did so, they scoured the landscape. Chunks of ice that broke from a glacier's leading edge were forced beneath the advancing ice and pushed into the substrate by its great weight. When the ice retreated, these poorly drained depressions filled with water to form wetlands known as potholes. Today the potholes of midwestern North America offer some of the best waterfowl breeding habitat on the continent (van der Valk 1989).

The term mangrove refers to trees that are rooted in the intertidal or subtidal zone; the type of community dominated by these trees is also referred to as mangrove, or mangrove swamp. Mangrove vegetation grows along the tropical coastlines of Asia, Australia, and the Americas in habitats that in temperate zones would contain salt marshes. Many of the organisms that colonize mangrove roots, including sponges, corals, algae, oysters, and barnacles, are sessile (attached to a substrate) for much of their life cycle. These, in turn, provide resources for diverse fishes, worms, crabs, lobsters, shrimps, and octopi. The decomposition of mangrove leaves adds nutrients to the marine community, and guano deposited by high densities of nesting seabirds, such as cormorants, frigatebirds, pelicans, herons, and egrets, further enriches the water. In addition, the roots, trunks, branches, and leaves of mangroves provide habitat for insects, raccoons, snakes, lizards, bats, and arboreal crabs (Simberloff 1983).

13.2.1   Variations in ecosystem productivity

The primary productivity of an ecosystem is the amount of biomass (measured as dry weight) manufactured by its producers. The amount of matter actually produced is termed gross primary production, but since plants consume some of their productivity in respiration, the amount left over, or net primary production (NPP), is what we can measure. For terrestrial plants, NPP includes biomass below as well as above ground, but in practice roots are often ignored and NPP values are based only on stems, leaves, flowers, fruits, and seeds.

By focusing on the flows of energy and matter in ecosystems, ecologists can gain insights into the conservation significance of different ecosystems and their vulnerability to anthropogenic disturbances. Net primary production is compared for some different ecosystems in Table 13.2. The values in this table demonstrate that ecosystems differ markedly in NPP. Production is influenced both by climate, especially moisture and temperature, and by nutrient availability. Moist tropical forests, estuaries, reefs, swamps, and marshes have high productivities, comparable to or exceeding croplands. This is because in these ecosystems high levels of nutrients are available under climatic conditions that allow them to be utilized.

In contrast, the productivity of subarctic and subalpine forests, deserts, and the open ocean is relatively low. In terrestrial ecosystems, temperature and moisture are critical determinants of productivity. Productivity is low in arid environments, such as deserts, and the cold environments at high latitudes (arctic tundra) and altitudes (alpine tundra). The scarcity of available water limits productivity in deserts, whereas productivity at high latitudes and elevations is limited by the short growing season. In these environments, vegetative cover is sparse, and plants grow slowly because of high rates of desiccation, extreme temperatures, and infertile soils. In the deep ocean, productivity is low for a different reason. Marine organisms sink to the bottom when they die. On the cold, dark ocean floor, nutrients are inaccessible, unless they are brought to the surface by upwellings. Regions characterized by these vertical currents, such as arctic and antarctic seas, are relatively productive compared to most of the open ocean.

Productivity is closely related to economic importance, and sometimes to ecological importance as well. We have already alluded to the tremendous ecological importance of moist tropical forests, which are highly productive. Estuaries are another type of ecosystem that is both highly productive and economically valuable. They produce more fish biomass per cubic meter of water than either freshwater or marine environments. This is partly due to the fact that many species of fish and invertebrates spend part of their life cycle in estuaries. About two-thirds of the biomass of commercially caught fish and shellfish is thought to come from species that depend on estuaries at some point in their life. In fact, "protein yields from managed estuarine regions equal or exceed protein yields from managed terrestrial systems producing beef" (Woodwell *et al.* 1973:236). (Of course, ecosystems with low productivity may be ecologically important also. Although on an annual basis the net primary productivity of arctic tundra is low, this ecosystem supports prodigious numbers of breeding migratory waterfowl in summer.)

Ecosystem productivity often affects sensitivity to disturbance. When biomass in an environment of low productivity is removed, it is usually replaced quite slowly. During the time that a disturbed area is denuded of vegetation, exposed soils may be eroded by wind or water. Consequently, dunes and deserts as well as alpine, arctic, subalpine, and subarctic communities are slow to return to predisturbance conditions after vegetation is removed by fire, grazing, trampling, burrowing, cultivation, or vehicular traffic.

Ecosystems with low productivity also tend to be quite sensitive to disturbance from exotic species. Introduced organisms such as salt cedars, burros, and mountain goats have had negative impacts on arid and alpine environments. There are two reasons for this sensitivity. First, if an invading species uses resources that are unavailable to native species in a situation where productivity is low, then the alien can reduce the availability of that resource to natives, which will have a profound effect upon ecosystem structure and function. For example, salt cedar taps water resources that are unavailable to native plants and alters desert hydrology as a result. Because water is a limiting resource in arid environments, salt cedar or any other introduced organism that depletes water can cause significant ecological changes. Second, if the activities of an alien species remove the sparse vegetative cover of a relatively unproductive ecosystem, then the invader is likely to have a major impact because recovery will proceed extremely slowly. This is the case with mountain goats introduced into the Olympic Mountains and feral burros in the arid Southwest (see Box 13.10).

Box 13.10   Some effects of introduced species on ecosystems with low productivity

Salt cedar in American deserts: Alteration of hydro-logical cycles in an arid ecosystem

Tamarisk, or salt cedar, is an alien shrub from Eurasia that has invaded arid parts of the western United States. It grows rapidly and forms dense thickets that displace native vegetation. The success of salt cedar is partly due to its ability to reach water sources that are unavailable to native plants. Salt cedar extends deep roots to the water table and uses prodigious amounts of water. As a result, invasion by salt cedar is accompanied by major changes in local hydrological cycles. Desert oases and species dependent upon them, such as desert pupfish, yellow-billed cuckoos, and desert fan palms, are seriously affected by tamarisk encroachment (Horton 1964, 1977; Johnson 1986).

Mountain goats in the Olympic Mountains: Accelerated erosion rates

The activities of mountain goats introduced to the Olympic Mountains have caused erosion to accelerate in the goats' alpine habitat. Mountain goats are native to the Cascade Range in Washington, but not to the Olympic Mountains because natural barriers prevented colonization of the Olympic Peninsula. In the 1920s, goats from British Columbia and Alaska were deliberately transplanted to the Olympic Mountains. In a short period of time, this nucleus increased dramatically to between 500 and 700 animals. Most of the goats remained within Olympic National Park, where hunting is prohibited. This successful introduction threatens the park's fragile alpine communities, and several endemic plant species they support. Wallowing goats remove a stabilizing layer of mosses and lichens from the soil surface. Bare areas created by the goats are subject to erosion. In spite of the fact that the goats pose a serious threat to native plant communities, efforts to control populations of these non-native species in the park are controversial because members of the public get upset at the idea of Park Service personnel killing appealing mammals (Wright 1992; Scheffer 1993).

Burros in the American Southwest: Increased utilization of sparse vegetation

In the sixteenth century, Spanish explorers brought the domestic burro, a descendant of the wild African ass, to the New World. Burros are well adapted to arid rangelands, and the development of large feral populations in the southwestern U.S.A. was followed by significant damage to this ecologically vulnerable region (Carothers *et al.* 1976). Burros are more efficient than native ungulates at exploiting scarce resources in arid ecosystems. This is due to adaptations of their digestive system. Burros, like other members of the horse family, digest plant foods primarily in the cecum, a blind outpocket of the posterior portion of the digestive system. This type of digestion is inefficient, but it is rapid and allows for partial digestion of high-fiber foods. In contrast, bighorn sheep, like cattle and deer, are ruminants characterized by slow but more thorough digestion of plant materials. As a consequence of their digestive abilities, burros are able to utilize high-fiber foods that native ungulates cannot digest. Thus, burros remove more biomass than native species, a serious matter in arid ecosystems where productivity is low because of limited water. The introduction of burros in the southwestern U.S.A. caused a change in the type of herbivory experienced by native plant species, which led to changes in the composition of plant communities and their associated fauna, as some species increased in abundance while others decreased.

Although ecosystems with low productivity tend to be sensitive to disturbance and invasion, productive ecosystems do not necessarily fare any better. Moist tropical forests and coral reefs are examples of ecosystems that are highly productive and yet sensitive for other reasons.

13.2.2   Variations in soil fertility

Soils are relatively infertile in environments where climatic factors limit plant growth, because where vegetation grows slowly little organic matter is incorporated into the soil. The low fertility of desert, tundra, and dune soils operates as a feedback mechanism to further retard plant growth. Thus, low soil fertility contributes to the sensitivity of these ecosystems to disturbance.

The connection between infertile soils and sensitivity to disturbance is not limited to ecosystems in which productivity is low, however. Moist tropical forests are very sensitive to disturbances that remove biomass, such as logging, grazing, and farming, yet they are extremely productive (see Table 13.2). How can this paradox be explained? Most of the biomass in moist tropical forests is concentrated above ground, in contrast to temperate forests where considerable biomass is located in the soil (see Box 13.6). The plants of moist tropical forests are characterized by a host of adaptations for trapping nutrients above ground, because once nutrients enter the soil they are likely to be leached out by the extremely high rainfall that is characteristic of the moist tropics. For instance, when leaves fall and decay, the released nutrients are rapidly absorbed by shallow root systems. Consequently, nutrients are quickly returned to the aboveground component of the ecosystem instead of entering the soil. As a result, the soils of moist tropical forests are surprisingly low in organic matter and nutrients. So when the aboveground biomass is removed from a tropical moist forest, the loss of nutrients is great, and the soil that remains contains relatively little organic matter to nourish regenerating plants. The low organic matter content of these soils is one reason for the slow recovery times of tropical moist forests.

Some tropical rainforest soils are high in iron and aluminum. This characteristic also makes them vulnerable to disturbance. Rocks weather rapidly in the warm, wet tropics, and acids are released as the vegetation decays. Under these chemical conditions the elements iron and aluminum form insoluble oxides. The net result of these effects is a clayey soil that is low in organic matter. When vegetative cover is removed from this type of soil and its surface is exposed to sun and oxygen, the soil becomes baked and forms a hard, brick-like substance, termed laterite, which is not easily penetrated by roots. This process is virtually irreversible.

13.2.3   Variations in adaptations to disturbance

The unique evolutionary history of a community influences its sensitivity to disturbance and to invasion. Sometimes because of the circumstances of its evolutionary, geological, and biogeographic history, a group of species may never have come in contact with certain types of antagonists. These isolated species are extremely vulnerable when unfamiliar antagonists are introduced. This is especially evident on oceanic islands, where some types of antagonistic interactions were lacking in native communities prior to introductions of exotic species. When people arrive on oceanic islands, they usually bring a variety of plant and animal invaders. Polynesians and subsequently Europeans introduced numerous species to the Hawaiian archipelago, including mosquitoes and the microorganisms they transmit (which were devastating to native birds), rats, pigs, mongoose, and domestic and wild ungulates. These exotic predators, diseases, and herbivores had severe impacts on native Hawaiian communities. For example, because the plants of Hawaii lack evolutionary experience with large herbivorous mammals, they are defenseless in the face of introduced cattle and sheep.

The threat to biodiversity from alien introductions is likely to be especially serious in communities that contain many endemic species, because when an endemic organism is locally extirpated it also becomes globally extinct. Communities that evolved in isolation, such as those on oceanic islands and isolated mountaintops, are characterized by large numbers of endemics. The introduction of exotic species to these areas frequently results in high rates of extinction among native organisms.

If a region's disturbance regime changes suddenly, then organisms that are adapted to the new disturbance regime will prosper (see Box 13.11).

Box 13.11   What happens when new disturbances and new, disturbance-adapted species arrive simultaneously?

During the last 150 years, a combination of habitat availability and transportation allowed cheatgrass and other immigrant weeds from Eurasia to virtually replace native bunchgrasses in many areas of the Intermountain West. The grasses of central Eurasia and the Intermountain West evolved in similar climates but with dissimilar natural disturbance regimes. In the Intermountain West, large herds of bison and other massive herbivores were absent; therefore, the grasses of this region did not evolve adaptations to cope with the disturbances that large groups of large ungulates create by grazing, wallowing, and trampling. The native grasses' lack of evolutionary experience with large-scale biotic disturbances left them vulnerable to any anthropogenic changes that remove vegetation on a similar scale.

The steppes of the Intermountain West were once dominated by bunchgrasses such as bluebunch wheatgrass and Idaho fescue. These grasses form clumps separated by soil covered with a thin crust of mosses and lichens. When agriculture and livestock came to the Intermountain West in the nineteenth century, the native grasses were unable to re-establish themselves rapidly on the large areas of bare soil that were created. Native vegetation grew back only slowly after these new land uses disturbed the soil. In contrast to the bunchgrasses of North America, cheatgrass and other Eurasian grasses are well adapted to germinate on areas of bare soil that become available after a disturbance, because these grasses evolved with centuries of agriculture in the steppes of Eurasia.

Thus, the sudden introduction of agriculture and livestock grazing into the Intermountain West in the nineteenth century disturbed the native bunchgrasses on an unprecedented scale. At the same time the railroads brought in grain shipments contaminated by the seeds of alien grasses. Disturbed areas were quickly colonized by Eurasian grasses that were adapted to frequent disturbances. This made for an unbeatable combination – an increase in disturbed habitat and the simultaneous arrival of exotic colonists – which set the stage for massive plant invasions (Mack 1981, 1986; Mack and Thompson 1982).

13.3   Geographic context

We saw in Chapter 10 that preservationist managers apply insights from landscape ecology to assess the effects of patch size and isolation. The flux-of-nature viewpoint suggests that managers should take this approach a step further by asking: How can the lands between reserves be managed to maximize their positive contributions to species richness and ecosystem function?

Reserves are surrounded by lands that are utilized by people in a variety of ways. Although the lands in this matrix between reserves produce commodities, nevertheless, they may also be managed so as to maximize their contribution to nearby reserves. A matrix that is ecosystem-friendly can contribute to the maintenance of biological diversity and ecosystem integrity in at least two ways.

13.3.1   Maximizing potential for recovery from disturbance

Landscape patterns have a lot to do with how quickly a disturbed area returns to its predisturbance state. For example, the size of a disturbance affects physical properties, such as the temperature and humidity of the disturbed habitat, and biotic features, like the availability of sources of colonists. When vegetation is completely removed in a patch of forest by a treefall, a fire, or logging, recovery proceeds inward from the margins of the surrounding vegetation. One reason for this is the extreme physical conditions that prevail at the center. High temperature and low moisture at the center of a disturbance impede germination and therefore retard the regeneration of plant cover. Another reason why recovery is greatest at the margins of a clearing stems from the organisms that are present there.

The 1980 eruption of Mount St. Helen's in the Cascade Range allowed scientists to test predictions about recovery rates of natural vegetation. Vast areas were covered with mud and ash, yet recovery proceeded much more rapidly than scientists had expected. This was in part because of biological legacies (organic materials and surviving organisms left behind after a disturbance) that remained after the eruption (Franklin 1990, 1995). In some places, vegetation was totally obliterated by events associated with the eruption. Flows of volcanic material, debris, and mud buried all plants, and by the end of the first growing season after the eruption, hardly any plants had returned to these areas. In areas where vegetation survived, however, recovery was surprisingly rapid. Where snowpack afforded protection, plants were able to resprout after the eruption and emerge through volcanic deposits up to 9 cm thick. Individual trees or clumps of vegetation acted as foci for regeneration (del Moral and Bliss 1987; McKee *et al.* 1987).

Plants that are left on a site enhance the potential for revegetation because they provide shade, moderate temperatures, and offer a source of animals and seeds to repopulate the disturbed patch. If no remnants of vegetation remain in the midst of a clearing, regeneration must proceed inward from its edges. Seeds are blown into the cleared patch or are brought onsite by birds or bats, but research on the regeneration of deforested areas in the Amazon has shown that most fruit-eating birds at such sites will not move into large, unvegetated openings. For this reason, birds bring few tree seeds to the centers of large clearings, although seeds are deposited at the margins. By leaving a few trees standing in the center of a cleared area, recovery can be speeded up considerably, because birds will visit these trees and deposit seeds beneath them (Uhl 1988). Green trees left on cutover sites also act as refuges for forest invertebrates. Many of these have poor dispersal abilities, so if they are eliminated from a cutover patch, their chances of recolonizing it at a later date are low. If green trees are left, however, they can maintain populations in these biological legacies until the trees return.

Thus, forest cuts can be designed so that they facilitate regeneration of a harvested area by leaving standing green trees. Insights about the role of biological legacies in ecosystem recovery can be used to develop practices that promote rather than impede regeneration after a disturbance in highly modified lands. In the Pacific Northwest, for example, the Interagency Scientific Committee charged with developing a strategy for forest management to conserve the northern spotted owl recommended that land between forest reserves should be managed so as to maintain 50% of the area in trees that are at least 4.3 cm in diameter and provide 40% crown cover (Franklin 1993).

13.3.2 Maximizing movement of organisms through the matrix

Passage through the matrix is critical for processes such as interpatch colonization. Homogeneous landscapes, such as large monocultures, allow few organisms to move through them. Practices that leave some uncultivated cover and maintain habitat diversity – such as minimum tillage systems, shelterbelts, fencerows, and buffer strips of vegetation along stream banks – make the matrix more friendly to many dispersing and migrating organisms. In some cases, however, it is not desirable to facilitate movement of organisms into reserves. This is true when the matrix is home to predators and parasites, or to non-native species that will colonize natural disturbances in the reserve (Janzen 1983).

13.4 Examples

The three examples below demonstrate how the concepts discussed above can be applied to managing sustainable ecosystems.

13.4.1 Maintaining and restoring structures and functions of late-successional forests in the Sierra Nevada ecosystem

In 1992 the U.S. Congress charged the Sierra Nevada Ecosystem Project with developing recommendations for management strategies that would maintain forest health and allow for sustainable use of resources in California's 63 000 km² Sierra Nevada ecosystem (Figure 13.3). An ambitious set of interdisciplinary studies was initiated to compare historical and current conditions in the Sierra Nevada ecosystem. Information was compiled on past climates

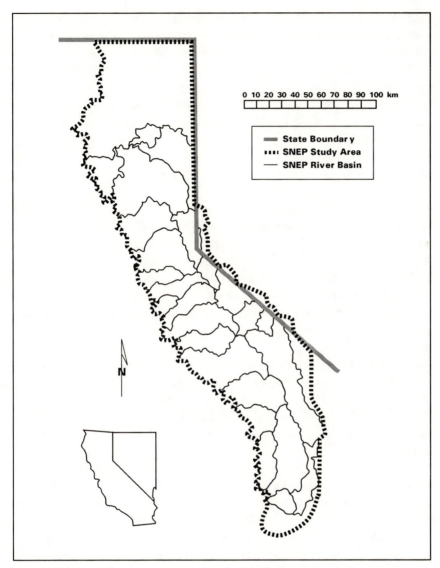

Figure 13.3. Map showing extent and location of the Sierra Nevada Ecosystem Project (SNEP). (After Sierra Nevada Ecosystem Project 1996.)

and disturbance regimes, hydrology, human uses, terrestrial and aquatic vertebrates and invertebrates, and vegetation. These studies indicated that fire was generally more frequent throughout much of the Sierra Nevada ecosystem before 1850. Current conditions encourage high-severity fires. More fuel is available, both on the forest floor and in the form of ladder fuels (shrubs and small trees that can carry a fire to the forest canopy). In the past, fire severity varied, depending on local climate, topography, elevation, vegetation, soil, and human practices. Fires of low or moderate intensity burned some areas, and high-intensity fires burned others. With fire suppression, however, the range of fire severities has narrowed. Low- to moderate-intensity fires have been virtually eliminated, but it was not possible to prevent large, severe fires (Erman and Jones 1996; McKelvey *et al.* 1996; Skinner and Chang 1996). Instead of a mosaic produced by a range of fire intensities, most sites either have not burned recently or have experienced severe fires that consumed most of the vegetation. As a result, late-successional and old-growth forests have become more homogeneous.

Most people think of old forests as having a dense, closed canopy, but this is not the case in the Sierra Nevada. (This conception is based on familiar images of old-growth forests of the Pacific Northwest.) In the Sierra Nevada, high-quality late-successional and old-growth forests are structurally complex and varied (Figure 13.4). Some patches have a dense, closed canopy, but in others, gaps and areas of partially closed canopy are common (Franklin and Fites-Kaufmann 1996; Franklin *et al.* 1996). Wildfires of low to moderate intensity were one of the main forces that created and maintained this heterogeneity.

Researchers used specific structural criteria, such as the density and size of large-diameter trees and snags, to assess the present extent and condition of late-successional/old-growth (LS/OG) forests in the Sierra Nevada Ecosystem. The structural criteria were used as indicators of the degree to which important LS/OG functions were being fulfilled. The research team mapped large, relatively uniform landscape units, termed polygons, and ranked each polygon for each criterion on a scale of 0 to 5, with 5 being the best. Polygon scores were used to quantitatively assess each polygon in terms of its contribution to LS/OG functions. Only 8.2% of the mapped polygons had rankings of 4 or 5, and polygons from which timber had been harvested had fairly low rankings. On the basis of these findings, the research team concluded that forests with high-quality LS/OG structures (and, presumably, functions) are much less extensive than they were prior to western settlement (Franklin and Fites-Kaufmann 1996).

The report of the Sierra Nevada Ecosystem Project concludes with recommendations for how LS/OG forests could be maintained and restored. Since the major process that has been altered is the disturbance regime, the authors recommend restoring low- to moderate-intensity fires. To do this, it will be necessary to reduce the current high fuel loads. This could be done by mechanical means, such as thinning and creating fuel breaks, but the use of

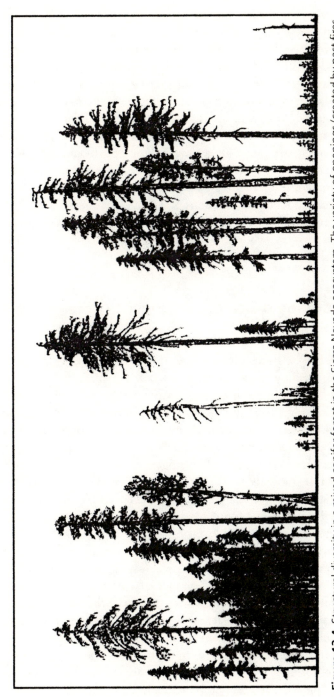

Figure 13.4. Structural diversity in a mixed conifer forest in the Sierra Nevada ecosystem. The variety of openings (created by past fires of varying intensities) creates both vertical and horizontal heterogeneity. (After Franklin and Fites-Kaufmann 1996.)

mechanical equipment disturbs the soil surface, compacts soils, and damages tree trunks and roots.

Franklin *et al.* (1996) suggest that the most effective way to maintain and restore LS/OG forests in the Sierra Nevada ecosystem would be through a two-pronged strategy that includes setting aside large reserves as well as managing the matrix between the reserves. Areas of Late-Successional Emphasis could be designated. Within these large reserves, LS/OG forests would be maintained using the least intrusive and most natural methods. The primary tool for restoring a more varied fire regime in these areas would be prescribed burning. Outside of these reserves, mechanical means of reducing fuel loads would be permitted, and structural features such as large-diameter trees and snags, which contribute to the functions of LS/OG forests, would be maintained (Franklin *et al.* 1996). These actions should restore a heterogeneous fire regime, as well as a spatially diverse mosaic of patches created by fires of different intensities.

13.4.2   Restoring structural heterogeneity in the Negev Desert

The Negev Desert of Israel (Evenari *et al.* 1971) illustrates how attention to structural heterogeneity at the appropriate spatial scale allows managers to identify and manage critical functions and structures. In this arid ecosystem, the capture of dust by plants and the incorporation of that dust into soil is a controlling process that influences many other aspects of ecosystem structure and function. As dust accumulates, soil mounds form around patches of vegetation. This in turn results in spatial heterogeneity on a scale of meters. The habitat between the mounds consists of soil with a crust of microorganisms such as cyanobacteria, bacteria, algae, mosses, and lichens. Scattered throughout this crusted matrix are soil mounds and their associated plants (Shachak and Pickett 1999).

Soil texture differs in the crusted soil and on the mounds. The surface of the mounds consists of loose soil particles, whereas in the matrix between the mounds, crusted soil particles are bound together, resulting in a harder surface. Water does not move easily between these particles of cemented soil. Instead, water moves downslope across the ground surface until it comes in contact with the more porous soil of a mound, where it infiltrates into the ground. The vegetation on the mounds produces litter when it dies. As water moves through this litter into the soil, nutrients are leached out of the vegetation and transported into the soil. In addition, desert animals concentrate their activities near the mounds, and their urine and feces enrich the soil. As a result of these processes, the soil in the mounds is relatively moist and rich in nutrients, and it provides favorable sites for seed germination. Thus, whether we are considering resources, organisms, or structures, the mounds and the crusted soil between them provide two radically different environments. Species richness and productivity are relatively high in the mounds and lower in the crusted soil.

Manipulating the environment to enhance mound formation is an unobtrusive way to increase biodiversity and productivity in this desert ecosystem. The rate of mound formation depends on surface relief. Small depressions provide pits where water and nutrients accumulate; vegetation patches develop in these pits. Because water is scarce in the desert, the pits where water is concentrated serve as foci for plant growth. Thus, a heterogeneous surface is critical for mound formation.

Several years ago, a team of hydrologists, ecologists, and managers set out to develop a management strategy for the Negev. The area has been subjected to uncontrolled grazing for several thousand years, which has caused a decrease in vegetative cover and allowed soil erosion to increase. The management objective of the Jewish National Fund, the agency in charge of wild and pastoral lands in Israel, was to create an ecological park with increased productivity and species diversity.

The scientists charged with developing a management plan for the proposed park began by studying the relationship between habitat patches and ecosystem properties in a nearby park, which served as a reference area for comparison to the managed site. They hypothesized that productivity and diversity in this environment are controlled by the microtopography of the ground surface and predicted that increasing the number of pits would enhance productivity and diversity. When the investigators created experimental pits in the landscape to allow for the storage of nutrients and runoff, they found that diversity and productivity did increase, as they had predicted, because as they suspected, the surface pits were a key structural feature (Boeken and Shachak 1994).

In this example, a subtle alteration of patchiness altered the flux of resources and affected productivity and species diversity. As scientists' understanding of the system grows, their conceptual model of a how ecosystem processes operate in the Negev Desert can be readjusted further. The conservation approach illustrated by this example is neither strictly utilitarian nor strictly preservationist. It certainly is not a hands-off approach, but on the other hand it is rather different from conventional utilitarian management too. Utilitarian managers would have tried to restore the degraded range of the

Negev by planting, watering, and fertilizing. The new approach also manipulates the desert ecosystem, but the way in which it does so incorporates a more sophisticated understanding of ecosystem dynamics. One advantage of this strategy is that if it turns out managers are wrong about what is going on in this system, the impacts of creating small surface pits will probably be a lot less severe than the impacts of wholesale planting, watering, and fertilizing. This approach also differs from typical preservationist management. Instead of identifying featured species of interest and directly trying to increase their populations, the scientists chose to identify key processes and structures and to manipulate them with the goal of promoting conditions favorable to the desired outcome.

One of the interesting things about this example is the relatively small spatial scale at which critical ecosystem processes operate in the Negev. Disturbances and other processes occur at multiple spatial and temporal scales, and there is a danger that we may overlook important phenomena if we focus on inappropriate scales. In this case, even a scale of tens of meters would be too coarse to elucidate the processes generating significant spatial heterogeneity.

13.4.3   Restoring variations in river flow on the Roanoke River

Just as the structure and function of Sierra Nevada forests depend upon variations in fire intensity, the structure and function of a river's ecosystem are intimately tied to variations in the amount of water flowing through the river's channel. Hydrologist Brian Richter and his colleagues at The Nature Conservancy have developed a method of using historical variability in streamflow to assess the degree of alteration to the flow regimes of regulated rivers. This approach, termed the range of variability approach, is illustrated by the Roanoke River in North Carolina. Daily U.S. Geological Survey streamflow measurements going back to 1913 are available for this river. Beginning in 1950, several dams were constructed on the Roanoke for the purposes of flood control and hydropower generation. A comparison of conditions before and after damming of the river reveals, not surprisingly, that the timing and magnitude of flooding have been altered by the dams (Figure 13.5). On the basis of this information, scientists recommended that between April 1 and June 15 daily flows on the Roanoke should be kept within the 25th and 75th percentiles of pre-impoundment variation in flow rates. Since they made this recommendation, the monthly mean flow for April has fallen within the

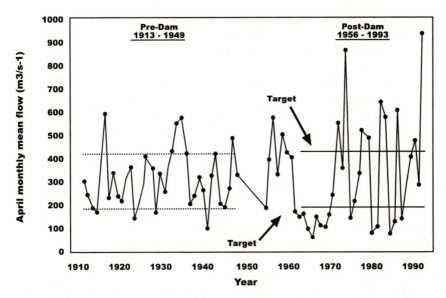

Figure 13.5. Monthly mean flows of the Roanoke River for April. The flow regime during the post-dam period differs markedly from the pre-dam regime. (After Richter *et al.* 1997.)

recommended range of variability in three years out of five. Preliminary data from follow-up monitoring suggest that these changes have had positive effects on striped bass recruitment, which appears to be influenced by the magnitude and rates of change in flow levels during the spawning period. One would expect that these modifications of the flow regime would also benefit other species, but additional data are needed to test this hypothesis (Richter *et al.* 1996, 1997).

Now that we have some understanding of how sustainable-ecosystem management can conserve processes and contexts, we need to consider ways that people can be integrated into conservation efforts. The next chapter does this.

References

Allan, J. D. and A. S. Flecker (1993). Biodiversity conservation in running waters. *BioScience* **43**:32–43.

Berry, K. H. (1980). A review of the effects of off-road vehicles on birds and other vertebrates. In *Management of Western Forests and Grasslands for Nongame Birds*, pp. 451–467. U.S. Department of Agriculture Forest Service, General Technical Report GTR-INT-86, Ogden, UT.

Boeken, B. and M. Shachak (1994). Desert plant communities in human-made patches: implications for management. *Ecological Applications* **4**:702–716.

Botkin, D. B. (1990). *Discordant Harmonies: A New Ecology for the Twenty-First Century.* New York, NY: Oxford University Press.

Brinson, M. M., B. L. Swift, R. C. Plantico, and S. J. Barclay (1981). *Riparian Ecosystems: Their Ecology and Status.* U.S. Department of the Interior Fish and Wildlife Service, Biological Services Program. Kearneysville, WV: Eastern Energy Land Use Team [and] National Water Resources Analysis Group .

Brown, J. H. and A. C. Gibson (1983). *Biogeography.* St. Louis, MO: C. V. Mosby.

Carlton, J. T. (1998). Apostrophe to the ocean. *Conservation Biology* **12**:1165–1167.

Carothers, S. W., M. E. Stitt, and R. R. Johnson (1976). Feral asses on public lands: an analysis of biotic impact, legal considerations, and management alternatives. *Transactions of the North American Wildlife and Natural Resources Conference* **41**:396–404.

Daubenmire, R. (1978). *Plant Geography.* New York, NY: Academic Press.

del Moral, R. and L. C. Bliss (1987). Initial vegetation recovery on subalpine slopes of Mount St. Helens, Washington. In *Mount St. Helens 1980: Botanical Consequences of the Explosive Eruptions,* ed. D. E. Bilderback, pp. 148–167. Berkeley, CA: University of California Press.

Erman, D. C. and R. Jones (1996). Fire frequency analysis of Sierra forests. In *Sierra Nevada Ecosystem Project, Final Report to Congress,* vol. 2, *Assessments and Scientific Basis for Management Options,* Wildland Resources Center Report no. 37, pp. 1139–1146. Davis, CA: University of California.

Evenari, M., L. Shanan, and N. Tadmor (1971). *The Negev: The Challenge of a Desert.* Cambridge, MA: Harvard University Press.

Ewel, J. J., M. M. Mazzarino, and C. W. Berish (1991). Tropical soil fertility changes under monocultures and successional communities of different structures. *Ecological Applications* **1**:289–302.

Franklin, J. F. (1990). Biological legacies: a critical management concept from Mount St. Helens. *Transactions of the North American Wildlife and Natural Resources Conference* **25**:216–219.

Franklin, J. F. (1993). Fundamentals of ecosystem management in the Pacific Northwest. In *Defining Sustainable Forestry,* ed. G. H. Aplet, N. Johnson, J. T. Olson, and V. A. Sample, pp. 127–144. Washington, DC: Island Press.

Franklin, J. F. (1995). Sustainability of managed temperate forest ecosystems. In *Defining and Measuring Sustainability: The Biogeophysical Foundations,* ed. M. Munasinghe and W. Shearer, pp. 355–385. Washington, DC: The World Bank.

Franklin, J. F. and J. Fites-Kaufmann (1996). Assessment of late-successional forests of the Sierra Nevada. In *Sierra Nevada Ecosystem Project, Final Report to Congress,* vol. 2, *Assessments and Scientific Basis for Management Options,* Wildland Resources Center Report no. 37, pp. 627–661. Davis, CA: University of California.

Franklin, J. F., D. Graber, K. N. Johnson, J. Fites-Kaufmann, K. Menning, D. Parsons, J. Sessions, T. A. Spies, J. Tappeiner, and D. Thornburgh (1996). Alternative approaches to conservation of late-successional forests in the Sierra Nevada and their evaluation. In *Sierra Nevada Ecosystem Project, Final Report to Congress*, Wildland Resources Center, Addendum, pp. 53–69. Davis, CA: University of California.

Glynn, W. P. and W. H. de Weerdt (1991). Elimination of two reef-building hydro-corals following the 1982–1983 El Niño warming event. *Science* **253**:69–71.

Harwell, M. A. (1997). Ecosystem management in South Florida. *BioScience* **47**:499–512.

Holling, C. S. and G. K. Meffe (1996). Command and control and the pathology of natural resource management. *Conservation Biology* **10**:328–337.

Horton, J. S. (1964). Notes on the introduction of deciduous tamarisk. U.S. Department of Agriculture Forest Service, Research Note RM-16, Fort Collins, CO.

Horton, J. S. (1977). The development and perpetuation of the permanent tamarisk type in the phreatophyte zone of the Southwest. In *Importance, Preservation, and Management of the Riparian Habitat: A Symposium*, pp. 124–127. U.S. Department of Agriculture Forest Service, General Technical Report RM-GTR-43, Fort Collins, CO.

Janzen, D. H. (1983). No park is an island: increase in interference from outside as park size decreases. *Oikos* **41**:403–410.

Janzen, D. H. (1988). Tropical dry forests: the most endangered tropical ecosystem. In *Biodiversity*, ed. E. O. Wilson, pp. 130–137. Washington, DC: National Academy Press.

Johnson, S. (1986). Alien plants drain western waters. *Nature Conservancy News* **36**(5):24–25.

Landres, P. B., P. Morgan, and F. J. Swanson (1999). Evaluating the utility of natural variability concepts in managing ecological systems. *Ecological Applications* **9**:1179–1188.

Likens, G. E. and F. H. Bormann (1972). Nutrient cycling in ecosystems. In *Ecosystem Structure and Function, Proceedings of the 31st Annual Biology Colloquium*, ed. J. A. Wiens, pp. 25–67. Corvallis, OR: Oregon State University Press.

Likens, G. E., F. H. Bormann, N. M. Johnson, and R. S. Pierce (1967). The calcium, magnesium, potassium, and sodium budgets for a small forested ecosystem. *Ecology* **48**:772–785.

Likens, G. E., F. H. Bormann, and N. M. Johnson (1969). Nitrification: importance to nutrient losses from a cutover forested ecosystem. *Science* **163**:1205–1206.

Mack, R. N. (1981). Invasion of *Bromus tectorum* L. into western North America: an ecological chronicle. *AgroEcosystems* **7**:145–165.

Mack, R. N. (1986). Alien plant invasion into the Intermountain West: a case history. In *Ecology of Biological Invasions of North America and Hawaii*, ed. H. A. Mooney and J. A. Drake, pp. 191–213. New York, NY: Springer-Verlag.

Mack, R. N. and J. N. Thompson (1982). Evolution in steppe with few large hooved mammals. *American Naturalist* **119**:757–773.

McKee, A., J. E. Means, W. H. Moir, and J. F. Franklin (1987). First-year recovery of upland and riparian vegetation in the devastated area around Mount St. Helens. In *Mount St. Helens 1980: Botanical Consequences of the Explosive Eruptions*, ed. D. E. Bilderback, pp. 168–187. Berkeley, CA: University of California Press.

McKelvey, K. S., C. N. Skinner, C. Chang, D. C. Erman, S. J. Husari, D. J. Parsons, J. W. van Wagtendonk, and C. P. Weatherspoon (1996). An overview of fire in the Sierra Nevada. In *Sierra Nevada Ecosystem Project, Final Report to Congress,* vol. 2, *Assessments and Scientific Basis for Management Options*, Wildland Resources Center Report no. 37, pp. 1033–1140. Davis, CA: University of California.

Menge, B. A. (1979). Coexistence between the seastars *Asterias vulgaris* and *A. forbesi* in a heterogeneous environment: a nonequilibrium explanation. *Oecologia* **41**:245–272.

Mitsch, W. J. and J. G. Gosselink (2000). *Wetlands*, 3rd edn. New York, NY: John Wiley.

Morgan, P., G. H. Aplet, J. B. Haufler, H. C. Humphries, M. M. Moore, and W. D. Wilson (1994). Historical range of variability: a useful tool for evaluating ecosystem change. *Journal of Sustainable Forestry* **2**:87–112.

O'Connell, M. A., J. G. Hallett, and S. G. West (1993). *Wildlife Use of Riparian Habitats: A Literature Review*. Washington Department of Natural Resources, Timber, Fish, and Wildlife, TFW-WL1–93–001, Olympia, WA.

Pickett, S. T. A., V. T. Parker, and P. Fiedler (1992). The new paradigm in ecology: implications for conservation biology above the species level. In *Conservation Biology: The Theory and Practice of Nature Conservation, Preservation, and Management*, ed. P. L. Fiedler and S. K. Jain, pp. 65–88. New York, NY: Chapman and Hall.

Poff, N. L., J. D. Allan, M. B. Bain, J. R. Karr, B. D. Richter, R. E. Sparks, and J. C. Stromberg (1997). The natural flow regime. *BioScience* **47**:769–784.

Reaka-Kudla, M. L. (1997). The global biodiversity of coral reefs: a comparison with rain forests. In *Biodiversity II*, ed. M. L. Reaka-Kudla, D. E. Wilson, and E. O. Wilson, pp. 83–108, Washington, DC: Joseph Henry Press.

Richter, B. D., J. V. Baumgartner, J. Powell, and D. P. Braun (1996). A method for assessing hydrologic alteration within ecosystems. *Conservation Biology* **10**:1163–1174.

Richter, B. D., J. V. Baumgartner, J. Powell, and D. P. Braun (1997). How much water does a river need? *Freshwater Biology* **37**:231–239.

Roberts, L. (1991). Greenhouse role in reef stress unproven. *Science* **253**:258–259.

Scheffer, V. B. (1993). The Olympic goat controversy: a perspective. *Conservation Biology* **7**:916–919.

Shachak, M. and S. T. A. Pickett (1999). Linking ecological understanding and application: patchiness in a dryland system. In *The Ecological Basis of Conservation: Heterogeneity, Ecosystems, and Biodiversity*, ed. S. T. A. Pickett, R. S. Ostfeld, M. Shachak, and G. E. Likens, pp. 108–119. New York, NY: Chapman and Hall.

Sierra Nevada Ecosystem Project (1996). *Sierra Nevada Ecosystem Project, Final Report to Congress*, vol. 2, *Assessments and Scientific Basis for Management Options*, Wildland Resources Center Report no. 37. Davis, CA: University of California.

Simberloff, D. S. (1983). Mangroves. In *Costa Rican Natural History*, ed. D. H. Janzen, pp. 273–276. Chicago, IL: University of Chicago Press.

Skinner, C. N. and C. Chang (1996). Fire regimes past and present. In *Sierra Nevada Ecosystem Project, Final Report to Congress*, vol. 2, *Assessments and Scientific Basis for Management Options*, Wildland Resources Center Report no. 37, pp. 1041–1069. Davis, CA: University of California.

Swanson, F. J., J. A. Jones, D. O. Wallin, and J. H. Cissel (1993). Natural variability: implications for ecosystem management. In *Ecosystem Management: Principles and Applications*, vol. 2, *Eastside Forest Ecosystem Health Assessment*, ed. M. E. Jensen and P. S. Bourgeron, pp. 89–103. U.S. Department of Agriculture Forest Service, Pacific Northwest Research Station, Forestry Sciences Laboratory, Wenatchee, WA.

Uhl, C. (1983). You *can* keep a good forest down. *Natural History* 92(4):70–79.

Uhl, C. (1988). Restoration of degraded lands in the Amazon Basin. In *Biodiversity*, ed. E. O. Wilson, pp. 326–332. Washington, DC: National Academy Press.

van der Valk, A. (ed.) (1989). *Northern Prairie Wetlands*. Ames, IA: Iowa State University.

Watling, L. and E. A. Norse (1998). Disturbance of the seabed by mobile fishing gear: a comparison to forest clearcutting. *Conservation Biology* 12:1180–1197.

Weaver, J. E. and F. E. Clements (1938). *Plant Ecology*, 2nd edn. New York, NY: McGraw-Hill.

Whittaker, R. H. (1975). *Communities and Ecosystems*, 2nd edn. New York, NY: Macmillan.

Whittaker, R. H. and G. E. Likens (1975). The biosphere and man. In *Primary Productivity of the Biosphere*, ed. H. Lieth and R. H. Whittaker, pp. 305–328. New York, NY: Springer-Verlag.

Woodwell, G. M., P. H. Rich, and C. A. S. Hall (1973). Carbon in estuaries. In *Carbon and the Biosphere*, ed. G. M. Woodwell and E. V. Pecan, pp. 221–239. U.S. Atomic Energy Commission, Technical Information Center, Office of Information Services, Springfield, VA.

Wright, R. G. (1992). *Wildlife Research and Management in the National Parks*. Urbana, IL: University of Illinois Press.

14

Techniques – including people in the conservation process

Euroamerican traditions of utilitarian and preservationist resource management both seek to manage people in a human-dominated landscape within which parcels are set off as "natural." We have seen that resource managers are becoming increasingly aware of political, ecological, and ethical reasons for including the activities of people explicitly in management plans, however. A number of innovative strategies for doing this have emerged within the past two decades.

In Chapter 10 we examined one type of economic conservation incentive, the debt-for-nature swap. In such an arrangement, parties from the developed world pay a developing country (by getting its debt reduced) to protect lands from exploitation. This strategy of having outsiders with a stake in resource preservation pay to support limits on resource use has been successful in some instances, but long-term resource conservation also requires the integration of resource use and conservation.

The examples described in this chapter interweave three themes. First, traditional modes of resource use as well as novel forms of exploitation can, under some circumstances, be carried on within areas that are considered preserves. Second, resource uses outside preserves should support, rather than undermine, ecosystem conservation. Third, resource conservation must be linked to tangible benefits for local people. This is especially important if the people who benefit from conservation live far from the place where that conservation takes place. Before looking at these themes in more detail, however, it is necessary to revisit the concept of sustainable use.

14.1   Sustainable use versus sustained yield

We hear a lot about "sustainability" and "sustainable development" these days. The concept of sustainable use has been around for over a century. In the developed world it has been narrowly applied, while in developing countries there are some local examples of sustainable use, but in many places a host of political and economic factors have undermined traditional uses of resources that were apparently sustainable or worked against adopting sustainable practices.

Sustainable use is usually defined as use that meets the needs of the present without compromising the ability of future generations to meet their needs. When resources are used sustainably, harvest does not exceed productivity, and resource use does not compromise ecosystem functioning or cause a loss of biodiversity. If use of a resource is truly sustainable, then it can be continued indefinitely without degrading ecosystem productivity. Nonsustainable uses erode the resource base and adversely affect biodiversity either directly or indirectly. They can result from overexploitation or from activities that interfere with ecosystem structure and function by modifying habitats, introducing pollutants, or introducing exotic species.

Sustainable use is possible because matter and energy are continually recycled. Some of the solar energy input to the biosphere is trapped by photosynthesizing plants and converted into biomass, passed up the food chain, and subsequently decomposed. Atmospheric nitrogen is captured by nitrogen-fixing bacteria and cyanobacteria and converted into organic forms. Water is recycled in a hydrological cycle. If resource use does not seriously interfere with these cyclical processes, then sustainable harvest is possible (though it is not guaranteed).

A caution about terminology is appropriate here. The term sustainability has been used in so many different ways that it is in danger of becoming useless. In particular, there is a tendency to confuse sustainability in the ecological sense (the ability of an ecosystem to put up with a given level of use indefinitely) with economic viability (the ability of a given level of harvest to continue to produce profits or rewards over a long period of time). Resource use is ecologically sustainable if it does not cause a loss of ecosystem function or biological diversity. Economic viability, on the other hand, depends upon whether an enterprise is profitable enough under current market conditions to persist into the foreseeable future. Ecological sustainability and economic viability need to be evaluated separately. Of course, in the long run, nothing can be economically viable if it is not also ecologically sustainable, because if the ecosystem supporting an economic enterprise collapses, then clearly that enterprise will collapse as well. But in the short term, we should distinguish between economic considerations and ecological ones. Otherwise, we may erroneously conclude that something is ecologically sustainable just because it is profitable.

Setting levels at which species can be harvested without compromising their ability to replace themselves is the central goal of utilitarian resource management. Professional managers of wildlife, rangelands, fisheries, and forests have sought to harvest renewable resources sustainably since the days of Gifford Pinchot. The focus of this type of management for sustained yield, however, is maximization of productivity in a few species (such as deer, muskrats, whales, forage plants, or trees) to produce products for recreation or market. This approach looks only at reproduction in the harvested species. In fact, the concept of compensatory mortality in wildlife management ("If we don't kill them something else will") explicitly ignores the consequences of harvest on interacting species; it is assumed that the removal of a harvestable surplus is benign (Chapter 4). A similar philosophy guided forest and range management for many years. The sustainability of timber harvest and grazing were judged in terms of the ability of trees or range plants to grow back after their removal rather than in terms of ecosystem functioning. The objective of sustained yield was not always met, either because estimates of allowable harvest were too optimistic or because recommendations made by scientists were ignored. Nevertheless, the philosophy of conservation as a means of ensuring continuous production pervaded the disciplines of wildlife management, forestry, and range management in the developed world for many decades.

By narrowing its focus to reproduction in a small number of species harvested for sport or commercial gain, the utilitarian approach to resource management overlooked the effects of harvest on interacting species that were not considered economically valuable. The consequences of lost food for a predator or a scavenger or of lost dead wood for a cavity-nesting bird were not addressed. If we kill a muskrat that would otherwise die of disease, this may not have any effect on the level of the muskrat population, but that is not to say that the removal will have no ecological effects at all. Clearly the muskrat's death can affect interacting species in myriad ways. Similarly, if we harvest a large, old tree, we remove a source of habitat for many species. As resource managers developed a more sophisticated understanding of ecosystems, the limitations of management for narrowly defined maximum sustained yield became evident.

14.2   Returning profits from biodiversity-based products to local communities

It is well known that because of their high biodiversity, tropical forests are a promising place to search for new medicines and crops. Many valuable products have been derived from tropical forest organisms, but until recently most of the profits from these discoveries went to companies that marketed them in the developed world, and the country of origin got nothing. For example, the leaves of the Madagascar rosy periwinkle plant yield the chemicals vincristine and vinblastine. Chemists at Lilly Research Laboratories discovered that these chemicals have antitumor properties. Between 1963 and 1985, the sale of these drugs grossed approximately $100 million, but none of this revenue went back to Madagascar (Ehrlich and Ehrlich 1985; Farnsworth 1988). Similarly, corticosteroids developed from a Mexican yam reaped large profits for drug companies, but relatively little revenue returned to Mexico (Joyce 1992).

If the nations that are the source of valuable products reap economic benefits from them, however, they have more incentive to save the habitats that provide those products (Roberts 1992). To respond to this concern, some pharmaceutical companies have begun paying royalties to source nations, but unfortunately the time-lag between searching for and marketing drugs involves decades, so these arrangements still fall short of providing immediate incentives for conservation. This problem was addressed by a novel arrangement between a private organization in Costa Rica, the Instituto Nacional de Biodiversidad (INBio) or National Institute of Biodiversity, and the pharmaceutical giant Merck & Co., Inc. Under the terms of a two-year agreement signed in 1991, prospecting for molecules with unique biological activity would be treated like prospecting for timber or minerals. Merck agreed to pay INBio $1 million to search for substances with pharmaceutical potential. In addition, if the search yielded any commercially viable products, the company agreed to pay royalties on any products it marketed as a result of its chemical prospecting. Under this arrangement, Costa Rica not only would share in profits from the pharmaceutical company's discoveries, it also would receive payment up-front for the right to search for sources of new products. Furthermore, 10% of this money and 50% of any royalties received were to be used for biodiversity conservation (Gershon 1992; Joyce 1992).

The agreement between INBio and Merck was a significant departure from business as usual. It provided a mechanism whereby companies that benefit from a tropical nation's biodiversity pay for using that biodiversity. Further, by paying for the right to search for new products regardless of whether the company turns up anything it considers worthy of development, Merck acknowledged that such chemical prospecting is a use of tropical forests for which users should pay. The agreement was designed to enhance the source country's control over uses of its biological products. In addition, by providing equipment and training, it set the stage for an even greater degree of involvement and control from within Costa Rica in the future (Reid *et al.* 1993).

This issue – fair return to the country of origin of financial rewards from biodiversity prospecting – concerned many delegates to the United Nations Conference on Environment and Development held in Rio de Janeiro in June 1992, especially those from poor countries. Conference participants drafted a Convention on Biological Diversity, which sought to guarantee fair and equitable sharing of the profits derived from utilizing biodiversity by requiring that the nation where a species or a gene originated would reap financial benefits if that species or gene were developed into a profitable product. Some developed nations refused to sign the treaty, however, because of concerns that these profit-sharing provisions did not give biotechnology firms enough control over the results of their research. By late in the year 2000, 179 nations, including many from the developed world but not the U.S.A., had become signatories to the convention.

14.3   Integrating economic development and conservation

People living next to protected areas bear the costs of conflicts with wildlife that moves outside park borders. Depending on the severity of the problem, they may lose their crops, livestock, or lives. The fact that even the largest protected areas do not provide for all the needs of large, mobile mammals exacerbates this situation, because wide-ranging predators and migratory ungulates leave park boundaries often. For herbivores, this is especially likely to be true if pressure from hunters or predators has diminished. In this situation, the externalities of conservation (see Chapter 7) are borne by local people, whereas the benefits are enjoyed by outsiders (Chapter 11).

When local people benefit from biodiversity protection, however, they are more likely to support it. One approach to including people in the conservation process seeks to integrate economic development and resource management in poor countries. Since poverty creates economic pressures on species and habitats, it should be possible to increase support for conservation programs by creating mechanisms by which conservation programs generate income for local people. Programs that integrate economic incentives and local control in ways that conserve wildlife and meet the needs of local communities will generally receive more support than purely protectionist approaches to wildlife conservation. These programs are sometimes referred to as Integrated Conservation and Development Programs, or ICDPs. An ICDP typically involves a partnership between: (1) a national agency in charge of managing forests, wildlife, or parks, (2) a foreign donor agency, (3) a nongovernmental organization, and (4) an organization representing the local community (Alpert 1996). ICDPs are in place in quite a few African nations, especially those where populations of large mammals are viewed as pests in communities adjoining preserves, as well as in Asia and Latin America.

Diverting a portion of the funds generated from conservation-related activities such as tourism to local communities is one of the most common ways of linking conservation and development. Another approach is to promote private wildlife-related businesses such as game ranching on private land (Kiss 1990). In addition to the income from fees, these enterprises provide jobs for local people as scouts, wardens, guides, and maintenance workers, as well as in service positions, and the money that visitors spend in the area provides additional income. These funds may be spent by individual households or on community projects such as schools, clinics, or wells (Alpert 1996).

Tourism in which visitors are attracted by opportunities to observe wildlife or experience special natural places is sometimes termed ecotourism. (This type of tourism is not restricted to the developing world. Economic analyses of the Greater Yellowstone Ecosystem, GYE, point to ecotourism as the primary employment sector in the GYE economy (Power 1991).) In the developing world, tourists are willing to spend considerable amounts of money to see certain habitats and species, such as tropical rainforest, large mammals of the African savanna, or nonhuman primates (Box 14.1).

Box 14.1   Tourism in the Virungas of Rwanda

Ecotourism has a number of features that make it an attractive form of economic development, but it also has some significant drawbacks. Rwanda's Parc National des Volcans illustrates both the potential gains from and the instability of ecotourism. The Virungas of central Africa are located in three nations: Rwanda, Zaïre, and Uganda. In Rwanda, parts of the volcanoes form the Parc National des Volcans. Within the park, the upper slopes of the Virungas support bamboo thickets, high-elevation rainforests, shrub thickets, and alpine communities. This vegetation provides important ecological benefits for villagers living at lower elevations. During the rainy season, the dense mountain vegetation prevents water from running downslope and allows it to soak into the soil. Then during the dry season, subsurface water is gradually released. In this way, the mountaintop plant communities even out the distribution of water during the year to reduce flooding, erosion, and siltation in the wet season and to increase the availability of water in the dry season. The Virunga forests also support many endemic species of birds as well as numerous rare mammals, including the mountain gorilla.

By the late 1970s, the mountain gorilla was "one of the most studied and best-known species in Africa" (Vedder and Weber 1990:84), but this knowledge had been obtained by and was communicated to outsiders only (see Box 11.1). Through films, books, and articles, western scientists had disseminated their findings to people in the developed world but not to the park's neighbors. Amy Vedder and William Weber, who studied the economic and social context of gorilla conservation, pointed out that "No Rwandan scientist had ever seen, let alone studied gorillas, no university students had been trained to fill this void, no references were made to the gorilla or its habitat in primary or secondary school curricula, and no effort had been made in the broad area of public education" (Vedder and Weber 1990:85). It is not surprising, therefore, that Rwandans knew little about mountain gorillas and had little interest in conserving them.

To address the problem of gorilla conservation in a context that would serve local needs as well as national and international ones, a group of non-governmental organizations developed the Mountain Gorilla Project. Project personnel recommended tourism as a means of providing an economic incentive for gorilla conservation. The proposed program involved charging tourists for the privilege of viewing wild gorillas during closely supervised tours. Initially, this suggestion met with skepticism. Park staff did not expect that many visitors would be willing to pay high prices to slog through cold, wet mud for an opportunity to view gorillas. They were wrong.

During the 1980s, tourism flourished in the Parc National des Volcans. Tourists paid $200 each to view one of several family groups of mountain gorillas that had become habituated to the presence of people. This enterprise provided jobs for local people as guides, and in addition tourists spent money in hotels and restaurants and bought local products during their visits. By the mid 1980s, tourism had become the fastest-growing sector of the Rwandan economy and one of the nation's principal sources of foreign exchange. The Mountain Gorilla Project also instituted a program to educate people living near the park about gorillas and about the ecological services provided by the park. This was accomplished through films, slide shows, talks, radio programs, and seminars. A number of tangible, positive results followed. Poaching within the park declined, the gorilla population grew, and local support for gorilla conservation increased. By 1990, the Mountain Gorilla Project had made substantial progress in conserving gorillas and their habitat.

The program was not without its problems, however. Most of its economic benefits were reaped by the central government rather than by local people. Furthermore, in 1988 the government abolished its two-tiered fee system, thus charging Rwandans the same $200 fee as foreign tourists. This effectively prevented local visitors from taking the tours. Finally, even after a decade, information and expertise had not been fully transferred to Rwandan personnel, so park employees remained somewhat dependent upon foreign experts (Vedder and Weber 1990).

These difficulties paled in the face of the problems Rwandan tourism encountered during the following decade, when ethnic tensions erupted in civil war. Widespread violence made travel unsafe, and several western tourists were murdered. By July 1999, the situation had improved somewhat, and the Parc National des Volcans reopened. Gorilla viewing (with the aid of military escorts) resumed, but clearly the prospects for tourism in Rwanda do not look bright for the near future.

Safari hunting and trophy hunting are particularly lucrative forms of tourism. Trophy hunters pay tens of thousands of dollars per person for the opportunity to hunt rare species such as the blue sheep and argali of Tibet (Marshall 1990). The numbers of animals killed by legal trophy hunting are small and carefully controlled, so that the hunted species are not likely to be depleted. Like other forms of tourism, trophy hunting provides income that serves as an incentive for wildlife conservation. It is also a relatively low-impact form of tourism because it brings only a small number of visitors into remote areas.

Trophy hunting and other forms of tourism have the potential to involve local people in a positive way in conservation efforts. This approach has some potential problems, however. First, the income from tourism in poor nations does not always benefit local communities; it often goes to the central government or to foreign hotel owners or tour operators instead. Second, poorly planned tours can damage sensitive ecosystems and harm organisms (by interfering with normal behavior and reproduction, attracting predators, or introducing weeds or diseases). Third, tourism can lead to cultural disruption. To aid prospective tourists who are concerned about these issues, some guidebooks now rate tourist hotels in terms of whether or not their income benefits local economies and their practices protect the environment and respect local culture.

An additional drawback to tourism is that its economic viability is tied to an infusion of foreign-generated income, which is sensitive to weather and to economic, political, and social circumstances. When conditions are unfavorable, tourist revenue declines and those who depend on it suffer. For example, trophy hunting expeditions to China declined after the Tiananmen Square uprising, and Rwanda's civil war interrupted tours to view gorilla groups in their natural habitat (see Box 14.1). It is questionable whether enterprises that are so volatile and dependent upon external funds are capable of providing sustained benefits for conservation.

In addition to these specific projects associated with tourism, critics have expressed concern about some more fundamental issues associated with ICDPs. For one thing, economic development and conservation are not always compatible. Tourism benefits conservation if it requires areas of relatively undisturbed habitat. (In some countries, ecotourism has become a powerful force opposing development. For example, there have been violent confrontations between hotel owners and oil company employees in the Amazon.) On the other hand, ill-advised roads and dams associated with economic development projects can make the sustainable use of ecosystems less rather than more likely (Alpert 1996).

An even more fundamental problem is that in ICDPs the national government retains ultimate authority for managing wildlife. This implies that local people are not capable of, entitled to, or willing to manage their environment in a sustainable fashion. A related concern is that focusing on monetary rewards suggests that local people have only utilitarian interests in preserving wild things and minimizes the aesthetic, intellectual, and cultural importance of the natural environment for local people (Bell 1987).

14.4   Basing resource management in local communities

A special type of ICDP called community-based conservation seeks to address this problem (Western and Wright 1994). It starts from the premise that local people have the collective capacity to manage natural resources. In community-based conservation programs, local communities are responsible for managing their natural resources (Getz *et al.* 1999; Newmark and Hough 2000). The best-known of these is the Communal Area Management Programme for Indigenous Resources (CAMPFIRE). Since the early 1980s, CAMPFIRE has sought to increase wildlife populations while raising village incomes in poor areas of Zimbabwe. In 1975 the national government gave authority for wildlife management to local landowners. Villagers now manage local resources, often through contracts with safari operators and other tourist concessions. The income from these arrangements is divided among village households. As a result of this program, income from sport hunting of elephant, leopard, lion, and buffalo populations has increased, and the number of problem elephants killed has declined dramatically.

Reading through the literature on conservation, it is easy to get the impression that most conservation projects are initiated by people in the developed world. Yet, indigenous people from the poles to the tropics have initiated some highly effective and innovative conservation projects. These reflect a strong desire for self-determination as well as a knowledge of local resources and an appreciation of the importance of managing them sustainably. Two of these are described in Box 14.2.

Box 14.2   Examples of conservation programs initiated by indigenous people

The Kuna-Yala Indigenous Reserve

In 1925 the Kuna-Yala of Panama's northern coast began a revolutionary uprising which led to their being granted territory within Panama. Their homeland extends from the continental divide to the Caribbean Sea and encompasses wetlands, moist tropical forests, coral reefs, mangroves, and coastal lagoons. Most of the approximately 30 000 Kuna live in coral island communities a short distance from the coast. There are also a few coastal settlements and inland villages. The Kuna diet is primarily seafood, supplemented by crops, pigs, chickens, and some hunting and gathering. Their territory supports populations of several endangered cats, plus the harpy eagle, giant anteater, Baird's tapir, crocodilians, and marine turtles. Many species of migratory birds winter in the area, and it also probably supports a number of plants not yet described by scientists.

In the 1980s, when the U.S. Agency for International Development (USAID) funded construction of a road through their lands, the Kuna became concerned about potential ecological and cultural impacts, particularly deforestation and colonization by settlers. The Kuna General Council proposed an Ecological Programme for the Management of the Forest Areas of Kuna-Yala. Its goals were full protection for 100000 hectares and sustainable management of the Kuna territory's forests. The national government supported the project, as did several nongovernmental organizations. In 1994 the government established the proposed reserve. Traditional use of the environment by the Kuna continues, accompanied by tourism, indigenous medicine, environmental education, agroforestry, and research on forest resources.

The Kuna exercise a high degree of control over reserve management and research (Houseal *et al.* 1985; Wright *et al.* 1985; Clay 1991; Gregg 1991; González (no date)). Because it affords a high level of environmental and cultural protection, the Kuna-Yala Indigenous Reserve has become "a cause celebre among environmentalists and indigenous rights activists." But its unusual integration of indigenous land uses and biodiversity protection may be difficult to copy in other settings. Jason Clay, an anthropologist with the nongovernmental organization Cultural Survival, suggests that the Kuna situation is unique in several ways that could prevent it from serving as a blueprint for similar projects elsewhere (Clay 1991:261). (Since each situation involving cultural and natural resources is unique, this is not really surprising.) First, due to their primary reliance on seafood, the Kuna never seriously depleted forest resources. Second, the Kuna have been able to maintain an unusual amount of autonomy and control over their resources, because their culture is very cohesive and because they have title to their lands.

Inupiat harvest of the bowhead whale

The Inupiat of Alaska have hunted bowhead whales for many generations. At one time, they depended on the bowhead for meat, blubber, bones, skin, and baleen. Today, however, the situation has been changed by several modern developments. Bowheads are listed as an endangered species, and their harvest is strictly regulated by the International Whaling Commission (IWC) (International Whaling Commission 1982) (see Chapter 9).

In 1977 scientists concluded that there were only 1000 to 2000 bowheads left. Reliable data on the sizes of whale populations are difficult to obtain, however, and the Inupiat took issue with the claim that bowhead populations could not sustain a limited harvest. They argued that the IWC's census figures were too low because they omitted whales migrating beneath the ice. To back up their contention, they verified the presence of whales under the ice with hydrophones. In 1985 the Inupiat estimated the bowhead population at 4400 animals. When they presented the IWC with these data, the commission reversed its ban on native hunting of bow-heads. Native whalers also formed the Alaskan Eskimo Whaling Commission, which now sends representatives to IWC meetings (Blair 1985). Like the Kuna, the Inupiat are managing resources to maintain both cultural values and biodiversity.

14.5   Locating reserves in a compatible landscape

Integrated conservation and development programs strive to gain local support for protected areas, to go beyond the "fences and fines" approach to reserves (Newmark and Hough 2000:586). But, as we noted in Chapter 13, the lands between protected areas are equally important if not more so. Several innovative approaches to integrating resource uses in the larger landscape have been developed.

14.5.1   Integrating traditional resource uses and conservation: Biosphere reserves

We noted in Chapter 12 that people are unlikely to support conservation when they are evicted from their homes in order to create protected areas. One way of addressing this problem is through the creation of reserves within which people are allowed to harvest resources. This approach has several advantages. It can preempt more destructive forms of development and preserve tradi-tional knowledge, and in agricultural ecosystems it can preserve the genetic diversity of valuable varieties of domesticated plants and animals.

The idea of regulated resource extraction within preserves is not new. National forests, national wildlife refuges, and lands administered by the Bureau of Land Management are three types of reserves within the U.S.A. from which people are allowed to harvest natural products. What is new, however, is the creation of reserves within which people live and support themselves. These are sometimes referred to as extractive reserves. Although the term "extractive reserve" literally means any reserve from which resources are harvested, it is generally used to refer specifically to protected areas, usually in tropical forests, within which settled people harvest products other than timber. The harvest of nontimber forest products from extractive reserves is discussed below.

In 1971 the United Nations Education, Scientific, and Cultural Organization (UNESCO) established the Man and the Biosphere Program (MAB) to conserve "natural areas and the genetic material they contain" (quoted in Dyer and Holland 1988:635). One way that MAB seeks to accomplish this goal is through an international network of multipurpose reserves, termed biosphere reserves. The goals of the biosphere reserve program are to conserve landscapes with long-established sustainable use patterns and to identify ways of using natural resources without causing environmental degradation.

Most national parks feature outstanding or rare physical and biological phenomena, but biosphere reserves are supposed to showcase typical, rather than spectacular, examples of ecosystems. Environmental changes are monitored within biosphere reserves, and biologists and social scientists study interactions between people and their environment, including the effects of human activities on ecosystems, the responses of people to alterations in their environment, and the restoration of degraded ecosystems. As of January 2000, 91 nations had designated 368 biosphere reserves.

Biosphere reserves usually involve hierarchies of protection, with resources being more strictly controlled in some areas than in others (von Droste zu Hulshoff and Gregg 1985). Ideally, biosphere management plans consider the relationship of lands in a protected area and the surrounding matrix, and land uses are adjusted or regulated accordingly. The regulation of land uses is not an innovation; the idea of restrictions on land uses within habitats of special importance underlies the creation of any reserve, and the regulation of land uses in order to group compatible activities together is the basis of zoning. The new elements in hierarchies of protection are the degree to which ecological considerations have been incorporated into the planning process and the landscape scale used in planning.

UNESCO recommends that biosphere reserves be organized around a core, surrounded by a series of concentric zones. The core should protect ecologically significant sites; within it resource extraction should be prohibited, and only activities such as monitoring and research should be allowed (Figure 14.1). Normally, the core should be large enough to contain viable

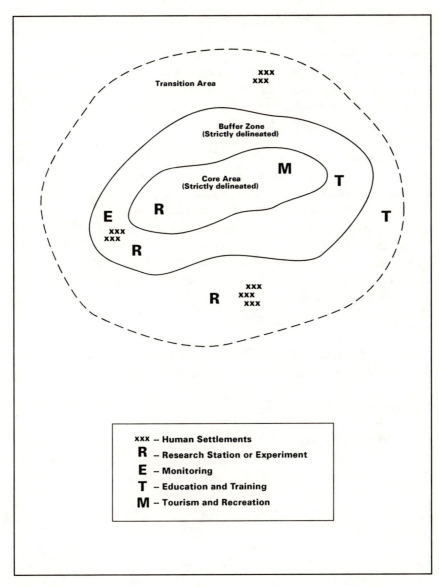

Figure 14.1. The recommended zonation for a biosphere reserve. (After Gregg 1991.)

populations of wide-ranging species. (In regions that have been densely settled for centuries, where large natural areas no longer exist, biosphere reserves may consist of small nature reserves embedded in a larger landscape supporting traditional land uses.) Previously designated wilderness areas, nature sanctuaries, and sacred lands are often incorporated into biosphere reserve core areas (Halffter 1985). The core area can be surrounded by areas in which even more intensive resource use takes place. In the recommended plan, a buffer zone surrounds the core area. Activities that are compatible with protection of the core area, such as nature tourism, research and education, restoration, and traditional land uses, are to be permitted in buffer zones. Cores and buffers should be clearly delineated by a legal boundary and strictly protected. Surrounding the buffer there should be a transition area, which is conceived of as a zone of cooperation that ties the reserve to the surrounding region. People live in the transition zone and pursue economic activities such as forestry, farming, and fishing that are compatible with conservation of the core and buffer. The boundaries of transition zones are typically flexible (Tangley 1988; Gregg 1991). Some examples of MAB biosphere reserves are described in Box 14.3.

Box 14.3   Examples of biosphere reserves in UNESCO's Man and the Biosphere Program

Cévennes

The Cévennes Bioreserve in the hills of southeastern France is a 323 000-ha landscape mosaic of deciduous forests, fields, pastures, and settlements. The species diversity of the area is unusually high for western Europe. It supports over 1800 species of vascular plants and 50 species of mammals. People have lived in the region continuously since Roman times, but their effects on the environment have not always been benign. In the eighteenth and nineteenth centuries, overgrazing by sheep and the clearing of extensive areas of beech and other hardwoods for agriculture and charcoal production resulted in widespread erosion. To counteract this problem, large areas of pine were planted. This, combined with succession on abandoned agricultural lands, homogenized parts of the landscape. Because of dwindling markets for chestnuts, sheepskins, and silk, traditional agroforestry practices have declined, and the region's rural population is decreasing. The Parc National de Cévennes, created in 1970, has statutory control over 90 000 ha. The larger biosphere reserve includes the park (which forms the core and buffer area) plus an adjacent zone of cooperation. The reserve emphasizes traditional rural culture, through measures such as a program for the preservation of traditional rural music and incentives for using traditional architectural forms. These programs attempt to enhance the quality of life for the region's residents and minimize migration out of the area. The rural lifestyles that are preserved also attract tourists. In addition, the park protects an endangered population of Aubrac cattle, a breed that is adapted to mountain conditions (Collin 1985; Gregg 1991).

Rio Platano

The Rio Platano Biosphere Reserve consists of 500 000 ha of moist tropical forest, sandy beaches, saltwater lagoons, mangroves, coastal pine savanna, rivers, and oxbow lakes in the Mosquitia region of northern Honduras. Elevation ranges from sea level at the Caribbean coast to over 1300 m. This steep topography is one reason for the diversity of habitats contained within the reserve. Burial sites and petroglyphs are evidence of the region's long history of use by native people. Before Europeans arrived in the area, Miskito Indians from South America immigrated and displaced the native tribes. Eventually the original Miskitos, other Indian tribes, and escaped African slaves formed a blended culture known today as "Miskito." Most of the people living in the reserve are descended from this group, but a small group of Paya Indians remains, as well as some Garifunos (Afro-Caribbeans) and mestizos. The reserve's inhabitants support themselves with shifting agriculture and grazing, supplemented by fishing, hunting, and gathering. In addition, the biosphere reserve provides habitat for nearly 400 species of birds, over 100 amphibians and reptiles, and nearly 40 species of mammals (such as spider monkeys, jaguar, jaguarundi, southern river otter, and kinkajou), many of which are globally rare or endangered. The biggest threats to the region's ecological and cultural resources have been habitat destruction caused by settlers colonizing the region and loss of indigenous skills, knowledge, and identity. In 1960 the Honduran government established the Ciudad Blanca Archaeological Reserve; 20 years later this designation was superseded by the creation of the Rio Platano Biosphere Reserve. The biosphere reserve designation confers some protection on its inhabitants, who do not have legal title to their lands, but illegal exploitation of forest resources remains a problem. Recently, many young men have begun working in the lobster and fish industries. This has increased their incomes, but it is also associated with an increase in alcoholism (Glick and Betancourt 1983).

Waterton Lakes

Waterton Lakes National Park in Alberta, Canada, was designated a biosphere reserve in 1979. The biosphere reserve consists of the 53000-ha national park plus an adjacent zone of cooperation. Together with Glacier National Park across the U.S. border, it also forms the Waterton/Glacier International Peace Park. Fostering cooperation with neighboring ranchers is a major objective of this biosphere reserve. A Biosphere Reserve Management Committee chaired by local ranchers was formed early in the project. Ranchers and park personnel share a common concern about knapweed, a highly invasive exotic weed that is unpalatable to livestock and displaces native vegetation. The management committee developed a cooperative program for controlling and monitoring knapweed, and then moved on to more controversial resource management issues, such as conflicts between elk that leave the park and livestock. Eventually more permissive provincial hunting regulations were adopted, and more effective fencing was installed to control the elk. Through continuing research, education, and interpretive programs, management personnel and local resource users are striving to devise mutually acceptable solutions to resource management problems (Lieff 1985; Gregg 1991).

Chernozem

The Central Chernozem State Biosphere Reserve, located in central Russia about 560 km south of Moscow, includes temperate deciduous forest and meadow steppe ecosystems as well as ploughed areas (see Box 13.4). The extensive, unmodified Chernozem soils protected by this reserve are of particular interest. These deep, black, fertile steppe soils were formed in the continental interiors of North America and Eurasia. In the mid-nineteenth century, much of the Ukrainian steppe was cultivated, and Russian scientists developed theories that emphasized the importance of studying natural ecological processes in unmodified reference areas. As a result, they became interested in setting aside some areas of unmodified vegetation. Within the Central Chernozem State Biosphere Reserve, researchers study the effects of agriculture on Chernozem soils (Sokolov 1981, 1985). But with Russia's current economic, political, and military problems, reserve research and management have become low priorities. Troops that withdrew from previously occupied territories have settled near the reserve. Although the effects of these settlements have not been studied, populations of birds of prey have reportedly declined in the reserve, and some observers fear that these declines indicate other ecological problems (Maleshin 1997).

Virginia Coast

The Virginia Coast Biosphere Reserve in the U.S.A. consists of barrier islands along the Atlantic seaboard plus associated coastal waters and salt marshes. These habitats support economically valuable fish and shellfish populations and colonial nesting birds. The islands, bays, and salt marshes constitute a 13600-ha core area, all of which is controlled by either The Nature Conservancy, the state of Virginia, or the U.S. Fish and Wildlife Service. Adjacent lands comprise a buffer area within which traditional uses deemed compatible and sustainable are encouraged. These include farming and harvesting seafood. To implement this part of the project, The Nature Conservancy bought several seaside farms and resold them with conservation easements (voluntary, legally binding limitations on development). The emphasis is on compatible development, tourism, and protection of water quality. Finally, a transition zone includes farms, villages, businesses, light industry, and recreation facilities. As a result of this integrated planning, water quality has been protected, and this in turn has benefited farming, fisheries, and tourism and allowed for the preservation of rural lifestyles. In addition, long-term, multidisciplinary ecological research is being carried out to monitor the effects of land use on the coastal ecosystem, and outreach programs have been designed to educate students, visitors, and community members about these studies (Badger 1990).

14.5.2   Developing alternative resource uses: Nontimber forest products

Recently, the collection and marketing of a variety of nontimber products – such as nuts, rubber, and resins – from tropical forests has been hailed as a method of conserving biodiversity while providing economic and social gains for local users. This stems from the idea that nontimber forest products can be harvested sustainably and generate profits that are at least as high as more destructive land uses (Peters *et al.* 1989). Many nontimber products can be obtained without cutting down trees, so it is possible that they can be harvested indefinitely (like maple syrup from temperate zone forests). But harvesting nontimber products has not always been good for forests or for people. For example, the New World rubber boom at the turn of the century generated profits for rubber companies but brought about environmental degradation and poverty in parts of Amazonia (Taylor 1997).

The harvest of nontimber forest products works as a conservation strategy only if it is economically viable and ecologically sustainable. The economic benefits accruing from extraction of nontimber products depend upon a host of ecological, socioeconomic, and political factors (Pendleton 1992). It is necessary to evaluate the economic, ecological, and social impacts of nontimber forest product harvests on a case-by-case basis. A comparison of Petén, Guatemala and West Kalimantan, Indonesia illustrates this (Salafsky *et al.* 1993). In both these settings, indigenous people have harvested nontimber forest products for centuries and probably also traded some surpluses at local markets. More recently, these products have been harvested for sale to international markets as well.

To assess the sustainability of this endeavor, researchers lived for several weeks in villages in each of the two study areas, going with workers on harvesting trips and interviewing people involved in harvest, processing, and trade. In the Maya Biosphere Reserve within the Department of Petén, Guatemala, the principal harvested products are: (1) chicle, which is used in the manufacture of chewing gum and glue, (2) xate, fronds from palms that are used in floral arrangements, and (3) allspice. These three products form the basis of an intensive export-oriented industry that employs over 7000 people as harvesters, contractors, or processors. In 1989 export income from these enterprises was estimated at $4 million to $7 million (Reining and Heinzman 1992; Salafsky *et al.* 1993).

The tropical forests of Petén seem to be well suited to economies based on the harvest of nontimber forest products. There are about 50 to 100 species of trees per hectare, a relatively low species richness for tropical forest (although still much higher than temperate zone forests!). Because there are relatively few tree species, many individuals of each species are present. This makes it easy for harvesters to get to trees of the desired species. Furthermore, the harvest of these species is spread out temporally and spatially, so that over-exploitation of any one species is avoided. Chicle must be harvested during the rainy season, from August to January; palm fronds are available through-out the year, but harvest peaks between March and June (partly because the demand for wedding floral arrangements in Europe and North America is greatest in spring); and allspice is available only in July and August. This stag-gered availability provides harvesters with year-round income and tends to protect the palm fronds from continuous harvest. Because the supply of non-timber forest products is predictable, stable markets have developed.

Chicle, xate, and allspice are usually harvested without killing reproduc-tively mature individuals, and perhaps without limiting recruitment of young plants into the population. Chicle is harvested by tapping. If properly tapped, individual trees can continue producing for decades. Xate harvest does not involve the removal of reproductive structures, and allspice fruits can be col-lected from areas where they are concentrated on the forest floor. Survival is low in the dense populations of allspice seedlings beneath the crown of the parent plant. In other words, seedling survival is density-dependent. Hence, harvesters can remove considerable numbers of fruits from beneath the parent tree without appreciably inhibiting reproductive potential. These char-acteristics contribute to the sustainability of nontimber forest product extrac-tion in Petén. In addition, chicle, palm fronds, and allspice are relatively easy to store and transport, and roads, airstrips, warehouses, and other compo-nents of the infrastructure necessary for processing and delivering these products have been available for some time because of an export-based chicle industry dating back to the nineteenth century. Thus, a variety of biological, economic, and sociological factors contribute to the sustainability and viabil-ity of nontimber forest product use in Petén.

In West Kalimantan, the situation is quite different. Species diversity is high in Indonesian tropical forests (150–225 species per hectare). Because many species are packed into a given area, the number of individuals of any one species is fairly low. And because the harvested species occur at low densities, it takes a long time to get to each plant and a long time to transport the harvested products. Under these conditions, harvesting nontimber forest products is inefficient. Furthermore, certain characteristics of the harvested species make it unlikely that they will be used in a sustainable fashion. Harvesters tend to concentrate on species that command a high price. Gaharu, the resin from diseased heartwood of trees in the genus *Aquilaria*, is one such product. The resin from a single tree can be sold for thousands of dollars. With such high profits, there is an incentive to maximize short-term gains and deplete the resource. In addition, some of the nontimber products of Indonesian forests, such as medang (bark from trees in the genus *Litsea*), which is used in mosquito coils, are harvested by killing the source tree.

To make matters worse, many of the tree species in Indonesian forests are mast species; they produce fruit only once every three to five years, and many species produce fruit simultaneously. (Recall from Chapter 1 that this phenomenon is thought to be an adaptation that satiates seed predators, so that some seeds escape predation.) This unpredictable timing prevents stable markets from developing. When fruits do become available there is a glut on the market, and prices drop.

Thus, the harvest of nontimber forest products appears to be economically successful and ecologically sustainable in Petén but not in West Kalimantan. The economic success of the Petén enterprises stems in part from ecological characteristics such as the predictable availability of forest products, which contributes to market stability, and the relatively high density of harvested species, which makes them easy to locate and collect, as well as from social and economic factors. In Kalimantan, the ecology of the harvested species as well as economic and social factors seem to work against the sustainable harvest of nontimber forest products. Unfortunately, however, although this study evaluated circumstantial evidence pertaining to the ecological effects of harvesting nontimber forest products, no ecological data were obtained to test the hypothesis that harvests negatively impacted harvested species in West Kalimantan but not in Petén (Salafsky *et al.* 1993).

One of the most important economic factors affecting the prognosis for nontimber forestry in the tropics is land tenure. For the extraction of nontimber products to be worthwhile, a harvester must be reasonably sure that he or she will have access to the resource in the foreseeable future. In contrast, logging and ranching yield immediate rewards. Without a guarantee of access, it may make more economic sense to convert resources to profits "as quickly and as lucratively as possible" (Pendleton 1992:256). For this reason, nontimber forest products are not likely to replace more destructive enterprises unless local people are guaranteed long-term access to forest resources.

14.6   Evaluating attempts to include people in the conservation process

It is easy to say that we should conserve biodiversity and meet the needs of people at the same time, but how can we tell if we are really accomplishing these objectives? How do we know that such projects are not causing unacceptable harm to species and ecosystems? Or to cultures? And how do we assess whether or not they are really providing tangible economic benefits?

Programs that seek to integrate human needs and conservation generate a lot of enthusiasm, but critics suggest that in most cases clearcut benefits from these programs have not been demonstrated (Inamdar *et al.* 1999). For example, Agi Kiss of the World Bank's Protected Areas and Wildlife Services Project argues that the integrated-conservation-and-development-program "*hypothesis* has moved rapidly from an untested hypothesis to being regarded as 'best practice,' but without having demonstrated a significant measure of success" (Kiss 1998:347; emphasis added). This statement underscores an important point: resource managers should take as their starting-point the hypothesis that a particular program will accomplish certain objectives. This applies to utilitarian and preservationist managers, as well as to those who strive to practice sustainable-ecosystem management. All too often, however, the benefits of certain conservation actions are regarded as a foregone conclusion in the absence of empirical data supporting their effectiveness.

There are a number of reasons why it is hard to demonstrate tangible benefits from the types of programs discussed in this chapter. Sometimes ecological, economic, or social benefits are not obvious until several years after the start of a program. Funding for monitoring is often unavailable. Research scientists may be more interested in questions of theoretical and academic interest than questions with immediate practical applications. Some of the things we need to evaluate are difficult to measure. Nevertheless, it is clear that multidisciplinary efforts to monitor the effects of human-oriented conservation projects on the natural and human world are crucial (Table 14.1). Without such efforts, we cannot intelligently assess what we have accomplished and where we need to go from here (Kiss 1990).

14.7   Conclusions

Several conclusions can be drawn from the examples in this chapter. One is that secure land tenure is essential if people are to take care of natural resources where they live. A related theme is that local people, especially those who have lived sustainably in an area for centuries or longer, have irreplaceable knowledge about the ecosystems they inhabit. A third is that the cultural, political, economic, and ecological circumstances of each situation are unique. There are no shortcuts. There is no substitute for understanding the idiosyncrasies and complexities of the biological and cultural contexts of each case

Table 14.1. *Examples of objectives and criteria for evaluating the ecological, social, and economic effects of an Integrated Conservation and Development Program*

Type of objective	Example	Possible criteria for success
Ecological	Maintain populations of selected species	Population size
	Maintain biological diversity at the species level	Number of species
	Reduce illegal exploitation	Rate of poaching
	Maintain productive capacity of ecosystem	Rate of soil erosion Rate of biomass production
Social	Increase involvement of local people	Participation in meetings
	Improve social welfare of local communities	Number of clinics, schools, wells funded by income from wildlife-related activities
	Educate people about the ecosystem services provided by protected areas	Increase in knowledge of ecosystem services provided by protected areas
	Increase local support for conservation programs	Positive change in attitudes toward wildlife and protected areas
Economic	Increase economic benefits from wildlife	Household income from wildlife-related activities Jobs generated by wildlife-related activities
	Decrease costs due to wildlife	Deaths from wildlife attacks Crop losses from wildlife Livestock losses from predators Local livestock carrying capacity

where we seek to integrate use and protection of living natural resources. This endeavor requires interdisciplinary cooperation, institutional flexibility, and a lot of patience. Finally, although integrating the needs of people and wildlife is a laudable goal, good intentions are not good enough. Managers should test explicit hypotheses about the effects of conservation programs and use the resulting information to evaluate the benefits of those programs.

References

Alpert, P. (1996). Integrated conservation and development projects. *BioScience* **46**:845–855.

Badger, C. J. (1990). Eastern shore gold. *Nature Conservancy* **40**(4):6–15.

Bell, R. (1987). Conservation with a human face: conflict and reconciliation in African land use planning. In *Conservation in Africa: People, Policies, and Practice*, ed. D. Anderson and R. Grove, pp. 79–101. Cambridge: Cambridge University Press.

Blair, J. G. (1985). How hunters are saving the bowhead whale. *Technology Review* **88**:82–83.

Clay, J. (1991). Cultural survival and conservation: lessons from the past twenty years. In *Biodiversity: Culture, Conservation, and Ecodevelopment*, ed. M. L. Oldfield and J. B. Alcorn, pp. 248–273. Boulder, CO: Westview Press.

Collin, G. (1985). The Cévennes Biosphere Reserve: integrating traditional uses and ecosystem conservation. *Parks* **10**(2):12–14.

Dyer, M. I. and M. M. Holland (1988). Unesco's Man and the Biosphere Program. *BioScience* **38**:635–641.

Ehrlich, P. and A. Ehrlich (1985). *Extinction: The Causes and Consequences of the Disappearance of Species*, 2nd Ballantine printing. New York, NY: Ballantine Books.

Farnsworth, N. R. (1988). Screening plants for new medicines. In *Biodiversity*, ed. E. O. Wilson, pp. 83–97. Washington, DC: National Academy Press.

Gershon, D. (1992). If biological diversity has a price, who sets it and who should benefit? *Nature* **359**:565.

Getz, W. M., L. Fortmann, D. Cumming, J. du Toit, J. Hilty, R. Martin, M. Murphree, N. Owen-Smith, A. M. Starfield, and M. I. Westphal (1999). Sustaining natural and human capital: villagers and scientists. *Science* **283**:1855–1857.

Glick, D. and J. Betancourt (1983). The Rio Platano Biosphere Reserve: unique resource, unique alternative. *Ambio* **12**:168–173.

González, O. (No date). Kuna-Yala, Panama: sustainability for comprehensive development. http://www.iucn.org/themes/ssp/panama.html.

Gregg, W. P., Jr. (1991). MAB biosphere reserves and conservation of traditional land use systems. In *Biodiversity: Culture, Conservation, and Ecodevelopment*, ed. M. L. Oldfield and J. B. Alcorn, pp. 274–294. Boulder, CO: Westview Press.

Halffter, G. (1985). Biosphere reserves: conservation of nature for man. *Parks* **10**(3):15–18.

Houseal, B., C. MacFarland, G. Archibold, and A. Chiari (1985). Indigenous cultures and protected areas in Central America. *Cultural Survival* **9**(1):10–20.

Inamdar, A., H. de Jode, K. Lindsay, and S. Cobb (1999). Capitalizing on nature: protected area management. *Science* **283**:1856–1857.

International Whaling Commission (1982). *Aboriginal/Subsistence Whaling (With Special Reference to the Alaska and Greenland Fisheries)*. Reports of the International Whaling Commission, Special Issue no. 4, Cambridge.

Joyce, C. (1992). Western medicine men return to the field. *BioScience* **42**:399–403.

Kiss, A. (ed.) (1990). *Living with Wildlife: Wildlife Resource Management with Local Participation in Africa.* Africa Technical Department Series, World Bank Technical Paper no. 130, Washington, DC.

Kiss, A. (1998). Kenyan wildlife conservation. *Science* **281**:347–348.

Lieff, B. (1985). Waterton Lakes biosphere reserve: developing a harmonious relationship. *Parks* **10**(3):9–11.

Maleshin, N. (1997). Russian reserves face an uncertain future. *Forum for Applied Research and Public Policy* **12**:120–122.

Marshall, E. (1990). Mountain sheep experts draw hunters' fire. *Science* **248**:437–438.

Newmark, W. D. and J. L. Hough (2000). Conserving wildlife in Africa: integrated conservation and development projects and beyond. *BioScience* **50**:585–592.

Pendleton, L. H. (1992). Trouble in paradise: practical obstacles to nontimber forestry in Latin America. In *Sustainable Harvest and Marketing of Rain Forest Products*, ed. M. Plotkin and L. Famolare, pp. 252–262. Washington, DC: Island Press.

Peters, C. M., A. H. Gentry, and R. O. Mendelsohn (1989). Valuation of an Amazonian rain forest. *Nature* **339**:655–656.

Power, T. M. (1991). Ecosystem preservation and the economy in the Greater Yellowstone area. *Conservation Biology* **5**:395–404.

Reid, W. V., S. A. Laird, C. A. Meyer, R. Gámez, A. Sittenfeld, D. H. Janzen, M. A. Gollin, and C. Juma (eds) (1993). *Biodiversity Prospecting: Using Genetic Resources for Sustainable Development.* Washington, DC: World Resources Institute.

Reining, C. and R. Heinzman (1992). Nontimber forest products in the Petén, Guatemala: why extractive reserves are critical for both conservation and development. In *Sustainable Harvest and Marketing of Rain Forest Products*, ed. M. Plotkin and L. Famolare, pp. 110–117. Washington, DC: Island Press.

Roberts, L. (1992). Chemical prospecting: hope for vanishing ecosystems? *Science* **256**:1142–1143.

Salafsky, N., B. L. Dugelby, and J. W. Terborgh (1993). Can extractive reserves save the rainforest? An ecological and socioeconomic comparison of nontimber forest product extraction systems in Petén, Guatemala, and West Kalimantan, Indonesia. *Conservation Biology* **7**:39–52.

Sokolov, V. (1981). The biosphere reserve concept in the USSR. *Ambio* **10**:97–101.

Sokolov, V. (1985). The system of biosphere reserves in the USSR. *Parks* **10**(3):6–8.

Tangley, L. (1988). A new era for biosphere reserves. *BioScience* **38**:148–155.

Taylor, D. A. (1997). Alternative products from woodlands. *Environment* **39**:6–11,33–36.

Vedder, A. and W. Weber (1990). Rwanda: the mountain gorilla project (Volcanoes National Park). In *Living with Wildlife: Wildlife Resource Management with Local Participation in Africa*, ed. A. Kiss, pp. 83–90. Africa Technical Department Series, World Bank Technical Paper no. 130, Washington, DC.

von Droste zu Hulshoff, B. and W. P. Gregg, Jr. (1985). The international network of biosphere reserves: demonstrating the value of conservation in sustaining society. *Parks* **10**(3):2–5.

Western, D. and R. M. Wright (eds) (1994). *Natural Connections: Perspectives in Community-Based Conservation.* Washington, DC: Island Press.

Wright, R. M., B. Houseal, and C. De Leon (1985). Kuna Yala: indigenous biosphere reserve in the making? *Parks* **10**(3):25–27.

Postscript

Each of the three types of natural resource management considered in this book – utilitarian, preservationist, and sustainable-ecosystem management – can make unique contributions to solving practical problems. Each has advantages and disadvantages, and each is appropriate in certain situations. In many cases elements from more than one approach can be blended.

I believe that the flux-of-nature viewpoint is a valuable contribution and that managing ecosystems to preserve their complexity is an exciting new development. But I also believe that as we continue to search for responsible ways to manage living natural resources, a large dose of humility is appropriate. Science, whether theoretical or applied, is an ongoing process. Just as the flux-of-nature viewpoint encompasses certain observations that did not fit comfortably into equilibrium explanations, it is likely that this perspective has limitations that are not obvious at present. Although our understanding of the natural world is more detailed than it used to be, there is still a lot we do not know. Management should err on the side of caution, therefore. There will always be surprises.

There are no easy answers or shortcuts. There is no substitute for understanding the details of context. We cannot design useful nature reserves without understanding whether organisms can move through the matrix that separates them (geographical context). We cannot restore ecosystems without understanding the key processes and structures specific to each case (ecological context). We cannot devise effective conservation programs without understanding who has access to resources, what forms of ownership are in place, and who gains and who pays the costs of conservation (political and economic contexts). We also need to understand traditional knowledge of and institutions for managing the natural world (cultural and historical contexts). Whether we are talking about ecological questions or social issues, there is no substitute for understanding the intricate specifics of each unique situation. As the logo for the Sian Ka'an Biosphere Reserve conveys so strikingly, the threads of history, biology, and culture are inextricably linked.

Logo for the Sian Ka'an Biosphere Reserve, used with permission of the Direction of the Sian Ka'an Biosphere Reserve, National Commission of Protected Natural Areas, Mexico.

Appendix: Scientific names of organisms mentioned in the text

Common name	Scientific name
Fungi	
chestnut blight	*Endothia parasitica*
Plants	
acacia	*Acacia* spp. (mostly *Acacia tortilis*)
allspice	*Pimenta dioica*
American chestnut	*Castanea dentata*
balsamo tree	*Myroxylon pereirae*
beech	*Fagus* spp.
big sagebrush	*Artemisia tridentata*
big-leaf maple	*Acer macrophyllum*
black spruce	*Picea mariana*
bloodroot	*Sanguinaria canadensis*
blue grama	*Bouteloua gracilis*
blueberry	*Vaccinium* spp.
bluebunch wheatgrass	*Pseudoroegneria spicata* (formerly *Agropyron spicatum*)
bracken fern	*Pteridium aquilinum*
Brazil nut	*Bertholletia excela*
buffalo grass	*Buchloe dactyloides*
California cordgrass	*Spartina foliosa*
camas	*Camassia* spp.
Cascade Oregongrape	*Berberis nervosa*
cassava (manioc)	*Manihot utilissima*

Common name	Scientific name
cattail	*Typha* spp.
ceanothus	*Ceanothus velutinous*
cedar	*Thuja* spp., *Juniperus* spp.
cheatgrass	*Bromus tectorum*
chestnut (see American chestnut)	
chicle tree	*Manilkara zapota*
cocoa	*Theobroma* spp.
coconut (coconut palm)	*Cocos nucifera*
coffee	*Coffea arabica*
common thistle	*Cirsium vulgare*
cordgrass (see California cordgrass)	
cottonwood	*Populus* spp.
dandelion	*Taraxacum* spp.
desert fan palm	*Washingtonia* spp.
Douglas-fir	*Pseudotsuga menziesii*
eastern hemlock	*Tsuga canadensis*
eucalypts	*Eucalyptus* spp.
fir	*Abies* spp.
fireweed	*Epilobium angustifolium*
giant sequoia	*Sequoiadendron giganteum*
guanacaste tree	*Enterolobium cyclocarpum*
heather	*Calluna vulgaris*
hickory	*Carya* spp.
Idaho fescue	*Festuca idahoensis*
indigo	*Indigofera tinctoria, I. suffruticosa*
jack pine	*Pinus banksiana*
Jeffrey pine	*Pinus jeffreyi*
juniper	*Juniperus* spp.
knapweed	*Centaurea* spp.
loblolly pine	*Pinus taeda*
lodgepole pine	*Pinus contorta*
longleaf pine	*Pinus palustris*
Madagascar rosy periwinkle	*Catharanthus roseus*
mahogany	*Afzelia* spp.
maize	*Zea mays*
manioc (cassava)	*Manihot utilissima*
maple	*Acer* spp.
mesquite	*Prosopis* spp.
musk thistle	*Carduus* spp.
oak	*Quercus* spp.

Common name	Scientific name
Pacific blackberry	*Rubus ursinus*
Pacific yew	*Taxus brevifolia*
palm	*Chaemdorea* spp.
paper birch	*Betula papyrifera*
pine	*Pinus* spp.
pitcher plant	*Darlingtonia* spp., *Sarracenia* spp.
Pitcher's thistle	*Cirsium pitcheri*
ponderosa pine	*Pinus ponderosa*
quaking aspen	*Populus tremuloides*
red mangrove	*Rhizophora* spp.
red pine	*Pinus resinosa*
redwood	*Sequoia sempervirens*
rose	*Rosa* spp.
rubber tree	*Castilla* spp.
sagebrush	*Artemisia* spp.
salal	*Gaultheria shallon*
salt cedar (tamarisk)	*Tamarix* spp.
salt marsh bird's beak	*Cordylanthus maritimus* ssp. *maritimus*
sand reed	*Ammophila arundinacea*
sedge	*Carex* spp.
Sitka spruce	*Picea sitchensis*
southern beech	*Nothofagus* spp.
spruce	*Picea* spp.
sugar maple	*Acer saccharum*
tamarisk (salt cedar)	*Tamarix* spp.
teak	*Tectona grandis*
thistle	*Carduus* spp.; *Cirsium* spp.
thyme	*Thymus praecox*
Venus'-flytrap	*Dionaea muscipula*
vine maple	*Acer circinatum*
western hemlock	*Tsuga heterophylla*
western redcedar	*Thuja plicata*
western rhododendron	*Rhododendrom macrophyllum*
wild ginger	*Asarum canadense*
willow	*Salix* spp.
wire grass	*Aristeda longiseta*
wood groundsel	*Senecio sylvaticus*
yam	*Dioscorea* spp.

Common name	Scientific name

Animals
Invertebrates
abalone	*Haliotis* spp.
Bay checkerspot butterfly	*Euphydryas editha bayensis*
Clear Lake gnat	*Chaoborus astictopus*
large blue butterfly	*Maculina arion*
lobster	*Homarus* spp.
monarch butterfly	*Danaus plexippus*
octopus	*Octopus* spp.
red ant	*Myrmica* spp.
tsetse-fly	*Glossina* spp.

Vertebrates
Fishes
alewife	*Alosa pseudoharengus*
desert pupfish	*Cyprinodon macularius*
lake trout	*Salvelinus namaycush*
salmon	*Oncorhynchus* spp., *Salmo* spp.
sea lamprey	*Petromyzon marinus*
snail darter	*Percina tanasi*
striped bass	*Morone saxatilis*

Reptiles
alligator (see American alligator)	
American alligator	*Alligator mississippiensis*
rattlesnake	*Crotalus* spp., *Sistrurus* spp.

Birds
American black duck	*Anas rubripes*
American coot	*Fulica americana*
bald eagle	*Haliaeetus leucocephalus*
black duck (see American black duck)	
black-headed gull	*Larus ridibundus*
black robin	*Turdus infuscatus*
bluebird	*Sialia* spp.
blue-winged teal	*Anas discors*
Brewer's sparrow	*Spizella breweri*
brown-headed cowbird	*Molothrus ater*
California condor	*Gymnogyps californianus*

Common name	Scientific name
California least tern	*Sterna albifrons browni*
Canada goose	*Branta canadensis*
Chatham Island tit	*Petroica macrocephala chathamensis*
chickadee	*Poecile* spp.
chicken	*Gallus gallus*
chough	*Pyrrhocorax pyrrhocorax*
chukar	*Alectoris chukar*
condor (see California condor)	
Cooper's hawk	*Accipiter cooperii*
cormorant	*Phalacrocorax* spp.
cowbird (see brown-headed cowbird)	
crow	*Corvus* spp.
dodo	*Raphus cucullatus*
duck hawk (see peregrine falcon)	
frigatebird	*Fregata* spp.
gadwall	*Anas strepera*
gannet	*Sula bassana*
golden eagle	*Aquila chrysaetos*
goldfinch	*Carduelis* spp.
goshawk	*Accipiter gentilis*
greywing francolin	*Francolinus africanus*
great-crested grebe	*Podiceps cristatus*
greater prairie-chicken	*Tympanuchus cupido*
greater sage-grouse	*Centrocercus urophasianus*
green-winged teal	*Anas crecca*
harpy eagle	*Harpia harpyja*
heath hen	*Tympanuchus cupido cupido*
Kirtland's warbler	*Dendroica kirtlandii*
light-footed clapper rail	*Rallus longirostris levipes*
magpie	*Pica* spp.
mallard	*Anas platyrhynchos*
marbled murrelet	*Brachyramphus marmoratus*
merlin	*Falco columbarius*
mourning dove	*Zenaida macroura*
northern bobwhite	*Colinus virginianus*
northern pintail	*Anas acuta*
northern shoveler	*Anas clypeata*
northern spotted owl	*Strix occidentalis caurina*
nuthatch	*Sitta* spp.
osprey	*Pandion haliaetus*

Common name	Scientific name
passenger pigeon	*Ectopistes migratorius*
pelican	*Pelecanus* spp.
peregrine falcon	*Falco peregrinus*
pheasant (see ring-necked pheasant)	
pigeon (see rock dove)	
pigeon hawk (see merlin)	
pintail (see northern pintail)	
prairie-chicken (see greater prairie-chicken)	
red-backed shrike	*Lanius collurio*
red grouse	*Lagopus lagopus scoticus*
red kite	*Milvus milvus*
red-winged blackbird	*Agelaius phoeniceus*
ring-necked pheasant	*Phasianus colchicus*
rock dove	*Columba livia*
rook	*Corvus frugilegus*
sage-grouse (see greater sage-grouse)	
sage sparrow	*Amphispiza belli*
sage thrasher	*Oreoscoptes montanus*
sandhill crane	*Grus canadensis*
sharp-shinned hawk	*Accipiter striatus*
snipe	*Gallinago* spp., *Lymnocryptes*
spotted owl	*Strix occidentalis*
starling	*Sturnus vulgaris*
street pigeon (see rock dove)	
turkey, domestic	*Meleagris gallopavo*
western grebe	*Aechmophorus occidentalis*
white-tailed eagle	*Haliaeetus albicilla*
whooping crane	*Grus americana*
wood duck	*Aix sponsa*
woodcock	*Scolopax* spp.
woodpigeon	*Columba palumbus*
yellow-billed cuckoo	*Coccyzus americanus*
yellow-headed blackbird	*Xanthocephalus xanthocephalus*

Mammals

African ass	*Equus asinus*
African buffalo	*Syncerus caffer*
African elephant	*Loxodonta africana*
American beaver	*Castor canadensis*

Common name	Scientific name
American mink	*Mustela vison*
argali	*Ovis ammon*
aurochs	*Bos taurus*
Baird's tapir	*Tapirus baridii*
beaver (see American beaver, giant beaver)	
bighorn sheep (see mountain sheep)	
bison	*Bison bison*
black bear	*Ursus americanus*
black-footed ferret	*Mustela nigripes*
blue sheep	*Pseudois nayaur*
blue whale	*Balaenoptera musculus*
bobcat	*Lynx rufus*
bowhead whale	*Balaena mysticetus*
brown bear	*Ursus arctos*
buffalo (see African buffalo and bison)	
burro (see African ass)	
caribou	*Rangifer tarandus*
cat, domestic	*Felis catus*
cattle, domestic	*Bos taurus*
cheetah	*Acinonyx jubatus*
common hare	*Lepus* spp.
cougar (see mountain lion)	
coyote	*Canis latrans*
deer mouse	*Peromyscus maniculatus*
dingo	*Canis familiaris dingo*
dog, domestic	*Canis familiaris*
eastern mountain lion	*Felis concolor coryi*
eland	*Taurotragus* spp.
elephant (see African elephant)	
elephant seal	*Mirounga angustirostris*
elk	*Cervus elaphus*
European bison	*Bison bonasus*
European hare	*Oryctolagus cuniculus*
European mink	*Mustela lutreola*
European polecat	*Mustela putorius*
fin whale	*Balaenoptera physalus*
Florida panther	*Felis concolor coryi*
gazelle	*Gazella* spp.
giant anteater	*Myrmecophaga tridactyla*

Common name	Scientific name
giant beaver	*Castaoroides* spp.
giant panda	*Ailuropoda melanoleuca*
goat, domestic	*Capra hircus*
gorilla (see mountain gorilla)	
gray whale	*Eschrichtius robustus*
gray wolf	*Canis lupus*
grizzly bear	*Ursus horribilis*
ground squirrel	*Spermophilus* spp.
Himalayan thar	*Hemitragus jemlahicus*
horse, domestic	*Equus caballus*
house mouse	*Mus musculus*
humpback whale	*Megaptera novaeangliae*
jackrabbit	*Lepus* spp.
jaguar	*Panthera onca*
jaguarundi	*Herpailurus yaguarondi*
javelina	*Pecari tajacu*
kinkajou	*Potos flavus*
koala	*Phascolarctos cinereus*
leopard	*Panthera pardus*
lion	*Panthera leo*
llama	*Lama glama*
lynx	*Lynx* spp.
mammoth	*Mammuthus* spp.
mastodon	*Mammus* spp.
Mediterranean monk seal	*Monachus monachus*
mink (see American mink, European mink)	
minke whale	*Balaenoptera acutorostrata*
mongoose	*Herpestes* spp.
monk seal (see Mediterranean monk seal)	
moose	*Alces alces*
mountain goat	*Oreamnos americanus*
mountain gorilla	*Gorilla gorilla beringei*
mountain lion	*Felis concolor*
mountain sheep	*Ovis canadensis*
mule deer	*Odocoileus hemionus*
multimammate rat	*Mastomys natalensis*
muskrat	*Ondatra zibethica*
narwhal	*Monodon monoceros*
northern fur seal	*Callorhinus ursinus*

Common name	Scientific name
nutria	*Myocastor coypus*
otter	*Lutra* spp.
panda (see giant panda)	
pig, domestic	*Sus* spp.
pine marten	*Martes martes*
polar bear	*Ursus maritimus*
polecat (see European polecat)	
prairie dog	*Cynomys* spp.
pronghorn	*Antilocapra americana*
Przewalski's horse	*Equus caballus przewalski*
quagga	*Equus quagga*
rabbit (see European hare, common hare, jackrabbit)	
raccoon	*Procyon lotor*
red deer	*Cervus elaphus elaphus*
red wolf	*Canis rufus*
reindeer	*Rangifer tarandus*
rhinoceros (see white rhinoceros)	
right whale	*Eubalaena* spp.
sea otter	*Enhydra lutris*
sheep, domestic	*Ovis aries*
skunk	*Mustela* spp., *Spilogale* spp.
southern right whale	*Eubalaena australis*
southern river otter	*Lontra provocax*
sperm whale	*Physeter catodon*
spider monkey	*Ateles geoffroyi*
stoat	*Mustela erminea*
Tasmanian wolf	*Thylacinus cynocephalus*
tiger	*Panthera tigris*
weasel	*Mustela* spp.
white rhinoceros	*Ceratotherium simum*
white-tailed deer	*Odocoileus virginianus*
wild cat	*Felis silvestris*
wildebeest	*Connochaetes* spp.
wisent	*Bison bonasus*
wolf (see gray wolf)	
zebra	*Equus* spp.

Index

Pages on which terms are defined are shown in bold type. Pages on which items are referred to in figures or tables are shown in italics.

Printed in the United States
120909LV00004B/70-99/A

9 780521 788120